The Pursuit of Harmony

The Pursuit of Harmony

Kepler on Cosmos, Confession, and Community

AVIVA ROTHMAN

The University of Chicago Press   Chicago and London

The University of Chicago Press, Chicago 60637
The University of Chicago Press, Ltd., London
© 2017 by The University of Chicago
Published 2017
Printed in the United States of America
26 25 24 23 22 21 20 19 18 17   1 2 3 4 5

ISBN-13: 978-0-226-49697-9 (cloth)
ISBN-13: 978-0-226-49702-0 (e-book)
DOI: 10.7208/chicago/9780226497020.001.0001

Library of Congress Cataloging-in-Publication Data

Names: Rothman, Aviva, author.
Title: The pursuit of harmony : Kepler on cosmos, confession, and
    community / Aviva Rothman.
Description: Chicago : The University of Chicago Press, 2017. |
    Includes bibliographical references and index.
Identifiers: LCCN 2017013388 | ISBN 9780226496979 (cloth :
    alk. paper) | ISBN 9780226497020 (e-book)
Subjects: LCSH: Kepler, Johannes, 1571–1630. | Harmony
    (Aesthetics) | Religion and science. | Cosmology.
Classification: LCC QB36.K44 R68 2017 | DDC 113—dc23
LC record available at https://lccn.loc.gov/2017013388

Contents

On Kepler's Works
and Translations

Most of the material by Kepler that I have cited comes from

Johannes Kepler Gesammelte Werke [*KGW*]. Edited by Max
   Caspar et al. 22 vols. Munich: C. H. Beck, 1937–.

In the case of letters, I've given the volume number and
letter number (e.g., *KGW* 17:747), while in the case of other
works, I've given the volume number, page number, and
(when available) line number(s) (e.g., *KGW* 12:29.17–19).
I've also cited from the following publications:

Joannis Kepleri Astronomi Opera Omnia [*KOO*]. Edited by
   Christian Frisch. 8 vols. Frankfurt and Erlangen: Heyder
   und Zimmer, 1858–71.
Nova Kepleriana. Vol. 6, *Joh. Kepleri Notae ad Epistolam
   D. D. M. Hafenrefferi*. Edited by Max Caspar. Abhand-
   lungen der Bayerischen Akademie der Wissenschaften,
   Mathematisch-naturwissenschaftliche Abteilung, Neue
   Folge, Heft 14. Munich: Abhandlungen der Bayerischen
   Akademie der Wissenschaften, 1932.

All translations from the volumes listed above are my own
unless otherwise indicated. I've relied as well on the following translations of three of Kepler's books:

Harmony of the World. Translated by E. J. Aiton, A. M. Duncan, and J. V. Field. Philadelphia: American Philosophical Society, 1997.

New Astronomy. Translated by William H. Donahue. Cambridge: Cambridge University Press, 1993.

Optics: Paralipomena to Witelo and Optical Part of Astronomy. Translated by William H. Donahue. Santa Fe, NM: Green Lion Press, 2000.

In referring to Kepler's books, I've used a combination of original-language titles and English translations, depending on popular usage.

Kepler and the Harmonic Ideal

How might we create a more perfect world? What would it mean to establish the best kind of political order or the best kind of religious community—and to what degree is it possible? These were especially pressing questions in seventeenth-century Europe, as the political order seemed ever more unstable and religious communities ever more at odds. In answer, many men and women offered utopian visions of how things might be different, optimistic plans for churchly reconciliation, or arguments for the kinds of states that would be strong enough to withstand the wars that had raged in the previous century and that would rage yet again in this one.

Johannes Kepler (1571–1630) and his friends and correspondents, like so many others, agonized over the troubled times in which they lived. Even before the start of the Thirty Years' War, it was obvious to many that the world was on the brink of catastrophe. "There is no corner of the world where the seeds of war are not dispersed and growing," wrote one friend to Kepler in 1616.[1] By 1621, three years into the war, the ravages of the conflict seemed ever present. "The fires of civil war are raging in Germany," Kepler wrote then. "Everything in my neighborhood seems abandoned to flame and destruction."[2] So staggering were the effects of war that many believed that the end of the world was near—or at least the end of the world as they knew it. "Europe is our Europe no more," lamented a friend to Kepler in 1622.[3] Yet as Kepler continued to

I.1   Portrait of Kepler, artist unknown (1610)

pursue his own studies in the midst of the cacophonous devastation, he came to believe that those very studies—astronomical, mathematical, and musical—had revealed the answer to the problems plaguing his world, and the model by which the world might be created anew: harmony.

Kepler is popularly known today as one of the major figures of the Scientific Revolution. In Prague, he collaborated with Tycho Brahe, famed astronomer and imperial mathematician at the court of Rudolf II. This collaboration led to the publication of Kepler's *New Astronomy* in 1609, which includes the first two of his famous three laws of planetary motion, though he did not refer to them as such. When Tycho died, Kepler became his successor as imperial mathematician, continuing Tycho's project of the construction of new planetary tables, finally published in 1627 as the *Rudolfine Tables*. In the meantime, in 1619, Kepler published the *Harmony of the World*, which investigates the distances of the planets on the basis of the harmonies they orchestrate as they move through their orbits and contains Kepler's third planetary law. In addition to these major works, Kepler's written output was prodigious throughout his lifetime and encompassed an extensive array of subjects ranging from astronomy and astrology to optics and mathematics and from theology and politics to music and chronology, as well as a vast and lively correspondence that stretched across Europe. This flow of writing continued uninterrupted through perpetual financial hardship, Kepler's excommunication from the Lutheran Church, his defense of his mother during her witchcraft trial, and the turbulent period of the Thirty Years' War, which colored the last twelve years of his life.[4]

If Kepler has become a popular figurehead for the Scientific Revolution, he has also been remembered primarily for his astronomical discoveries. Kepler himself, however, was much more interested in harmony. In 1605, mere weeks after he had made the discovery that would forever enshrine him as a herald of modern science and that would form the centerpiece of his *New Astronomy*—that the orbit of Mars was elliptical—Kepler wrote a letter from Prague, where he served as imperial mathematician, to Christopher Heydon in England. "If only God would set me free from astronomy," he lamented, "so that I might turn to the care of my work on the harmony of the world."[5]

What did harmony mean to Kepler, and why was it so important? In the letter to Heydon, he undoubtedly was referring to the book that he saw as the culmination of all his studies, the *Harmony of the World*, finally published in 1619. Yet Kepler likely meant something much more sweeping as well. For though Kepler's book focused on harmony in a variety of aspects—mathematical, musical, astrological, astronomical, and cosmological—he saw the book, and his studies on harmony more generally, as serving broader social goals. In the same letter to Heydon in which he wrote of his longing to return to studies of harmony,

Ioannis Keppleri

HARMONICES
MVNDI
LIBRI V. Qvorvm

Primus GEOMETRICVS, De Figurarum Regularium, quæ Proportiónes Harmonicas conftituunt, ortu & demonftrationibus.

Secundus ARCHITECTONICVS, feu ex GEOMETRIA FIGVRATA, De Figurarum Regularium Congruentia in plano vel folido:

Tertius propriè HARMONICVS, De Proportionum Harmonicarum ortu ex Figuris; deque Natura & Differentiis rerum ad cantum pertinentium, contra Veteres:

Quartus METAPHYSICVS, PSYCHOLOGICVS & ASTROLOGICVS, De Harmoniarum mentali Effentiâ earumque generibus in Mundo; præfertim de Harmonia radiorum, ex corporibus cœleftibus in Terram defcendentibus, eiufque effectu in Natura feu Anima fublunari & Humana:

Quintus ASTRONOMICVS & METAPHYSICVS, De Harmoniis abfolutiffimis motuum cœleftium, ortuque Eccentricitatum ex proportionibus Harmonicis.

Appendix habet comparationem huius Operis cum Harmonices Cl. Ptolemæi libro III. cumque Roberti de Fluctibus, dicti Flud. Medici Oxonienfis fpeculationibus Harmonicis, operi de Macrocofmo & Microcofmo infertis.

Cum S.C.M^{ti}. Priuilegio ad annos XV.

Lincii Auftriæ,

Sumptibus GODOFREDI TAMPACHII Bibl. Francof.
Excudebat IOANNES PLANCVS.

Anno M. DC. XIX.

1.2 Title page of Kepler's *Harmony of the World* (1619)

Kepler noted that he had already decided to dedicate his book on harmony to King James I of England, the man whom Kepler believed was best suited to apply its lessons to the most pressing harmony of all: the harmony of church and state.

In speaking of harmony in this way, Kepler invoked the ancient tradition of Pythagorean harmony, a tradition that linked heaven and earth, God and man, and society and individual through the language of music. To Kepler and many of his contemporaries in the sixteenth and seventeenth centuries, as well as to their ancient and medieval predecessors, this was no mere decorative metaphor or poetic fiction; harmony was very real, an archetype embedded in the cosmos by its divine creator.[6] The ways that Kepler and his contemporaries understood that cosmos were changing, and Kepler was one of the first to follow in the footsteps of Copernicus and unmoor the earth from its traditional place at the center of the cosmic symphony. Yet Kepler's appeals to harmony survived the slow shattering of the crystalline spheres that had upheld the geocentric cosmos of old.

Harmony was the cause to which Kepler devoted his life; it was both the intellectual bedrock and the crucial goal for his seemingly disparate endeavors. To Kepler, moreover, the quest for harmony was not merely academic. As Kepler slowly sought harmony in his own work, the Holy Roman Empire was moving ever closer to a devastating religious and civil war, a war that ignited with the Defenestration of Prague a mere four days before Kepler completed his *Harmony of the World* and that was to wipe out one-third of the population of Germany. Despite this, Kepler persisted in pursuing his goal of harmony, through the discordant havoc of war, exile, his own excommunication, and a great deal of personal loss and hardship. The corpus of his work bears testimony to his desperate efforts to create a unified enterprise of his own, one that he hoped might mend his crumbling world.

This book will explore the ways that Kepler sought both to reveal the harmony in nature and to work toward a worldly harmony that might follow from it. Yet what precisely did that harmony reveal about the natural order, and what kind of world would best mirror it? Kepler's own answer to this question changed over time. At the start of his career, as we will see in chapter 1, Kepler hoped to use his theories— and, in particular, his vision of the true nature of God's harmonic cosmos—to unify Europe, by proving that one way of life, and one approach to religious truth, was clearly and inarguably better than another. Over time, however, as we will see in the subsequent chapters of this book, he came to emphasize harmony in a very different sense; following God's harmonic model came to mean, for Kepler, accepting the peaceful coexistence of diverse perspectives—in particular, diverse religious views—within one larger community. Kepler urged his own Lutheran Church toward more inclusiveness (chapter 2), put forth a vision of the Catholic Church that emphasized its place in the larger body of Christendom (chapter 3), fought steadfastly for the dissemination of Copernicanism in notably religious terms (chapter 4), and argued for the merit of multiple political configurations, so long as they protected the harmony of the state (chapter 5). Finally, in a world that seemed bent on the elimination of division through brutal warfare, Kepler argued for the importance of mathematics as a tool of toleration (chapter 6)—the impartial mathematician, he believed, was ideally situated to mediate between divided confessions and achieve a unique form of harmony.

Kepler's ultimate vision of harmony relied on much the same language and many of the same presuppositions as did earlier adherents of the harmonic tradition, but his ideas differed from those of his predecessors in certain crucial respects. I begin, therefore, by briefly reviewing traditional conceptions of harmony and then considering some of the ways that those traditional conceptions changed; Kepler and some of his contemporaries, I argue, ultimately linked these new visions of harmony to visions of a new world order. In making this claim, I am not offering an account of either revolution or continuity; this is, rather, a story where some things changed and some things stayed the same. The language of harmony had always allowed or constrained certain ways of understanding the world, and Kepler, along with his predecessors and contemporaries, marshaled that language both to comprehend the world as it was and to envision the world as it might be. Yet the language of harmony itself was also always shaped by the very specific contours of the world whose members gave voice to it. Kepler, in other words, used harmony to understand the cultural and social possibilities available to him, while at the very same time those cultural and social possibilities reciprocally affected the ways that he understood harmony. This was particularly so given that he lived at a time when the language of harmony itself was being debated on many fronts.[7]

The rest of this book will therefore detail Kepler's specific contextual influences: the devastation of war, the confessional disputes of his day, Kepler's own excommunication, his theological and metaphysical assumptions, the fledgling Copernican cause, the birth of new kinds of scientific communities, the problems of patronage and economy, the tensions of politics and political advising. Each chapter will begin by foregrounding and explicating a particular context that influenced Kepler or that sheds light on his work: the relationship between confession and metaphysics (chapter 1); personal conscience and toleration (chapter 2); theological accommodation (chapter 3); rhetoric, persuasion, and deception (chapter 4); political fiction (chapter 5); and the meaning of impartiality (chapter 6). From there, I will move to Kepler's own experiences and works and the interesting and inventive ways in which he understood, reshaped, or utilized the particular contexts and traditions I've foregrounded. On the one hand, I argue that Kepler was not a man who stood outside time, as the old heroes of the Scientific Revolution have often been portrayed. He was very much a product of the confessional age in the Holy Roman Empire. On the other hand, I emphasize that the very specific confluence of contextual factors surrounding him allowed him to marshal the intellectual categories available to him in ways that differed from those of very many of his contemporaries. This is ultimately a story of both continuity and revolution—of the ways that one man tried to craft something new while still clinging very much to the old. It is a story that should help us rethink the connections between science, religion, and politics at the very moment when modern science is said to have been born.[8] It should also help us rethink our conventional narratives about the rise of toleration, which often begin with Locke and presume both Enlightenment and secularization.[9]

Finally, it should be clear that there are other stories one might tell about the beliefs and practices I've described here—the ordered, mathematical cosmos, the ordered human body, the body politic, and the body of Christendom—that do not link them all to the tradition of harmony. What matters, for my purposes, is that *Kepler* linked them all to harmony, as did others who came before him, some of whom we will meet in these pages. Kepler opened the dedication of his *Harmony of the World* by introducing his subject as a "work on the harmony of the heavens, with [the] savor of Pythagoras and Plato."[10] He linked its purpose immediately to the "manifold dissonance in human affairs," invoked "God who . . . regulated all the melody of human life,"[11] and hoped that James I might improve that worldly dissonance as he had already done in England, "for what else is a kingdom but a harmony?"[12] He bemoaned his own excommunication by speaking of the "wounds to my person treated by what harmonies, by what physician?" and noted that his "cross-shaped wound [was] still swollen."[13] Ending his dedication, he hoped that "this enduring dissonance . . . will end in pure and abiding harmony" and suggested that this might be possible if James would "stir up in [himself] by the examples of the brilliance of concord in the visible works of God the zeal for concord and for peace in church and state."[14] The harmonic tradition linked all these concerns for Kepler—cosmos, individual, state, and church—and it is this linkage and its shifting meanings that I explore here.

The Harmonic Tradition

Theories of universal harmony inevitably begin with Pythagoras. According to myth, Pythagoras passed a blacksmith's shop one day and discovered that hammers of different weights produced different sounds, some consonant in combination and some dissonant.[15] As he later determined by experimenting with vibrating strings and their pitches, the reason for these differences lay in the numerical relationship between the various weights or lengths of string: the relative measurements of 2:1, 3:2, and 4:3 produced an octave, a fifth, and a fourth when struck or plucked together. Musical harmony could thus be linked directly to ordered numerical relationships. Though almost certainly false,[16] this myth—frequently repeated by classical and medieval authors—accurately captured the belief that musical harmony was founded on ordered relationships and that those relationships, since they were based on number, were accessible via reason. Though the Pythagoras of legend had discovered the theory of harmony empirically, ancient theorists insisted that the mathematical relationships governing harmony could be determined a priori and were necessarily limited to the set of ratios based on the first four integers—the Pythagorean tetrad, symbolizing harmony, whose sum equaled 10, symbolizing unity.

While Pythagoras himself may have linked musical harmony to cosmic harmony,[17] it was Plato who most famously publicized this linkage. In his *Timaeus*, Plato described the Demiurge's process of creation via the very proportions that Pythagoras had discovered in music. This resulted in a cosmos whose interplanetary distances could be represented on a musical scale and whose planetary motions produced beautiful harmonies that were orchestrated by the Demiurge, much as a musician played his instrument.[18] Likewise, in the Myth of Er, with which Plato concluded his *Republic*, the planetary spheres were depicted as concentric wheels turning around a spindle, on each of which sat a Siren singing. The combined singing of the Sirens produced, according

1.3 Pythagoras discovers harmony, from *Theorica Musicae* by Franchino Gaffurio (1492)

to Plato, an audible harmony governed by the mathematical motions of the spinning spheres.

This mathematical notion of harmony and the linkage between music and the heavens was underscored in the medieval theory of music, which was based in particular on the writings of Boethius. That theory placed music in the quadrivium, the four subjects that, along with the trivium of grammar, logic, and rhetoric, made up the seven liberal arts. The quadrivium of arithmetic, geometry, astronomy, and music established music as a science rather than an aesthetic taste or skill, one linked to the motions of the celestial bodies and governed by rules of mathematical order. Boethius's *De institutione musica* was the most authoritative musical text known to scholars of medieval Europe, and it formed the backbone of the musical education in the university.[19] Boethius famously identified three divisions: *musica instrumentalis*, which encompassed singing and instrumental performance; *musica humana*, the music of the body and soul; and *musica mundana*, the music of the spheres. This threefold division was a formalization of the Pythagorean notion of harmony; *musica* represented an archetypal harmony that linked the sounds produced by voices or keyboards with the movements of the planets via the appreciation for consonance built into the soul of man. Boethius was likewise responsible for transmitting the idea of the monochord, first described in Ptolemy's *Harmonics*, to the Latin West.[20] An instrument that divided a string according to various mathematical ratios via movable bridges, the monochord became both a means to investigate the mathematical properties of music and a visual image with which to represent the harmonies that underpinned the cosmos. To the medieval world after Boethius, to speak about proportions was necessarily to invoke some notion of music,[21] and discussions of each of the subjects of the quadrivium referred, whether explicitly or implicitly, to ideas of harmony.

Harmony implied both the idea of mathematical order and also, from the very start, the idea of uniting elements that were different; harmony was, by definition, the *concordia discors*, the discordant made consonant. The language of harmony invoked not only beauty, proportion, and number but also the reconciling of opposites. Harmony was represented, for Heraclites, by the bow and the lyre; the greater the tension between the two, the greater the resulting harmony.[22] It was alternatively represented by the lute and the arrow—both similarly shaped, both attributes of Apollo, yet the one an instrument of music and the other of war. Harmony and strife were opposite sides of the same coin: on one side, differences were positively reconciled, and on the other, they were not.[23] In one version of the myth of the goddess Harmonia she is the daughter of Ares and Aphrodite, the gods of war and love; in another she is the mother of the Muses.[24] The power of harmonious unity was that it embraced diversity; this was true in all invocations of harmony. Yet the embrace of diversity did not mean that harmony was unbounded by strict guidelines, nor that *any* configuration might be harmonious. Quite the opposite was understood to be the case; the dominant references to harmony in the ancient and medieval worlds emphasized rigid definitions and hierarchies. Though harmony might embrace difference, in order to *be* a harmony, rather than a discordant jumble of conflicting elements, only certain particular orders and configurations were allowed. This was true both for the mathematically determined musical harmonies and for the strictly enforced social and political hierarchies that followed from them. The Demiurge in Plato's *Timaeus* was able to create harmony only by imposing a mathematical order on the chaos around him; it was this ordered notion of harmony, Plato suggested, that human beings were supposed to emulate.

Harmony, in this way, was both mathematical and moral; it linked music not only to the ordering of the cosmos but also to the ordering of human society. Plato had made this linkage clear by ending his *Republic*, a vision of the ideal state, with the Myth of Er, a vision of the musical cosmos. In his *Timaeus* Plato had further emphasized the psychological effects of music on individuals, and in the *Republic* he forbade all innovation in music, because such innovation would inevitably alter the foundations of political society.[25] Music affected politics, and the theory of harmony represented both musical order and the ideal ordering of the state. In his *De republica*, Cicero too linked the well-ordered state with the notion of harmony. Like Plato before him, he ended his *De republica* with a myth, in this case the Dream of Scipio, in which he described the ways that the celestial motions produced "a great and pleasing sound" based on "carefully proportioned intervals."[26] And within the text itself, he argued that musical harmony was akin to the harmony of the state, as both were characterized by the hierarchical division of the individual elements that composed them: "from the just apportionment of the highest, middle, and lower classes, the state is maintained in concord and peace by the harmonic subordination of its discordant elements. And thus, that which is by musicians called harmony in song, answers and corresponds to what we call concord in the state."[27]

The linkage between harmony and the state continued unbroken through the centuries, as united, well-governed countries were understood to be "in tune" or "well-tempered."[28] Shakespeare often invoked the language of harmony in his plays, arguing, for example, that "government, though high and low and lower, / Put into parts, doth keep in one consent, / Congreeing in a full and natural close, like music."[29] Humanist Louis Le Roy likewise described political society as "composed of degrees or estates, as it were parts, which estates must be held in concord by a due proportion of each to other, even as the harmony in music."[30] The sixteenth-century political theorist Jean Bodin—whom we will meet again in chapter 5—followed in this tradition when he developed a particular mathematical harmonic series that would serve as a model for the ideal state. "If, therefore, choice be had of such proportions as make a sweet consent in the perpetuall course of numbers," he wrote, "the Commonwealth shall so be everlasting."[31]

Harmony, as we saw in the three divisions of Boethius, underpinned *musica mundana, humana*, and *instrumentalis*—the cosmos, man, and instrumental music. Furthering this Boethian division, the four components of the musical tetrad were linked to the four cosmic elements, which were then paralleled to the four humors of the human body, as elaborated by Galen.[32] The body was in harmony if its humors were well balanced and in tune. If they were not, then disharmony—that is, illness—would inevitably result. The images in Robert Fludd's *Utriusque Cosmi* of 1617 are among the more famous visual depictions of the ways that the theory of harmony linked the supralunar and sublunar worlds, and the macrocosm of the heavens to the microcosm of man, even into the seventeenth century. On the title page, Fludd depicted a series of concentric circles representing first the Ptolemaic cosmos, next the four elements, and finally the microcosm of man and the four Galenic humors. In a later image in the same text, Fludd depicted the cosmos as a divine monochord, which reached from the earth through the elements and up to the celestial spheres, with each space representing a musical interval. At the top of the image, the hand of God reaches down to tune the monochord and preserve its eternal harmony. Fludd's representation of the musical macrocosm and microcosm echoed those of countless before him who understood harmony as the principle that linked all creation, and it would be further echoed after him in works like Athanasius Kircher's *Musurgia Universalis* of 1650. Even as late as 1687, John Dryden could write that

From harmony, from heavenly harmony,
This universal frame began,
. . . .
Through all the compass of the notes it ran,
The diapason closing full in man.[33]

For Fludd and Dryden, as for earlier adherents of the theory of harmony, the body, as microcosm, was ordered much the way the musical cosmos was ordered; harmonic proportion governed all creation.

Further, since harmony was linked to the state—since, that is, the

I.4 The macrocosm and microcosm, from the frontispiece of Robert Fludd's *Utriusque Cosmi* (1617)

ideal form of the social order was expected to mirror the divinely established natural order—it is no surprise that the language of human
harmony, especially that referring to the human body, was used to describe communal harmony. At the same time as the state was described
as a harmony in the musical sense, the idea of the harmonious, ordered body was extended to the state via the metaphor of the body
politic. John of Salisbury invoked both the musical metaphor and the

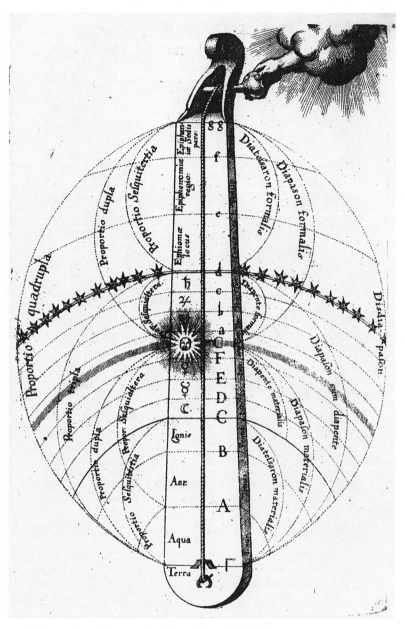

1.5   God tunes the divine monochord, from Robert Fludd's *Utriusque Cosmi* (1617)

metaphor of the body to describe the correctly ordered state. As he explained in his *Policraticus* of 1159, just as musicians "manage by great diligence to curb the fault of a wayward string and restore it to harmony with the others," so too should princes ensure "that subjects are made to be of a single mind in a household and the works of peace and charity create one perfect and great harmony out of pursuits which appear discordant."[34] From musical harmony he then turned to bodily harmony and explained that

the position of the head in the republic is occupied . . . by a prince subject only to God and to those who act in His place on earth, inasmuch as in the human body the head is stimulated and ruled by the soul. The place of the heart is occupied by the senate. . . . The duties of the ears, eyes, and mouth are claimed by the judges and governors of provinces. The hands coincide with officials and soldiers. Those who always assist the prince are comparable to the flanks. Treasurers and record keepers . . . resemble the shape of the stomach and intestines. . . . Furthermore, the feet coincide with peasants perpetually bound to the soil.[35]

Like the musical metaphor, the bodily metaphor makes clear that political harmony entailed both unity and a clear and well-established hierarchy. Each part must do only its own work and must obey the part controlling it—otherwise, discord and political illness would inevitably follow. Thomas Aquinas made precisely the same point linking the ordered body to the ordered state when he wrote: "Among members of the body there is one which moves all the rest, namely the heart: in the soul there is one faculty which is preeminent, namely reason. . . . It is [likewise] necessarily true in the case of human affairs that that community is best which is ruled by one."[36] James I, to whom Kepler would dedicate his work on harmony, similarly asserted: "Kings are compared to the head of this Microcosme of the body of man."[37] And in claiming absolute monarchy as the ideal form of the state, Jean Bodin argued that the wise king would create "a pleasant harmonie of all the subjects among themselves" and would unite them under his rule, which was "figured even in the nature of man himselfe, being the verie true image of a well ordered Commonweale . . . which still hath but one head, and all the rest of the members aptly fitted thereunto."[38]

Thomas Hobbes also famously embraced the tradition of the body politic in his *Leviathan*. The frontispiece of the text showed the monarch, as head of state, quite literally formed out of the individual bodies of the citizens. Within the text, Hobbes consistently invoked the metaphor of the body politic, explaining that the *sovereignty* is an artificial *soul,* as giving life and motion to the whole body; the *magistrates* and other *officers* of judicature and execution, artificial *joints; reward* and *punishment* . . . are the *nerves.* . . ; the *strength, salus populi* (the people's *safety*), its *business; counsellors* . . . are the *memory; equity* and *laws,* an artificial *reason* and *will; concord, health; sedition, sickness;* and *civil war, death.* Lastly, the pact and covenants . . . resemble that *fiat,* or the *let us make man,* pronounced by God in the Creation.[39]

A harmonious state, as Hobbes made clear, was a healthy state, and unrest was equivalent to illness; as he noted at one point, unlawful systems and assemblies were analogous to "wens, biles, and apostems, engendered by the unnatural conflux of evil humours."[40] Harmony, for Hobbes, ultimately meant unity, and the more unity was lacking, the more the body politic was at risk of demise. Political discord and disregard of the sovereign could, in the words of Hobbes, theoretically destroy even the "soul of the Common-wealth" itself.[41] "Take away in any kind of state, the obedience (and consequently the concord of the people)," he wrote, "and they shall not only not flourish, but in short time be dissolved."[42]

Just as the body was linked to the state via the notion of harmony, the body was invoked as well in understandings of the ideal Christian community, through the idea of the *corpus Christianum.* Some theologians simply adopted the notion of the political body and replaced the sovereign with the pope as head of church.[43] But beyond this, the church itself was a kind of body politic, as Paul had argued in Corinthians, writing that "just as the body is one and has many members, and all the members of the body, though many, are one body, so it is with Christ. For by Spirit we are all baptized into one body. . . . you are the body of Christ individually and members of it."[44] The body of Christ unified the universal community of Christians, both metaphorically and via the sacrament of Communion, the mystical consumption of that very body. Indeed, the very term "communion" signified the notion of harmonious unity and was often used for the church itself. Milton invoked these ideas when he wrote that "it is from the union and communion with the Father and with Christ, and among the members of Christ's body themselves, that there comes into being that mystic body, the Invisible Church, the head of which is Christ."[45]

The Eucharist as the sacrament of Communion was understood to physically represent harmony, for, as Erasmus explained, it consisted of "the mystic bread, brought together out of many grains into one flour, and the draught of wine fused into one liquid from many clus-

1.6 Frontispiece of *Leviathan*, by Thomas Hobbes (1651)

ters of grapes."[46] By extension, Erasmus linked the monastic life—the preeminent form of Christian communal living—to the Pythagorean harmonic ideal when he explained that Pythagoras "also instituted a kind of sharing [*communionem*] of life and property in this way, the very thing Christ wants to happen among Christians."[47] Likewise, in the larger Christian community, partaking of the sacrament of Communion was a ritual of social unity, a way to enact one's membership in the body of Christ and the social body of Christendom.[48] In a 1408 description of the sacrament, the Eucharist is described as the *"medium congruentissimum,"* the instrument of harmony that holds the community together.[49] The presence of Christ in the Mass, in this view, mimicked the communal function of the sacrament: "as Christ unites the members to the Head by means of his precious Passion, so we shall be united in faith, hope and charity by the daily celebration of this sacrament of remembrance."[50] In a sixteenth-century English prayer, the believer preparing for Communion implored that "I may be worthy to be incorporated into your body, which is the Church. May I be one of your members, and may you be my head, and that I may remain in You, and You in me, so that in the resurrection my lowly body may be conformed to Your glorious body."[51]

While the taking of Communion was sometimes described as a daily celebration, for the vast majority of people this was not the case; most people took communion only once a year, on Easter Sunday, alongside the rest of the community. This meant that the ritual not only represented social cohesion but enacted it; people came together for Communion and usually beforehand listened to a sermon by their priest emphasizing the importance of social unity.[52] And even when people could not partake of Communion, it still acted as a focus for communal activity. The annual Corpus Christi procession, one of the major civic events of the year, was a way to visually represent the bonds linking the community via the sacrament of the Eucharist.[53] Members of the community displayed banners and forms of social iconography along the procession route, and the celebration often included performances in which the entire community participated.

As taking Communion implied communal harmony, the refusal to partake—or the denial of the right to partake—implied communal discord. Some parishioners refused to take Communion if they were in the midst of a dispute with a fellow community member. Because such disputes threatened the larger community, the clergy struggled to heal communal rifts and to allow all members of the *corpus Christianum* to partake of Communion in harmony.[54] By contrast, when members of the community deliberately broke their ties with the rest and engaged in sin, they were excommunicated—literally, denied the ability to partake of Communion and cast out of the larger Christian community. Their sin, much like discord in the body politic, was often framed in terms of illness. To prevent disease from spreading throughout the rest of the *corpus Christianum*, cancers had to be removed, and sinners expelled—at least temporarily—from the Communion.[55] When disputes were resolved, the Eucharist was sometimes invoked as a symbol of reunion. In Italy, the peaceful resolution of a vendetta was marked by the reception of the Eucharist, and the newly reunited parties acknowledged as they took Communion that they deserved "the vendetta of God" were they to destroy their newfound harmony.[56]

During the confessional era—when, as many saw it, the body of Christendom had been grievously broken—Communion continued to signify communal harmony and belonging, but it became as well a primary sign of *proper* belonging. As in earlier days when individual members of the community were denied Communion for their sins, Communion in the post-Reformation era became a marker for membership in the right community, and the people with whom one took Communion came to matter a great deal. In 1537 Calvin warned his followers in France that they should avoid taking Catholic Communion at all costs; such a practice would mean that they had participated and signified their belonging in a community of idolaters, whose Mass was an "abominable sacrilege and Babylonical pollution."[57] Pierre Viret explicitly argued in 1558 that a central role of the sacrament was to denote proper communal allegiance: God, he argued, had instituted Communion "in order to separate us in the matter of religion from the whole assembly and from all persons who follow a doctrine and religion contrary to his."[58] He continued to insist that "whoever would be taken as a Christian and would participate in the true table of the Lord cannot at all communicate or assist in the Mass or such Supper of the papists if he does not wish to be at one time a participant in both the table of the Lord and the table of devils."[59] Taking Communion, as before, signified communal harmony via participation in the body of Christendom, but one needed to be ever more careful that it was the right and true body of Christendom rather than the devil in disguise.

Changes in the Harmonic Ideal

The harmonic tradition, in its musical, cosmological, social, and religious guises, remained powerful, for some, well into the seventeenth century, as many of the examples we've encountered above make clear. Yet the move from a geocentric to a heliocentric cosmos shook its foundations. Despite the persistent myth that Pythagoras himself may have granted the sun a central place in his system, the tradition of celestial harmony was anchored firmly to the geocentric Ptolemaic cosmos. The music of the spheres was linked not just to the planets but to the crystalline spheres themselves, those orbs whose solidity secured the planets and moved them in their cosmic dance across the sky. Further, a hierarchical chain tied the various layers of harmony together, from the macrocosm of the planets to the microcosm of man, and assumed an ordered cosmos in which one could descend by levels to the realm of man at the very center. For some, dissolving the spheres and moving the earth seemed to shake the entire harmonic edifice and called into question both the reality of the celestial harmonies and the proper role of man in the order of things. John Donne famously bemoaned the loss of the ordered and harmonious cosmos caused by the Copernican vision, and in his 1611 "Anatomy of the World" lamented a world in which "'Tis all in pieces, all coherence gone / All just supply and all Relation."[60]

Yet this despair was not Donne's only word on the subject, nor was harmony extinguished by the new cosmic order. We ought not to forget that the year before Donne wrote his famous poem, he argued for brotherly charity in a sermon that emphasized that "heaven and earth are as a musical instrument; if you touch a string below, the motion goes to the top. Any good done to Christ's poor members upon earth affects him in heaven."[61] The old harmonic vision may have seemed, at times, to wobble, but it could still be revived and reinforced. Copernicus himself had relied on the language of harmony to argue for the superiority of heliocentrism, and his follower Georg Joachim Rheticus had likewise insisted that the Copernican cosmos was *more* harmonious than the older world system. Rheticus argued that earlier astronomers would have had better luck had they more closely "imitate[d] the musicians who, when one string has either tightened or loosened, with great care and skill regulate and adjust the tone of all the other strings, until all together produce the desired harmony and no dissonance is heard in any."[62] Because the Copernican cosmos was unified and symmetrical in ways that the Ptolemaic was not, it was heralded by its promulgators as closer to the true Pythagorean vision.

Copernicus and Rheticus may have appealed to harmony only abstractly and largely rhetorically, but they did indicate that it was possible for the heavenly spheres to be "retuned—not untuned—by the new philosophy."[63] This, in fact, was one of the central tasks of Kepler's *Harmony of the World*—to rescue the theory of harmony in a post-Copernican cosmos, by describing the new planetary intervals that would yield harmonious proportions when the sun, rather than the earth, lay at the center of the world harmony. To do this, Kepler relied on what he perceived to be two factors that distinguished modern from ancient harmonies: polyphony and just intonation.[64] Though the origins of polyphony were debated in the sixteenth and seventeenth centuries, Kepler insisted—as did most of the scholars of his time— that polyphony was a modern innovation and that ancient music was monodic. A primary reason that the ancients could not have developed polyphonic music, Kepler believed, was their overly rigid method of deriving harmonic consonances via a priori numerical relationships— the very relationships that Pythagoras had discovered so long ago. Because the Pythagoreans had admitted only harmonies whose ratios could be formed from the tetrad, they had excluded thirds and sixths (which relied on the number 5) and considered those intervals disso-nant rather than consonant.

Kepler, along with other musical theorists of his day, believed that a theory of music that excluded as dissonant intervals that so clearly *sounded* consonant was untenable. The problem with earlier approaches to music, he argued in the *Harmony of the World*, was that "the Pythago-reans were so much given over to this form of philosophizing through numbers that they did not even stand by the judgment of their ears . . . but they marked out what was melodic and what was unmelodic, what was consonant and what was dissonant, from their numbers alone, do-ing violence to the natural prompting of hearing."[65] Kepler, by contrast, followed those who hoped to establish a theory of harmony that, while rooted in mathematics, "would satisfy the judgment of the ears in es-tablishing the number of the consonances, and the other melodic in-tervals, without trespassing beyond what the ears bear."[66]

This was especially important for Kepler because without a system of intonation that allowed for thirds and sixths, true polyphony was impossible. And it was polyphony, above all, that distinguished the superiority of modern music in Kepler's view. While most of Kepler's contemporaries saw in polyphony some good and some bad, and some argued that the simplicity of ancient music was to be emulated and preferred over the decadence of modern polyphony, Kepler contended that only in polyphonic music had man finally managed to imitate the cosmic harmonies. Writing to his patron Hans Georg Herwart von Ho-henburg in 1599, Kepler took a stance against the position of Ursus, a fellow astronomer who believed "the music of the ancients much no-bler than ours." This greatly surprised Kepler, he wrote, since "I shall never believe that the modulation of one simple voice [i.e., monody] is sweeter than four voices preserving unity in variety."[67] He later insisted in the *Harmony of the World* that "man, aping his creator, has at last found a method of singing in harmony which was unknown to the ancients, so that he might play . . . the perpetuity of the whole cosmic time in some brief fraction of an hour, by the artificial concert of several voices."[68] In contrast to the older, simpler cosmic harmonies, where one melody took precedence, Kepler felt that it had specifically pleased God to grant the planets a harmony in which multiple melodies could be heard simultaneously. Speaking again of polyphony, he wrote that "nature . . . brought you forth in these last centuries, you, the first true likenesses of the universe, and whispering through your ears, she has revealed her very self, as she exists in her deepest recesses, to the Mind of man."[69]

What did this mean, from the cosmological perspective? Ancient theories of the harmony of the spheres often did not offer detailed accounts of the musical relationship between the planets, aside from the assertion that such a relationship existed. When theorists did give specifics, they typically assigned a pitch to each individual planet according to the intervals between the planetary spheres.[70] These pitches corresponded to the Pythagorean system of harmony based on the tetrad, and together they formed a Pythagorean harmony, centered on the earth. In the *Harmony of the World*, by contrast, Kepler moved away not merely from the earth-centered system and the notion of Pythagorean harmony based on number but also from the linkage between harmony and planetary distance. He did this by emphasizing *geometry*, not number, as the archetype underpinning creation. The musical ratios, as Kepler elaborated in the *Harmony of the World*, corresponded in particular to the arcs of a circle cut off by regular and constructible polygons (the latter, for Kepler, meant that they could be drawn with only a compass and ruler).

That the harmonic archetype must be geometric and constructible was central to Kepler's entire metaphysics (as we'll examine more closely in chapter 1). That it was based on the circle was linked, for Kepler, to the idea of the sphere as the most perfect of shapes, one that represented both the Trinity and the cosmos itself.[71] Further, the construction of the circle from the sphere represented the linkage between God and man: a straight line, representing corporeal form, was rotated in the sphere, representing God, to create a circle, representing "the created mind, which is in charge of ruling the body." This, for Kepler, was "a confirmation from the harmonic proportions of the circle as the subject and source of their terms."[72] While repudiating the centrality of Pythagorean number, Kepler thus retained the Pythagorean linkage between harmony, cosmos, and man via the circle itself.

Kepler ultimately argued that inscribing certain geometric figures in a circle yielded a set of ratios, produced by comparing the arc subtended by one side of the polygon with the circumference to the arc subtended by the remaining sides with the circumference.[73] These ratios then corresponded not to the *distances* between the planets but rather to their speeds—and, in particular, to their angular velocities with reference to the sun, at the moments of perihelion and aphelion. These extreme speeds determined the scale of each planet by demarcating its highest and lowest notes. Moreover, the planets jointly produced polyphonic harmonies in their movements. While Kepler ultimately privileged polyphony, as we saw above, he created a system in which *both* monody and polyphony were present, the first in the motions of the individual planets and the second in their movements all together. While polyphony was superior because it represented the cosmos as a whole, Kepler insisted that monody, too, had a place in God's ultimate vision and contributed, in its own way, to the beauty of the whole: "different types of harmonies . . . must have been organized," he maintained, "so that the beauty of the world might be expressed in harmony through all possible forms of variation."[74]

A further implication of this approach to cosmic harmony was that when it came to the actual sounds produced by the planetary motions, they were, on the whole, dissonant. Dissonance itself had been embraced with increasing frequency in the musical theory of Kepler's contemporaries, for it was seen as providing an essential contribution to the ultimate beauty of the overall harmony. Context mattered, theorists began to insist with greater frequency, and it was not rational proportion alone that determined the quality of a harmony. As Girolamo Cardano argued, those who crafted harmonies must take into account the ways that experience changed perception; "better things," he wrote, "are always pleasing after worse ones . . . so light pleases after darkness, sweetness after bitterness, oil of roses after dill, and consonant tones

1.7   The music of the planets, from Kepler's *Harmony of the World* (1619)

after dissonances."[75] Kepler, too, emphasized dissonance as an essential ingredient in a true harmony; he compared the use of dissonance in musical harmony to the use of yeast, salt, or vinegar in cooking and noted that while "complete dishes are not made from them," they are still used to great effect for emphasis.[76] And given the specific intervals produced by each planet, moments of harmonic consonance between the majority of them would be incredibly rare. In Kepler's words, "harmonies of four planets now begin to be scattered over the centuries; and those of five planets over myriads of years. However, an agreement of all six is hedged about by very long gaps of ages; and I do not know whether it is altogether impossible for it to occur twice by a precise rotation, and it rather demonstrates that there was some beginning of time, from which every age of the world has descended."[77] In other words, the planets all played a perfect harmony at the very moment of Creation, and they might play one again at the end of days. In the interim, large-scale dissonance and smaller, more individual harmonies were all that could be expected.

Kepler's understanding of harmony thus privileged a number of components that were either absent or undervalued in most theories of harmony before the sixteenth and seventeenth centuries, as he understood them: polyphony, or the ability of multiple voices to express themselves; consonances that were true to experience rather than merely to a truth determined mathematically; variety of harmonic forms; and dissonance itself as both inevitable and central to the ultimate experience of harmony. And, as we've already seen, he emphasized at the very opening of his *Harmony of the World* that his new vision of harmony might yield important insights for those who hoped to achieve harmony of church and state. He likewise quoted Proclus in an epigram to book 4 of the text, emphasizing that the study of harmony "connects everything in the world, . . . restores friendship between things which are in conflict, and relations and mutual affection between those which are widely separated . . . [and] measures out the proper occasions for conducting affairs."[78] What kind of insights, then, followed from Kepler's vision of harmony? Kepler, of course, articulated his new visions of religious and political harmony in very specific contexts and to very specific ends throughout his life, and we will follow that story in detail over the course of this book. Yet, at the outset, we can point to some general ways in which Kepler may have drawn on his conception of cosmological and musical harmony in order to imagine and bolster these new visions of worldly harmony.

As we saw earlier, in the wake of the religious and civil wars that dominated so much of sixteenth- and seventeenth-century life, many thinkers, from Machiavelli to Bodin to Hobbes, leveraged the ancient tradition of harmony and the body politic to emphasize rigid hierarchy and centralized absolute power. Kepler, as we'll examine in some detail in chapter 5, rejected Bodin's direct linkage between theories of harmony and an absolutely unified state, in Bodin's case via monarchical order. He did so in the *Harmony of the World* itself, in an extended "political digression" devoted to criticizing Bodin's politics. Bodin had argued: "A Monarchie is natural . . . whether we behold this little world which hath but one bodie, and but one head for all the members . . . or if we looke to this great world which hath but one soveraigne God: or if we erect our eyes to heaven, we shall see but one sunne."[79] By contrast, Kepler linked the idea of harmony not to an absolute sovereign but to the public good when he wrote that "the public good has a certain correspondence with the way in which singing in harmonic parts is pleasing."[80] As there might be a wide range of ways in which the public good was achieved, Kepler refused to link the theory of harmony to a particular form of government. Instead, he argued that the ruler, "whether he be king, or the aristocracy, or the entire people,"[81] was following harmonic principles so long as he looked out for the public good. "This one supreme law, the mother of all laws—that anything on which the safety of the state depends is ordered to be sacred and lawful—is . . . consistent . . . with harmonic ratios."[82]

Likewise, the healthy body, as we saw earlier, was often linked to the healthy monarchical state, while the ill body was linked to problems in the political order. In fact, political problems were often compared not merely to ill bodies but also to monstrous ones. Hobbes described the uneasy tripartite division of power in England as just such a monster: "To what disease in the natural body of man, I may exactly compare this irregularity of a commonwealth, I know not. But I have seen a man, that had another man growing out of his side, with a head, armes, breast, and stomach, of his own: if he had another man growing out of his other side, the comparison might then have been exact."[83] Samuel Pufendorf would later criticize the structure of power in Germany in very similar terms: "There is now nothing left for us to say, but that Germany is an Irregular Body, like some mis-shapen Monster, if it be measured by the common Rules of Politicks and Civil Prudence."[84]

Yet the tradition of the body politic, too, could be co-opted to yield very different results depending on one's notion of harmony and of ideal forms of government. In 1580, for example, Montaigne wrote an essay about a "monstrous child" who had a complete twin, minus the

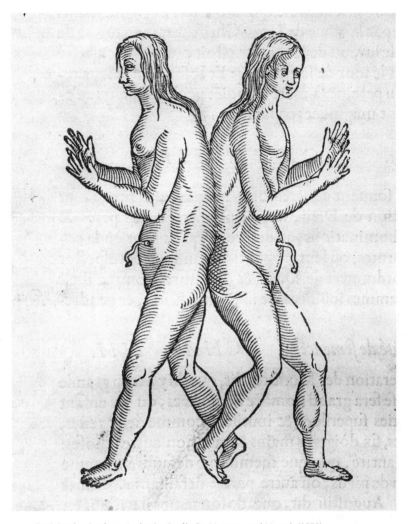

I.8   Conjoined twins from Ambroise Paré's *On Monsters and Marvels* (1573)

head, conjoined at his abdomen. Montaigne knew well that conjoined twins, traditional examples of monstrous birth, were usually prime examples of disharmony and disorder.[85] Yet Montaigne adopted the image of the monstrous conjoined twins to describe the body politic and argued that it might have *good* political resonances rather than the sickly and politically ominous resonance it would have traditionally conveyed.[86] "This double body and these several limbs, connected with a single head," he wrote, "might well furnish a favorable prognostic to the king that he will maintain under the union of his laws these various parts and factions of our state."[87] Montaigne, that is, specifically invoked a monstrous body to offer an image, not of political weakness or discord, but of potential political strength, a strength that relied on plurality and difference rather than unity and agreement. Montaigne argued that if the king could somehow bring the different factions in France together without forcing them to sacrifice their differences, the state might form a stronger harmonious whole.

Kepler, too, embraced a vision of society that valued diversity over homogeneous unity. When it came to music, he had written that "just as . . . individual consonances considered separately are pleasing on account of the fact that they are plainly not identical notes, but in a way figured and different notes . . . in the same way . . . the harmonious singing of parts . . . without any variety in them ceases to be pleasing altogether."[88] He invoked this notion of harmony when articulating his irenical vision of a religious community that embraced diversity and disagreement in its midst. Just as music was harmonious only if it contained many different notes, so too, Kepler believed, earthly communities needed to create a kind of cohesiveness that embraced difference rather than one that sought to do away with it.[89] Though Kepler identified as a Lutheran throughout his life and associated the Lutheran Church most closely with the truth, the reunified Christendom that he hoped to help create was not, in his view, to be equated with any one confession, even his own. Rather, it was to embrace them all, to offer some common ground on which all could agree, and then to allow for the fact that nobody would be able to agree on everything, particularly when it came to questions of theology. After all, Kepler wrote in his 1623 *Confession of Faith*, "Christ the Lord who spoke this word . . . neither was nor is Lutheran, nor Calvinist, nor Papist."[90] Kepler endeavored to highlight those points of commonality around which the different confessions might unite and hoped that his vision of harmony might provide the church with a true model to follow. But that model was one of unity within diversity, not without it.

Like Montaigne, Kepler too invoked the tradition of the body politic to elucidate this new kind of harmonious community, and he relied on the monstrous body to do it. When it came to the *corpus Christianum*, as when it came to the body politic, monstrous bodies typically signified great evil.[91] Martin Luther was represented in Johann Cochlaeus's famous image as a seven-headed beast, while the pope was similarly styled in Reformation propaganda. Likewise, Cornelius Gemma argued that in a period of such religious strife it was no wonder that mon-

1.9   Seven-headed Luther, by Johann Cochlaeus (1529)

strous births seemed to be happening with increased frequency,[92] and an anonymous pamphlet of 1613 emphasized that the birth of conjoined twins and other similar monstrous births "are brought forth to put us in mind of our iniquities . . . which are ever justly punished by the righteous lawe and justice of God."[93]

Kepler, by contrast, referred to one particular monstrous birth—conjoined twin girls born in Strasbourg in 1606—in order to offer up a model for political and religious *harmony*, rather than discord. To Kepler, writing both in his 1606 *De Stella Nova* and in his 1623 *Confession of Faith*, these twins might represent a ruler "who shows a way for two parties who have such different and dissimilar beliefs to grow together through brotherly love, with one heart (as the two girls have only one heart), and to let them have one direction (as the two girls

1.10   Seven-headed Papal Beast, anonymous (1530)

Schawet an das siben hewbtig tier
Gang eben der gstalt vnd manier
Wie Johannes gesehen hat
Ein tier an des meres gestat
Das hat siben vngleicher haubt
Eben wie diß pabstier gelaubt
Die waren all gekrönt bedewt
Die blatten der gaistlichen lewt
Das thier das het auch zehen horen
Deüt der gastlig gwalt vñ zumoren
Das thier trüg Gottes lesterung

Bedeüt jr verfleische zung
Das thier was aim pardel geleich
Bedeüt des Bapst mordische reich
Das auch hinricht durch tirdanney
Alles was jm entgegen sey
Auch so hat das thier peren füß
Deüt das das Euangeli süß
Ist von dem bastum vndertretten
Verschart/verdecket vñ zerknetten
Das thier het auch ains löwen mund
Bedeüt deß bapstum weiten schlund

Den doch gar nie erfüllen thetten
Ziples/pallium noch annaten
Bann/opffer/peicht/stifft zü Gotsdienst
Land vnd leüt Künigreich rent vñ zinß
Das es alles hat in sich verschlunden
Das thier entpfieng ain tödlich wunden
Deüt das Doctor Martin hat gschriben
Das bapstum tödlich wund gebliben
Mit dem oten des Herren mund
Gott geb das es gar gee zü grund
Amen.

have only one face and forehead), and to have them share a public way of speaking, with shared beliefs that they both hold in common (as the girls have only one mouth, throat, lungs, and stomach)."[94] The kind of body politic represented by these conjoined twins, according to Kepler, would epitomize the new kind of harmony, one in which, as he had earlier written about cosmic harmony, "the beauty of the world might be expressed . . . through all possible forms of variation."[95]

As we saw earlier, the ritual of Communion was the religious symbol of the idea of the communal body; participation in Communion signified that one was a member of a particular church, and exclusion from Communion meant that one was deviant and potentially injurious to that body. Kepler, of course, was excommunicated from the Lutheran Church (a story we will look at more closely in chapter 2), and in his attempts to regain admittance to Communion he articulated yet again a new conception of the *corpus Christianum* and its relationship to the ideal of harmony. He argued to the Jesuit Paul Guldin, as we will see in chapter 3, that against all traditional views to the contrary, he, as a believing Lutheran, should be allowed to partake of the Catholic Communion. He could do this, he argued, if the Catholic Church "clearly accepts my protestation and that of all my [family] that we do not agree to those things which we are persuaded to be in error, but only to the general and ultimate holy and catholic intention of the Mass."[96] Communion, in other words, would become a sign of agreement, not to a particular model of the *corpus Christianum*, but rather to a new, more expansive body of Christ that embraced all confessions equally and that allowed for dissent and plurality of opinion. Kepler similarly argued, as we will see in chapter 6, for a new method of calendar reform that might arrive at unity via dissent and plurality by allowing each confession to independently arrive at a method for calculating Easter that would result in all of them celebrating Easter on the very same day.

If Communion is the close religious analogue to the notion of the body politic, then the close religious analogue to the notion of harmony more generally is the idea of the Trinity. As Maria Antognazza has argued with respect to Leibniz, "if harmony is defined as diversity compensated by identity, the most perfect example of harmony is given precisely by the traditional doctrine of a Trinity of distinct persons in one single essence."[97] This too, of course, could be understood to have very different ramifications. For example, John Calvin, who saw the Trinity as an embodiment of harmony and as a model for the social order, understood that model in an older, more rigidly hierarchical sense.

In his *Institutes*, he described the triune nature of God hierarchically—the Father commanded the Son, and both commanded the Spirit—and viewed this as a model for the social community, in which people were born into natural roles, and conformity and obedience were expected of those on the lower rungs of the hierarchy: "Knowing that someone has been placed over us by the Lord's ordination," wrote Calvin, "we should render to him reverence, obedience, and gratefulness."[98] Kepler, by contrast—like Leibniz after him—saw the lesson of harmony, embodied by the Trinity, to be one of unity constituted via plurality rather than via conformity and obedience. Leibniz described harmony alternately as "unity in multiplicity" or "diversity compensated by unity," and since he saw the Trinity as so perfect a model for harmony, he spent a great deal of energy reflecting on its meaning and implications.[99] The same may be said of Kepler, and it was because of this very subject, and his disagreement with traditional Lutheran conceptions of the triune nature of God (as we will examine in more detail in chapter 1), that he was ultimately denied Communion by the Lutheran Church. Kepler's universe itself was Trinitarian, as we saw earlier, as was the figure of the sphere, which served as the model and source for all harmony. Indeed, Kepler noted that *because* he saw the Trinitarian harmony in the basic structure of the cosmos, he felt confident of his ultimate success in identifying the harmony of the planetary motions: "I dared to attempt this," he wrote in his first book, the *Mysterium Cosmographicum*, "because of the beautiful harmony of the things that are at rest—the Sun, the fixed stars, and the intermediate space—with God the Father, the Son, and the Holy Ghost. . . . Since the parts that are at rest are like this, I did not doubt that the moving parts would be so also."[100] In part because the Trinity was so central to Kepler's understanding of the cosmos, and so representative of the harmonic ideal, he could not accept a theological doctrine of the Trinity or its properties that he found philosophically insufficient. For Kepler, as for Leibniz, thinking about harmony meant thinking about God, in both the larger metaphysical and the more specifically doctrinal senses.

In sum, harmony continued to matter throughout the sixteenth and seventeenth centuries, and it continued to link musical notions to heaven and earth, man and state—yet the nature of those linkages and their implications fluctuated greatly. Kepler's own invocations of the harmonic ideal at times fractured the very order that the old harmonic tradition sought to uphold. For while the harmonic ideal had earlier signified a rigidly upheld world order whose contours were determined by a logic unconcerned with the particulars of context or perception, for Kepler, as for some of his contemporaries, harmony came to signify diversity more than unity and came to embrace multiple possible configurations rather than one absolute. Kepler, like the poet Alexander Pope a century after him, saw in harmony something that pointed the way to a world that might be improved, rather than harmed, by difference—a world where, in Pope's words, "Not, chaos-like together crush'd and bruis'd, But, as the world, harmoniously confused: When order in variety we see, And where, though all things differ, all agree."[101] Indeed, Kepler had emphasized that while the cosmos itself had once produced a perfect and complete harmony, it would not do so again until the end of days—and maybe not even then. God, it seemed, had meant for humans to be satisfied with the beauty of the smaller harmonies produced by individual groups of planets and to accommodate themselves to the dissonance of the whole. Might the same not be true for the church, once a great harmony but now broken into smaller confessions, united in themselves but dissonant overall? Kepler ended his *Harmony of the World* by invoking and elaborating on the words of the psalmist, who had called to God from a world that too seemed bleak and desolate, beset by conflict: "Great is our Lord, and great is His excellence. . . . Praise Him, heavenly harmonies, praise him, judges of the harmonies which have been disclosed; and you also, my soul, praise the Lord your Creator as long as I shall live."[102]

ONE

"The Study of Divine Things": Kepler as Astronomer-Priest

In the town of Leonberg in 1581, the ten-year-old Johannes Kepler first dreamed of devoting his life to God. What he really wanted was to be a prophet.[1] Even to this rather solitary dreamer, it was clear that the world desperately needed guidance, and as a prophet he would have direct access to God's plan for his wayward people. Yet he felt in his bones that he was too impure and knew that a life of prophecy was beyond him. If he could not speak to God, then, he would speak for him; he would become a Lutheran priest. Kepler pursued this dream for the next thirteen years, until, while he was completing his theology degree at the University of Tübingen, a letter arrived that was to change the course of his life. The Lutheran school in Graz required a new teacher of mathematics and requested that the faculty at Tübingen send along their best candidate. Kepler was their choice, and though he had no desire to teach mathematics (even though the subject was one of his great loves), he reluctantly agreed. After all, poor and dependent on the goodwill of his teachers, what else could he do?

In the years that followed, Kepler rose from his position as a lowly teacher and district mathematician to the post of imperial mathematician to the Holy Roman Emperor himself. Though this no doubt comforted the man still continually plagued by fears of poverty, what comforted him still further was the way he recast the position of astronomer so that it enabled him to fulfill his earlier dream. "I truly believe," he wrote, "that as astronomers we are priests of the Lord Most High with respect to the Book of Nature."[2] As an astronomer-priest, he believed that he could use his mathematical talents for the good of God and his church. And he ultimately claimed that as an astronomer-priest he had been able to fulfill the earlier dream of prophecy that had seemed impossibly elusive—he had been able to read the mind of God.

In this chapter, I focus on Kepler's early conception of the relationship between his mathematical and astronomical work (his science, in modern parlance) and his confessional identity (his religion). In particular, I consider the ways that Kepler hoped to use his mathematical astronomy in the service of his particular understanding of true religion. First, though, some contextualizing is necessary, for what place reason and philosophy (including both metaphysics and natural philosophy) should have in theology—indeed, whether they should have a place at all—was a contentious issue in the Lutheran Church of the seventeenth century. Delimiting the boundaries between various disciplines and choosing *what* metaphysics, if any, should inform one's theological views also formed a dividing line *between* confessions. Philosophy and science, that is, were often marshaled as weapons in the tense and volatile confessional battles of Kepler's day.

This was particularly true for the issue central to Kepler's own disagreement with the Lutheran doctrine of his age: the nature of Christ's presence in the Eucharist. Varying understandings of Christ's Eucharistic presence formed one of the central dividing lines—and the harshest points of contention—between the three major confessions in the post-Reformation era. The debate revolved not just around Communion itself but also around the larger question of how Christ's two natures—divine and human—related to one another. When Christ had pointed to the bread at the Last Supper and proclaimed "this is my body," what did that mean? Was his presence in the bread and wine at each and every subsequent Communion real or symbolic, human or divine? Catholics traditionally relied upon the Aristotelian distinction between substance (or essence) and accident to explain the real presence of Christ in the Eucharist. According to this perspective, known as transubstantiation, at the moment of the Mass the substance of the bread and the wine transformed into Christ's body and blood, while their accidents—their external appearance—remained the same.

Like his Catholic predecessors, Luther agreed that Christ's statement "hoc est corpus meum" implied his real presence in the Eucharist, though he took issue with the Aristotelian categories with which Catholics framed the debate. Christ's body and blood were, according to Luther, actually present in the bread and the wine, even if attempting to parse that presence in philosophical terms was a hopeless task. Luther did insist, however, that the presence of Christ's body and blood was not restricted to the bread and the wine at the moment of Mass alone. According to Luther: "Because we believe that Christ is God and man, and the two natures are one person, so that this person cannot be divided in two . . . it must follow that he . . . is and can be wherever God is, and that everything is full of Christ through and through, also according to his humanity—not according to the first, corporeal, limited manner, but according to the supernatural, divine manner."[3] On the basis of the doctrine of *communicatio idiomatum*, the communication of properties, Luther argued that all the consequences of Christ's divine nature applied equally to his human nature, since the two were one. As a result of this belief, Luther arrived at the doctrine known as ubiquity, referring to the omnipresence of Christ in both his divine and his human forms. This doctrine maintained that as God was omnipresent, so too was Christ's physical body to be found everywhere throughout the universe.

By extension, Luther did not assert a miraculous change of the substance of the bread and the wine into the body and blood, for Christ's body and blood were already there, as they were everywhere. The Mass was a powerful testament that Christ left behind for his followers, not a particular, localized miracle or transformation. Moreover, Luther argued that believers needed to refine their understanding of Christ's body and what its presence actually implied. Rather than a kind of pantheism, to which it steered dangerously close, Luther's doctrine of ubiquity maintained that Christ's body was not corporeal in the usual sense of the term, as it was not subject to any physical or natural limitations. Christ's body, according to Luther, was really present everywhere, but not locally so. The Eucharist did not link the body of Christ directly to the physical world, for the presence of Christ's body in the Eucharist could be understood in only a nonmaterial and nonlocal sense, as could the presence of Christ's body in the world more generally.

The Formula of Concord reinforced the Lutheran understanding of *communicatio idiomatum* and all its consequences for the doctrine of ubiquity. According to the Lutheran consensus articulated in the Formula, because of the fact that it has been personally united with the divine nature in Christ, the human nature in Christ . . . did receive in addition to and above its natural, essential, permanent properties also special, high, great, supernatural, inscrutable, ineffable, heavenly prerogatives and advantages in majesty, glory, power, and might . . . and accordingly, in the operations of the office of Christ, the human nature in Christ, in its measure and mode, is equally employed from and according to its natural, essential attributes . . . but chiefly from and according to the majesty, glory, power, and might which it has received through the personal union, glorification, and exaltation.[4]

The Formula further emphasized the omnipresence of Christ's body by quoting Luther's words about the ubiquitous presence of Christ and asserting that this was true "even according to His human nature."

Calvinists sharply disagreed with the Lutheran understanding of the Eucharist, and much of their dispute centered on the Lutheran doctrine of *communicatio idiomatum* and the omnipresence of the body of Christ. While the Calvinists believed that there was a real presence of Christ in the bread and the wine at the moment of the Mass, they argued that this presence was not physical but spiritual—it was the presence of Christ's spirit, descending to the Mass in order to elevate those partaking of it. Christ's body, by contrast, was in heaven; the moment of ascension, according to Calvin, was the moment when Christ's body was physically removed from earth. Calvinists argued that the doctrine of ubiquity was rationally unintelligible—what did it mean for a body to be everywhere and illocal, when to be a body meant to be bound in time and space? Calvinist theologians adhered to the general maxim that "finitum non est capax infiniti," the finite cannot contain the infinite, in order to argue that the Lutheran idea of Christ's body and of the omnipresence of Christ simply made no sense. Theodore Beza famously summed up the Calvinist position at the Colloquy of Poissy when he argued that in the Eucharist, Christ's body was "as far removed from the bread and wine as is heaven from earth."[5]

As should be clear from the Calvinist objection on the basis of rational intelligibility, reason—and by extension metaphysics and natural philosophy—became inextricably embroiled in post-Reformation Eucharistic debates. This was true even within the Catholic Church itself, as the Aristotelian physical system, with its distinction between substance and accident, was slowly replaced by the mechanical worldview and its emphasis on matter and motion alone, and the Catholic doctrine of transubstantiation became newly open to debate and reinterpretation. And it was particularly true in the debates between Lutherans and Calvinists, as Calvinists increasingly turned to reason to support their own confessional stance against Luther's claim that the finite could and did contain the infinite when it came to the physical body of Christ. Indeed, Luther had asserted that the very idea of rational intelligibility could not and should not be applied to the divine. He argued that "nothing is so small . . . that God is not still smaller. Nothing is so large that God is not still larger. Nothing is so short that God is not still shorter. Nothing is so long that God is not still longer. Nothing is so wide that God is not still wider. Nothing is so narrow that God is not still narrower."[6] The realm of God existed above and beyond the realm of reason, according to Luther.

In the face of Calvinist arguments against the Lutheran position of *communicatio idiomatum* and consubstantiation, Lutherans had two possible strategies. The first, adopted predominantly at the start of the movement, was to do as Luther had done: to deny the validity of the Calvinist critiques by denying the relevance of reason and metaphysics to the discussion in the first place. In his "Disputation against Scholastic Theology," Luther argued that scholastic theology represented the inappropriate use of reason; God and his work could be understood only through faith, with the scriptures as the only guide. "No syllogistic form holds for divine terms," Luther wrote.[7] Faith that relied on logic was not true faith but its opposite. True, Philip Melanchthon had found a place for natural philosophy within Luther's vision, but this was natural philosophy expressly concerned only with the discernment of God's presence and providence in the natural world, not a larger metaphysics that specifically bore on the nature of God's being.[8] Following Luther's rejection of Aristotelian scholasticism, Lutheran scholars and the universities that housed them rejected the discipline of metaphysics, along with any attempts to reconcile reason and faith, or philosophy and theology.[9] Christian mysteries were just that— mysteries, inaccessible via reason—and logic was not the method by which salvation was to be achieved. Lutheran universities in the sixteenth century slowly excluded metaphysics in favor of an emphasis on rhetoric, ethics, natural law, biblical exegesis, and natural sciences that were decoupled from any larger metaphysical considerations.

The second possible strategy in light of the Calvinist rational critique of Lutheran theology was to do exactly the opposite: to embrace metaphysics retroactively and to develop a Lutheran metaphysics that might refute the objections of the Calvinists. This strategy was adopted with increasingly frequency in the seventeenth century, a phenomenon that helps to explain, as Ian Hunter has, the reappearance of metaphysics and the new linkages between metaphysics, theology, and natural philosophy in seventeenth-century Lutheran universities. Lutheran Salomon Gessner, as Hunter notes, is just one example of this phenomenon. Gessner began his edition of a metaphysics textbook with the concern that since Calvinists were attacking the Lutheran conception of *communicatio idiomatum*, Lutherans needed to have actual tools to fight back, instead of just ignoring the attack. If his coreligionists might be taught to speak with greater subtlety and sophistication on questions of metaphysics, Gessner believed, they might have a better chance of defending their theological positions—key among them the doctrine of ubiquity—that the Calvinists attacked so disdainfully. Walter Sparn has likewise argued that the impetus behind studies of metaphysics in the seventeenth century was primarily theological and confessional; in Sparn's analysis, both the Lutheran and the Calvinist turns to philosophical inquiry were guided by theological beliefs and driven by confessional concerns.[10] As a consequence, the different confessions developed very different philosophies of God, nature, and being to uphold their particular doctrinal positions.

As a young man at the start of his career, Kepler, too, sought to harness metaphysics—in particular, the geometrical worldview that underpinned his personal metaphysics—to confessional theology, but the direction of the linkage he envisioned differed markedly from that of many of his contemporaries. As I will argue in this chapter, Kepler's personal metaphysics was itself theological from the start, as it began with a particular conception of God and proceeded to link God, man, and nature via the idea of geometry. Moreover, it was absolutely clear and certain; it could be derived a priori via reason; and it could be grasped as true instinctively even without reason. Instead of beginning with the truths of Lutheranism, then, and developing a metaphysics that might uphold those truths, Kepler began with the truths of geometry and considered what those truths revealed about the doctrinal positions of his own church. In doing so, Kepler came to argue, much as the Calvinists did, that the Lutheran notion of ubiquity was rationally incomprehensible. In fact, Kepler believed that his *Mysterium Cosmographicum*, which demonstrated just how clearly the world was modeled on geometry, had made the theological objections of the Calvinists stronger and more convincing.

Naïve idealist that he was, Kepler believed that he could convince others of this too, and in so doing help bring a little more unity to a Christian world that seemed to fracture more every day. Kepler hoped, that is, to use his work as an astronomer-priest—especially his very first book, the *Mysterium Cosmographicum*—to help resolve the theological disputes that divided the world of Christendom. And it was that larger Christian world, and not the more limited world of the Lutheran confession, that mattered most to Kepler. Though Kepler continued to identify as a Lutheran to the end of his days, he fought for a conception of the church that was far broader than the rigid confessional allegiances of many of his contemporaries. God's church, Kepler argued, could not be identified with any one confession alone. Though Kepler believed Lutheranism to be the confession that approached the truth most closely, he insisted that each confession contained elements of truth and that the idea of confession itself was problematic, as it implied a body of Christendom that was already broken. The true purpose of those who served the church, in Kepler's view, ought to be to guide it toward greater unity, not to tear it apart still further. At the start of his career, Kepler hoped to use his mathematical prowess in the service of this unity; geometry, he felt, might be the means by which disputing parties could come to agreement.

The *Mysterium Cosmographicum* and the Idea of Quantity

In October 1595 Johannes Kepler, teacher of mathematics and district mathematician in the Styrian city of Graz, wrote a jubilant letter to Michael Maestlin, his former professor of mathematics at the university in Tübingen. Kepler had just made a remarkable cosmological discovery, one that did much to brighten the drudgery of the previous months away from his place of greatest comfort, the university, and his preferred subject of study, theology. He documented his discovery and sent a copy of the manuscript to Maestlin, which he titled *Prodromus Dissertationum Cosmographicarum continens Mysterium Cosmographicum*.[11] "I truly desire," Kepler wrote to Maestlin, "that these things are published as quickly as possible for the glory of God, who wants to be known from the Book of Nature. . . . I wanted to be a theologian; for a long time I was distressed: behold, God is now celebrated too in my astronomical work."[12] Unable to devote himself to the Book of Scripture directly, Kepler had turned his focus to God's other book—the Book of Nature—which, he believed, also revealed God's providential plan. The astronomer who unfolded and clarified this plan, argued Kepler, performed a task analogous to that of the theologian—one illuminated God's words, while the other illuminated God's things.[13]

In the *Mysterium Cosmographicum*, Kepler believed he had done this

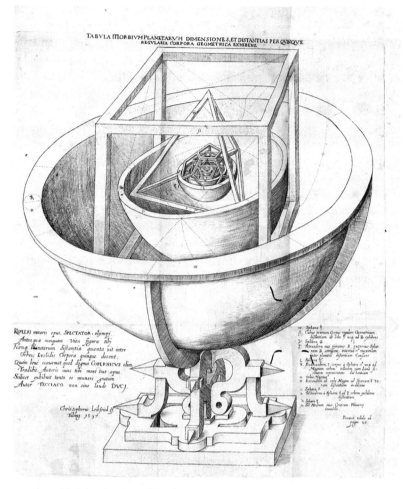

1.1 A model of the planetary orbits using nested Platonic solids, from Kepler's *Mysterium Cosmographicum* (1596)

by demonstrating the fundamental geometric structure underpinning the cosmos. A convinced Copernican, Kepler had long sought the reasons for the precise number of the planets and the distances between them. His solution, described in the *Mysterium Cosmographicum*, rested on the five Platonic solids (dodecahedron, tetrahedron, cube, icosahedron, octahedron). Kepler demonstrated that by nesting the Platonic solids one inside the other and then circumscribing circles around each one to represent the planetary orbits, one could arrive at the distances between the planets, ordered according to Copernican theory. Because there were only five Platonic solids, it was clear that there would be precisely six planets. Kepler further argued that the structuring of the cosmos according to the Platonic solids made perfect sense, for geometry was the tool with which God had created the universe and all things in it. To be intelligible was to be geometrical, Kepler contended, because the human mind was imprinted with the very geometrical archetypes that also structured the cosmos and was therefore uniquely suited to understand God's creations.[14]

Kepler was overjoyed by his successful geometrical articulation of the cosmos in the *Mysterium Cosmographicum*, and not just because it reinforced his Copernican views. The central role of geometry in his first book formed the foundation of Kepler's larger approach to the study of the natural world and at the same time formed a bridge between his conception of nature and his conception of the divine. In fact, Kepler opened the *Mysterium Cosmographicum* with the idea of geometry as just such a bridge, in a friendly note to the reader in which he claimed that in his book the reader would find explained "the world, and God's reason and plan for creating it, from where God [obtained] the numbers, what is the rule for such a great mass, why [God] made six orbits, why spaces fall in such ways between the spheres."[15] The book, Kepler suggested in grandiose tones, would explain a central fact about the arrangement of the cosmos, but it would also do something more: it would also allow its readers to understand the mind of God himself.

As it turns out, to Kepler this was not simply rhetorical grandstanding. To explain the ways in which his geometrical vision related to both nature and God, Kepler began the text by carefully constructing a vision of the cosmos built on the notion of what Kepler called quantity. In articulating the importance of quantity, Kepler contrasted quantity with numbers, the latter being abstract and imaginary, and the former being concrete and geometrical, linked to real, physical objects. In so doing, Kepler distanced himself from the Pythagorean school of thought, which, as Aristotle had noted in his *Metaphysics*, had believed abstract numbers to be the foundation of all reality.[16] Though Kepler at times identified himself with the Pythagorean school because of his mathematical vision of the cosmos, he differed by emphasizing geometric quantity rather than abstract number.[17] "It was matter which God created in the beginning," Kepler argued in his own beginning, "and if we know its definition, I think it will be tolerably clear why God created matter and not any other thing in the beginning. I say that what was put forward by God was quantity."[18]

What, then, was quantity? "The quantity of matter," Kepler continued, "to the extent that it is matter, is a certain form and the origin of its definition."[19] Quantity, that is, was number *combined* with physical form—it was counted numbers rather than *counting* numbers.[20] Number alone was meaningless unless combined with something physical and real: it was, as Kepler explained in a letter to Maestlin, "an accident of quantity, meaning worldly number. For before the world there was no number, except for the Trinity, which is God himself. Therefore, if the world was fashioned to the measure of numbers, it must have been to the measure of quantities. But neither in a line nor a surface is there number, but only infinity. Therefore, it is in bodies."[21]

Why did quantity assume such primacy in Kepler's worldview? The answer lies in Kepler's conception of God and his relationship to nature. Kepler began his *Mysterium Cosmographicum* with the theological premise, linked to Nicholas of Cusa, that geometry—in particular, the relationship between the idea of the curved and the idea of the straight—formed a method by which people might speak about God.[22] The distinction between the curved and the straight, for Cusa, was analogically equivalent to the distinction between the divine and the created; still further, the seeming impossibility of bridging that distinction—of squaring the circle—highlighted the unbridgeable gap between God and man. Further, Kepler also drew on Cusa in his geometrical framing of the Trinity; for both men, the Trinity could be represented as a sphere, with God the Father as the center, the Son as the surface, and the Holy Spirit "in the symmetry of the relationship between the point and the circumference."[23] With this opening to the *Mysterium Cosmographicum*, it seems at first that Kepler, much like Cusa, elevated geometry because it allowed people to speak analogically about God—it was, in other words, the language that allowed people to approximate the divine essence most closely.

Yet Kepler was actually saying something far more radical, something that contradicted Cusa's ultimate claim about the gap between God and man. For Cusa, the importance of geometry lay in its ability to analogically *approximate* the relationship between God and man, because only approximation was possible. Cusa argued that "between the infinite and the finite there is no possible proportion."[24] Geometry was an aid to understanding, and it brought people as close to God as possible—which was still impossibly far away, in Cusa's understanding of divine otherness. Kepler, by contrast, used geometry to try and close the gap that Cusa left open. "It is clear," he continued, "that by those laws which God himself in his goodness prescribes for himself, he could accept no other idea for the constitution of the world than that of his own essence. . . . . This image, this Idea, he wanted to imprint on the world, so that it might become the best and most beautiful possible; and so that it might become capable of accepting this Idea, he fashioned quantity."[25] Here, Kepler suggested that geometrical quantity was not merely a means to speak about God analogically—it was, rather, part of God's own essence.

With the publication of the second edition of the *Mysterium Cosmographicum* in 1621, Kepler made this idea still more explicit. "The Ideas of quantities," he clarified, "are and were coeternal with God, and God himself; and they are still a pattern in souls made in the image of God (also his essence)."[26] Geometry, Kepler believed, was the original archetype, the blueprint by which God created the universe, the mind of man, and all else. This idea, too, has its roots in the Platonism on which Cusa drew—after all, the Demiurge of the *Timaeus* used the eternal Forms to create the world, which was a reflection of those Forms. Plato, in his dialogues, often cited mathematical knowledge as the paradigm for all other kinds of knowledge and indicated that mathematical objects, like the Forms (and perhaps even full-fledged Forms themselves), were immaterial entities, formal models for the material world created in their image. But for Kepler, geometric forms were not divorced from matter but could be understood only in terms of material extension and dimension—they were ideas inextricably linked to physical objects.

Still further, for Kepler the geometric forms were more than just a blueprint; geometry, Kepler argued, was coeternal with God and reflected both the shape of creation and the nature of God himself. As he ultimately elaborated again in the *Harmony of the World*—the book which he believed completed the work of his "forerunner," the *Mysterium Cosmographicum*—"geometry, which before the origin of things was coeternal with the divine mind and is God himself (for what could there be in God which would not be God himself?), supplied God with patterns for the creation of the world, and passed over to Man along with the image of God."[27] Kepler certainly believed that by demonstrating that the underlying structure of the universe was geometrical, he had increased man's ability to understand and speak about God and his creation. Yet beginning with his work in the *Mysterium Cosmographicum*, Kepler also believed that he had done something that went much further than merely allowing people to speak about God's *works*: he had articulated a means by which they might speak directly about the very nature and essence of divinity.

The problem of how finite, temporal creatures could hope to understand the eternal and transcendent God is an old one.[28] Many Christian theologians (Cusa among them), relying both on biblical claims that "no one has seen God"[29] and on Plato's claim that "to discover the maker and father of this universe is indeed a hard task, and having found him it would be impossible to tell everyone about him,"[30] argued that it was impossible to speak directly about God, either because his nature was unknowable or because he did not have a nature, or genus, in the way that created things did.[31] Though it might be possible to speak about God indirectly, via his effects, or works, it was not possible to speak about God's essence. Origen, for example, had declared that God's nature "cannot be grasped or seen by the power of any human understanding."[32] Gregory of Nyssa had likewise believed that human nature was profoundly limited and had neither the capacity nor the language to understand or speak about the divine.[33] The only way to speak about God's essence or nature was negatively, by articulating the ways in which he was *not* like created things—this tradition is consequently often described as negative theology.

Kepler, with his long-standing interest in theology, was interested not only in describing God's creation but also in attempting to speak as directly as possible about the divine. To circumvent some of the problems of negative theology, he argued on the basis of the idea of quantity in the *Mysterium Cosmographicum* that God's structuring of the world according to archetypal geometrical forms revealed something direct and real about the very nature of God itself. As he explained in a letter to Maestlin about the conclusions of the *Mysterium Cosmographicum*,

as the eye was fashioned for understanding colors and the ear for understanding sounds, . . . the mind of man was fashioned not for understanding anything whatsoever but [specifically] for understanding quantities. And the closer something is to bare quantities—as it were, to its own origin—the more properly the mind perceives it; the farther it recedes from this, the more obscurity and errors there are. For by its own nature our mind carries its notions, built upon the category of quantity, with it toward the study of divine things: if it is deprived of them, it is able to assert nothing except by mere negations.[34]

Kepler here once again emphasized the importance of quantity, or geometric entities, in the construction of the cosmos and in the fashioning of the mind of man. God had used a geometrical blueprint to create the universe, and the language in which the blueprint was written was the language of quantities—physical measurements of geometric bodies. And that geometrical blueprint was not just incidentally chosen by God—it mirrored God himself. In fact, as Kepler wrote in both the *Mysterium Cosmographicum* and the *Harmony of the World*, geometry was coeternal with God. And since nothing coeternal with God could exist outside God, geometry ultimately *was* God.

Kepler echoed this sentiment in varying ways throughout all his works, including even his astronomical textbook, *The Epitome of Copernican Astronomy*.[35] By equating God with geometry, Kepler hoped to offer a positive definition of the nature of God himself.[36] Moreover, if the knowledge of the human mind was "of the same nature as that of God," as Kepler argued to Herwart von Hohenburg,[37] then this meant that unlike the claims of negative theologians, God had a nature, or a genus, and it was one that could be grasped by humans. In contrast to Descartes, who, in his *Meditations*, portrayed God as a "cause whose power surpasses the limits of human understanding,"[38] Kepler directly linked human understanding to the divine nature. While Gregory of Nyssa had claimed that there was no language capable of describing God, Kepler argued that there was, and that it was geometry. In Kepler's words to Maestlin, we can take "the category of quantity toward the study of divine things." Without geometry, Kepler agreed with the negative theologian that we can speak of God only "by mere negations"; armed with the notion of quantity, however, we might say something real and positive about God himself. And what we say might, as we will soon see, have very real implications for confessional practices that bear on God's nature—in particular, on the proper approach to the Eucharist.

Kepler's conception of the geometrical archetype and the ways that it linked the nature of God, the mind of man, and the structure of the cosmos had important implications for another central feature of Kepler's work: his belief that he could offer a real, physical description of the cosmos rather than just an approximation or a hypothetical arrangement that would adequately describe the phenomena. In the sixteenth and seventeenth centuries, astronomy was commonly understood as a discipline that did not offer—or seek to offer—a true picture of what actually happened in the heavens. Instead, astronomy's job was to offer a means for calculating planetary positions, positions that might not correspond to reality. This conception of astronomy was linked to two separate but related trends: on the one hand, the common disciplinary understanding of astronomy as distinct from natural philosophy and, on the other, the increasing influence of skepticism more generally. The disciplinary divisions between astronomy and natural philosophy went back to Aristotle's classification of the sciences and their respective methods and objects of inquiry. According to Aristotle, the mixed mathematical sciences, astronomy among them, were different from natural philosophy because they were not concerned with causes, or qualities, but simply with quantities. This was because, as Aristotle had argued in the *Posterior Analytics*, while natural philosophy could reason from causes to effects—while it could make a priori causal arguments, in other words—astronomy and its related disciplines could only consider nature a posteriori.[39] For this reason, astronomy could offer possible models for the motions of the planetary bodies, but any number of models might possibly correspond with the way things really worked, and there was no way for astronomical methods to distinguish between them. To mix mathematical astronomy with arguments about physical causes or to claim certain truth on the basis of astronomical claims was thus, in the eyes of most practitioners in the period, to make a basic category mistake.

It was this disciplinary distinction that allowed, for example, the widespread adoption of Copernican theory in Lutheran universities, despite the fact that Copernican astronomy seemed to contradict both accepted Aristotelian physics and traditional biblical interpretation. When Copernicus's book was first published, Andreas Osiander added a preface in which he presented Copernicus's theories in precisely this way, stating that "these hypotheses need not be true nor even probable. On the contrary, if they provide a calculus consistent with the observations, that alone is enough."[40] Osiander's approach was adopted by Philip Melanchthon in his reform of Lutheran universities and by other prominent astronomers, especially at the University of Wittenberg. In saying that astronomy did not seek to offer causal knowledge, proponents of this approach could then argue that Copernican theory could be acknowledged false in a causal sense but still accepted astronomically as a useful tool that might provide accurate results about the positions of the heavenly bodies. They might, that is, adopt the technical innovations of Copernican theory while discarding their physical implications.[41] Erasmus Reinhold, professor of mathematics at the University of Wittenberg from 1536 to 1553, was an early adoptee of Copernicus's technical innovations, which he used to construct a new set of widely used astronomical tables, the *Prutenic Tables*. Yet Reinhold and his successor, Caspar Peucer, argued against the theory that the earth moved.

Separate from but often linked to this disciplinary distinction between the goals of astronomy and the goals of natural philosophy was the skeptical attack on knowledge. Skeptical humanists, taking up a long tradition of Ciceronian skepticism, maintained that no elaborate system of knowledge constructed by humans could achieve incontrovertible truth. This was as true of natural philosophy as it was of astronomy, though it was often leveled at astronomy in particular in conjunction with the Aristotelian arguments against causal astronomical knowledge.[42] Osiander, for instance, argued not merely that astronomers should remain within their discipline but also that true knowledge in general was beyond the grasp of anyone. The astronomer should not bother aiming at truth, he declared in his preface to Copernicus's *De Revolutionibus*: "The philosopher will perhaps rather seek the semblance of the truth. But neither of them will understand or state anything certain, unless it has been divinely revealed to him."[43]

Kepler positioned himself against both these attacks on astronomical knowledge. In the first case, he argued that astronomy could speak to physical causes and that this was precisely his goal—he offered a physical astronomy, or celestial physics, that was based on a priori causes.[44] At the same time, he positioned himself against the skeptical approach to astronomical knowledge more generally, particularly in his *Apologia for Tycho against Ursus*.[45] In the *Apologia*, Kepler confidently rebuffed Ursus's skeptical critique of astronomical truth claims on the basis of both the many physical explanations he offered for Copernican theory and the very real progress he saw being made in the history of astronomy overall. In the *Mysterium Cosmographicum*, as in the *Apologia*, Kepler emphasized that his claims should be taken to represent the physical truth, and he argued against those "who rely on a model of accidental demonstration, which infers something true from false premises by the requirement of the syllogism."[46] In addition to denying the possibility that the true could follow from the false, Kepler argued in the *Mysterium Cosmographicum* for the physical truth of Copernicanism on its own grounds. Ptolemaic theory was incomplete, and Copernicanism far superior, for "the ancient hypotheses plainly do not offer an explanation for a number of central features," such as the number of spheres, the times of orbit, and the reason for their placement. "On all these matters," Kepler wrote, "since a most beautiful order is found in Copernicus, the cause must also necessarily be found in it. . . . Copernicus's beginning cannot be false, when it returns so consistently great a recounting of the phenomena, unknown to the ancients."[47]

Kepler appealed to the argument that Copernicus himself had made in his preface: that the beauty and symmetry of Copernicanism were themselves grounds for assent.[48] "These hypotheses of Copernicus not only do not sin against the nature of things but very much delight it," he wrote. "It loves simplicity, it loves unity. Nothing exists in it which is idle or superfluous; on the contrary, often one cause is designed to produce many effects."[49] Copernicus had united the cosmos into one simple system and had "freed nature from that onerous and useless furnishing of so many immense orbs."[50] Finally, Kepler insisted that his own arrival at Copernican theory a priori was a further truth of its veracity and was so unassailable that Copernican theory could be demonstrated from Kepler's discovery with no ambiguities for anyone, "even Aristotle himself, if he were alive."[51]

On top of all these claims, Kepler used his theory of quantity and its linkages to God to ground his realism still more securely. It is no accident that he began the *Mysterium Cosmographicum* with the description of quantity and its significance, for it was quantity that guaranteed, in Kepler's mind, that his theories represented the real, physical truth of the created universe. The skeptics might be right to argue that the senses deceive; as Kepler often noted, his poor eyes certainly deceived him much of the time, and the human eye in general was a flawed organ.[52] Reason was a reliable tool when used properly, but many people were unskilled in its use, and their conclusions were often faulty. Yet as Kepler explained in the *Harmony of the World* when he elaborated on his idea of quantity, because the human mind, the created world, and the divine nature were all fundamentally the same on a geometric level, the mind recognized geometrical forms without relying either on the senses or on reason. "To the human mind and to other minds," Kepler argued, "quantity is known by instinct. . . . Of itself it understands a straight line, of itself an equal distance from a given point, of itself it forms for itself from these an image of a circle."[53] Immediately thereafter, Kepler asserted that geometry was coeternal with God. Kepler suggested here that despite all the bumps along the way in fine-tuning a theory—bumps about which he was very clear in the *Mysterium Cosmographicum* and his later works—one's basic geometric *intuitions* about the structure of the world could be trusted on a fundamental level, because they were born out of the divine connection between matter and mind, based on quantity.

In many ways, this idea, like Kepler's emphasis on the geometric structure of the cosmos, hearkens back to Plato. In the *Meno*, Socrates famously asked an ignorant slave a series of questions about geometry, demonstrating first that the slave did not know any geometry, but later that he could be led to recollect it with the proper guidance. Socrates used this demonstration to prove that real knowledge (in the *Meno* he included both geometry and virtue) didn't come from the teacher—

from without—but rather existed already within every individual. Aristotle later would disagree and argue that people began with no knowledge and acquired it in the course of their lives via sensory investigation. In the *Harmony of the World*, Kepler explicitly allied himself with Plato and his later interpreter Proclus against Aristotle on precisely this point.[54] As to where this knowledge came from in the first place, Plato's answers differed. Though at first he indicated that this innate knowledge was recollected from previous lives lived, he later argued that the knowledge came from a time when souls were disembodied and lived among the gods, where they could contemplate the eternal Forms that were to be instantiated in the world in physical form. Since souls had previously contemplated these Forms in their true state, they could come to recollect them with the proper training or in flashes of insight.[55]

Kepler believed that though Proclus later conveyed this Platonic idea in pagan terms, he had ultimately recognized its Christian implications. The Platonic doctrine of recollection clearly signaled, to Kepler, that the Christian God had somehow imbued knowledge of fundamental truths—for Kepler, geometric objects—in the minds of his creatures. As Kepler wrote, Proclus had "even by his own Platonic philosophy perceived the son of God from far off by the natural light of his own mind,"[56] yet because of the time in which he lived, he refrained from explicitly mentioning Christ. Moreover, with Kepler's mention of natural light as the way in which human knowledge was rooted in the divine, Kepler referenced not only Plato and Proclus but also Augustine and his followers. Augustine had used Platonic ideas to articulate a theological doctrine of "natural light," and this doctrine was to prove of particular importance to Lutherans. As elaborated in particular by Melanchthon, the natural light doctrine held that certain truths, both moral and natural, were revealed directly by God, without need for any reasoning whatsoever. These truths were innate, placed in the soul by God and accessed intuitively and instinctively.[57] For Kepler, this natural light was specifically geometrical in nature, resting on the linkages between God, nature, and the human mind. As he explained in an early fragmentary work *On Quantities*, certain principles of mathematics were

understood by means of the common light of nature [and] do not need demonstration, and are at first associated with quantities, and then are applied to other things, insofar as they share something with quantities. Of those principles there are more in mathematics than in other contemplative sciences, because of the very nature of human understanding itself, which seems to be such, by the law of creation, that it cannot know anything perfectly except quantities or by means of quantities. This is why it is the case that the conclusions of mathematics are the most certain and unquestioned.[58]

The doctrine of natural light, for Kepler, pointed directly back to the notion of quantities.

Indeed, Kepler's own theory of the Platonic solids arose, as he described, in a moment of natural intuition: he had been drawing for his students the pattern of the conjunctions of Jupiter and Saturn, and the image on the board led him, in a flash of insight, to the idea of nested Platonic solids as an explanation for the planetary orbits. Kepler described this flash of insight as both "by chance" and "by divine inspiration,"[59] and it is clear that both terms were apt: Kepler's instinctive recognition of this "truth" was rooted, in his mind, in the fundamental role of geometric quantity. This may be why Kepler argued that he

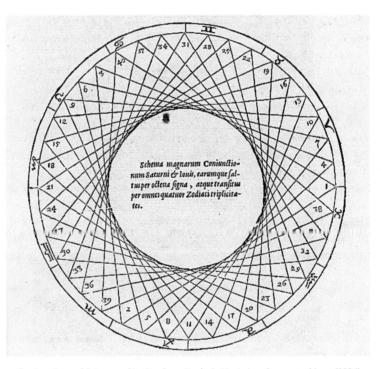

1.2   Conjunctions of Saturn and Jupiter from Kepler's *Mysterium Cosmographicum* (1596)

never abandoned the idea of the Platonic solids; he only modified this initial intuition in the harmonic theory that he ultimately arrived at in the *Harmony of the World*. As he later explained in the second edition of the *Mysterium Cosmographicum*, he held firmly to his initial theories because they were not really *his* at all. "It ought not to be regarded as a bare invention of my talent . . . since rather, as if it had been dictated to my pen, descended from a heavenly oracle . . . it was immediately recognized by all those with understanding as most genuine and true (as is the case with the manifest works of God)."[60]

Responses to the *Mysterium Cosmographicum*

When Kepler finally sent a draft copy of the *Mysterium Cosmographicum* to Michael Maestlin, his mathematics teacher, Maestlin responded to it with effusive praise. It was the strongly realist stance of the text that most delighted Maestlin, along with the book's strong grounding in geometry. Maestlin not only praised the book privately to Kepler but repeated his praise in an official endorsement of the manuscript written to Matthias Hafenreffer, professor of theology and prorector of the University of Tübingen. "I have read the work of the most learned master teacher Kepler," he wrote, "and I ask you to accept my judgment of it kindly. For whoever dared to think, much less to attempt, to teach a priori the number, order, magnitude, and motion of the heavenly spheres, either according to the standard [approach] or according to any other hypotheses, to explain them, and thus to produce them as though from the secret plan of God the creator?"[61] Like Kepler, Maestlin was one of the few, increasing in number, who believed that astronomy could offer true, a priori knowledge of the motions of the heavens. Maestlin supported Kepler's claims for the truth of his theories by indicating, like Kepler himself had done, his contempt for "the absurdity of the hypothesis . . . that the true can follow from the false. For although it can happen, it does so only by chance, and rarely at that."[62] Yet this aspect of the text was hardly likely to ingratiate it with the majority of the faculty at Tübingen, who adhered to the Wittenberg interpretation along with other Lutheran universities. Though Maestlin, who had long believed in the truth of Copernican theory himself, couldn't help but focus on this aspect of Kepler's work, he emphasized as well that even those who didn't accept the truth of Kepler's claims might still embrace the text's utility for the technical practice of astronomy.

"There is no doubt that those who collect observations are going to find these foundations, given a priori, of the greatest help in the reformation of the motion of the heavenly bodies."[63]

Along with his praise, Maestlin included some suggestions for revision of the text before its publication. In particular, he proposed that Kepler make it more clear and accessible to the popular reader. Kepler had written it for readers who were already familiar with geometry and with the technical details of Copernicus's astronomical theory, but Maestlin wanted it to reach an even wider audience. To accomplish this, he suggested that Kepler devote more space to an explanation of the properties of the regular geometric solids, the details of the Copernican theory, and the order and dimensions of the heavenly spheres that follow from it. "For the attentive reader," Maestlin wrote, "must not be stopped short by obscurity and enigmas but must be excited and encouraged by clarity and a plain and open discussion."[64]

Luckily for Kepler, Matthias Hafenreffer, with whom the decision for publication largely rested, was a sympathetic friend. Though a professor of theology, he was only ten years older than Kepler and had a reputation as an insightful scholar and an understanding mentor, one who was conciliatory and tolerant rather than harsh and doctrinaire.[65] He was also skilled in mathematics himself and was able to appreciate Kepler's broader interests and skills.[66] He and Kepler had formed a close bond in Kepler's years at Tübingen, and they remained in contact long after Kepler had left the halls of the university. Upon receipt of Maestlin's letter, Hafenreffer, representing the faculty of the university, accepted Maestlin's endorsement and praised Kepler's manuscript. The university Senate, he wrote to Kepler, "finds this discovery of yours to be as admirable as it is useful to all readers."[67] He did, however, agree with Maestlin's suggestions for improvement of the text and relayed to Kepler the Senate's order that he "set out as a preface both the hypotheses of Copernicus and the dimensions of regular bodies according to the recommendations of Maestlin . . . and remove all obscurity when possible."[68]

As to the means by which Kepler was to explain Copernican theory and remove all obscurity, Maestlin and Hafenreffer disagreed. Maestlin suggested that Kepler preface the manuscript with the *Narratio Prima*, the short synopsis of Copernican theory published by fellow Lutheran Georg Joachim Rheticus in 1540, three years before the publication of the *De Revolutionibus*. As he explained later, doing this would not only clearly set forth the details of Copernican theory but also situate the work more appropriately in the minds of his readers, providing them with a textual context within which to understand Kepler's work: "by this means," he asserted, "the interests of the mathematical republic of letters will be served better."[69] By contrast, Hafenreffer requested that Kepler explain both the geometry of the *Mysterium* and the ideas of Copernicanism in his own words in a short preface. The *Narratio* of Rheticus would be a poor substitute, wrote Hafenreffer, since it "is lengthier and contains certain things foreign to your *Prodromus*."[70] Though Hafenreffer did not say so explicitly at this juncture, his likely objection centered on the realist stance of the *Narratio*. Like Kepler and Maestlin, Rheticus, too, was one of the few Lutherans—and few astronomers more generally—to openly emphasize the physical reality of Copernicus's cosmos, and like Kepler he had argued in the *Narratio* for the beauty of the Copernican theory as a physical system, one characterized by unity and harmony.[71] So long as Kepler summarized Copernican theory in his own words, it could perhaps—if one read very generously and not very carefully—be seen as a continuation of the accepted tradition that Copernicanism was simply hypothetical. Yet the addition of the *Narratio* of Rheticus would alert readers from the outset to the fact that something more was intended—and possibly this was Maestlin's intention when he recommended its inclusion.

In Kepler's correspondence with both Maestlin and Hafenreffer about the ways that he might improve his manuscript, he expressed his desire to make one other central addition. Since he understood his Copernican arguments to be physically true, and since he also conceived of his work in the *Mysterium Cosmographicum* in fundamentally religious terms, he believed it necessary to reconcile any perceived conflicts between his cosmological claims and his religious belief. In particular, he felt that it was important to reconcile his belief in the truth of Copernicanism with his belief in the truth of scripture (particularly those scriptural passages that seemed, on the surface, to imply the motion of the sun and the earth's immobility), and he hoped to do so explicitly within the *Mysterium Cosmographicum* itself. As he wrote to Maestlin when he first revealed his plans for the book, "In the beginning I work with some theses about the scriptures: and I show how their authority may be preserved, and yet Copernicus, if he says things that are proper in other respects, can't be refuted by them."[72] That is, he intended to show how scriptural descriptions of the cosmos could be reconciled to the theories of Copernicus, while at the same time arguing that even if the two appeared to be at odds, natural phenomena proven to be true could not be disproved merely by reference to scriptural citations.

This may seem like a highly unorthodox position, in its preference for demonstrated scientific truth over accepted scriptural traditional. Yet in point of fact it is not that different from the later arguments of Cardinal Bellarmine, representative of the Catholic Church, in his discussion of Galileo's own attempts to reconcile scripture and nature. Bellarmine argued, as did proponents of the Wittenberg interpretation, that Copernicanism should be approached hypothetically. Yet he did concede that if any principle of science were to be demonstrated with certainty, scripture would have to be reinterpreted accordingly. He asserted, however, that Copernicanism had not yet reached that level of certainty.[73] Kepler believed that his a priori derivation of Copernican theory in the *Mysterium Cosmographicum*, combined with the basic geometrical intuitions that underpinned it (intuitions which themselves were of divine origin), had raised the Copernican claims to this kind of certainty, where even ostensible scriptural refutations would not be enough to contradict them.

At the start, however, neither Kepler nor Maestlin mentioned Kepler's plans for this section of the book in their correspondence with Hafenreffer and the university Senate, but merely outlined the dominant themes of the text, its Copernican premises, and why the discovery was noteworthy. Kepler evidently feared that the outspoken Copernicanism of the text would be enough to elicit objections from the Tübingen faculty; when no objections were raised at the outset, and Hafenreffer approved the text for publication, Kepler wrote to Maestlin, relieved that "no difficulty was put before my little book by the protectors of the Holy Scripture, as I had feared."[74]

Yet over the next few months, after the book had gone to press, it became clear that Kepler's relief was a bit premature. As Maestlin wrote to Kepler, his book

somewhat offends our theologians. . . . Doctor Hafenreffer time and again has assailed me (at least jokingly, although there seem to be serious [comments] mixed with jokes). He wants to dispute with me, all the while defending his Bible, etc. Likewise, not long ago at a public evening sermon, in the explanation of Genesis 1, [he said that] God did not hang the sun in the middle of the world like a lantern in the middle of a room. Indeed, I am accustomed to opposing these jokes with jokes, while they are jokes; if the matter had to be treated seriously, I would respond differently. The same Doctor Hafenreffer acknowledges [the *Mysterium Cosmographicum*] to be an excellent idea and skilled discovery but thinks that it is simply and totally opposed to sacred scripture and to truth itself. Indeed, with those who do not sufficiently grasp the principles of these matters (but who are otherwise most erudite and great men) it is preferable to act jokingly, as long as they accept jokes.[75]

Clearly, this description represents a shift from Hafenreffer's initial response to the book, in which he not only embraced Kepler's discovery wholeheartedly but also urged him to explain Copernican theory even more fully. In this description, by contrast, Hafenreffer appears threatened by the Copernican contents of the book; he speaks against Copernicanism in public sermons and argues that it flatly contradicts the words of scripture. What caused this change of mind-set in Hafenreffer? Why the move from endorsement and encouragement to the contention that the discovery was "simply and totally opposed to sacred scripture and to truth itself"?

Hafenreffer's move to distance himself from Kepler's book was first spurred by a discussion he appears to have had with Kepler when he visited Tübingen to set the stage for the printing of his manuscript. At that point, Kepler had apparently disclosed his plan to explicitly reconcile Copernican theory with scripture within the *Mysterium Cosmographicum* and had asked for Hafenreffer's advice. Hafenreffer's written reply, composed not in his official role as Tübingen theologian but instead in his role as Kepler's friend and mentor, was lengthy and heartfelt. Hafenreffer recounted Kepler's request for advice and noted that as "you desire my brotherly advice . . . I will clearly and candidly disclose to you, a most illustrious man and a most dear brother, what I think." He recalled Kepler's training at Tübingen, where he had surely been taught that the Copernican ideas were only hypothetical and ought not to be confused with the true sayings of scripture. He wrote: "From the moment that I first became aware of those hypotheses I have always felt it to be beyond doubt that one must distinguish openly between them and sacred scripture, which you observed when you were with us and even now you can rightly remember." For this reason, Hafenreffer continued, he recommended, in both his own name and that of his colleagues, that the proposed chapter on the reconciliation of scripture and Copernican theory "be omitted . . . except for some brief mention of the matter that is made immediately at the beginning."[76]

Hafenreffer argued that rather than include the proposed reconciliation (operating from the standpoint that as Copernicanism was physically true, it had to be reconciled to scripture), Kepler ought to adopt the standard Lutheran approach to Copernican theory, already made popular by astronomers at Wittenberg, which firmly emphasized the

Aristotelian disciplinary divisions between mathematics (and astronomy, one of the mixed mathematical sciences), physics (or natural philosophy), and theology. He wrote:

If there is some place for my council (as I firmly hope), you will act as a mathematician alone, unconcerned about whether [your objects of study] correspond to existing things or not. For I believe that a mathematician achieves his goal if he presents hypotheses to which the phenomena correspond as accurately as possible: and I think that you yourself would yield to someone who could offer better [hypotheses]. Nor does it follow that the truth of things conforms immediately to the hypotheses devised by each expert. I do not want to mention the irrefutable [proofs] that I could produce from the sacred scriptures. For I think that what is needed here is not debate but brotherly advice. And if you heed it (as I certainly am confident you will) and act as an abstract mathematician, I have no doubt that your thoughts will be judged very agreeable by many (as they certainly are by me).[77]

The traditional conception of mathematics as a discipline unconcerned with true causes would protect Kepler and prevent him from overstepping the boundaries of orthodoxy, Hafenreffer maintained. Since natural philosophers alone could discourse on the true nature of the heavens, while astronomers, practitioners of a mathematical discipline, could only describe the positions of heavenly bodies, Kepler's use of Copernican theory, from the standpoint of a mathematician, would not be viewed as dangerous or particularly controversial. Operating from the standpoint that the true could follow from the false, many of Kepler's conclusions could be utilized to improve astronomical calculations, while the basic Copernican premise could be discounted as a useful fiction.[78]

Though Hafenreffer urged Kepler to act as an abstract mathematician and ignore the relationship between Copernicanism and scripture, Kepler—far more deeply than Hafenreffer seemed to realize—was committed to the physical truth of the Copernican system. If Hafenreffer's appeal had ended here, then, it could have worked only had Kepler been worried enough about his overstepping of proper theological bounds. Hafenreffer, however, went one step further and argued for the omission of the proposed chapter on grounds that would have appealed much more strongly to Kepler. He urged Kepler to consider the cohesive bonds of community rather than simply the strict bounds of doctrine. His concern, he wrote, was not simply that Kepler himself would be contravening an accepted truth of the church but rather that since many Lutherans would perceive Kepler's actions that way, and

since some might even agree with him, Kepler's actions could only increase the strife and disagreement in an already contentious and fractured Lutheran Church. Hafenreffer implored Kepler "as a brother, that
you not attempt to propound or fight for that stated harmonization
publicly, for thus many good men would be offended, and not unjustly,
and the whole business could either be impeded or tainted with the
grave stain of dissension." With this plea, Hafenreffer appealed not to
Kepler's sense of orthodoxy but rather to Kepler's desire for harmony in
the church. Hafenreffer's own desire for churchly harmony was so important to him, he wrote, that potential harm to the church—by which
he meant the Lutheran Church in particular—would totally invalidate
any good that might have come from Kepler's discovery: "But if (and
may God, in his greatness, avert this) you want to publicly harmonize
those hypotheses of yours with the sacred scripture and to fight for
them, I fear that it is certain that this matter of yours may erupt in
dissension and battle. In which case I wish that I had never seen those
thoughts of yours, which in themselves and considered mathematically
are splendid and noble. As it is, already in the church of God there
has been more contention than is advisable for the weak." Hafenreffer
concluded his letter by assuring Kepler of his "most dear and sincere
brotherly love" and urging him to "act as a strict mathematician and
constantly foster tranquillity in the church, as I know was agreeable to
you in the past."[79]

Hafenreffer's own position makes a great deal of sense given his post
as a theologian at Tübingen. The Tübingen theologians had played a
central role in shaping the post-Reformation theological climate and
in fashioning the Lutheran movement into a strong and unified confession.[80] Jakob Andreae, Tübingen chancellor from 1561 to 1590, was
a pivotal figure in the attempt to create doctrinal accord between the
different branches of Lutheranism. Since Württemberg, the province
in which Tübingen was located, was straddled by Catholic Bavaria and
the Calvinist Palatinate, the need for Lutheran unity was pressing. Andreae had argued that the best way to achieve confessional unity was
to create a simple list of articles of faith with which the majority of
theologians could agree. He was instrumental in drafting the Formula
of Concord, which enumerated these articles and sharply distinguished
between Lutherans and their Catholic and Calvinist adversaries.[81] After
its completion in 1577, the Formula of Concord was adopted by two-
thirds of Lutheran Germany, including the province of Württemberg,
where all government and clerical officials, as well as all teachers and
university professors, were required to sign their assent.[82]

The Formula of Concord was therefore the symbol of the quest for Lutheran unity, a quest that emanated directly out of Württemberg and the University of Tübingen and that drew its strength from the perceived need for stability and agreement in the face of threats from Catholics and Calvinists, the enemies of the Lutheran Church. In light of this, it is clear that Hafenreffer's plea that Kepler not disturb Lutheran unity by raising the contentious issue of scripture and Copernicanism stemmed directly from a theological environment that prized Lutheran unity—and the particular doctrinal orthodoxy that undergirded it—above all else. Yet Hafenreffer focused not on the orthodox doctrines themselves but rather on the importance of unity, hoping that this alone would sway Kepler as no doctrinal arguments could.

As it turns out, Hafenreffer was right to assume that this plea for unity would appeal to Kepler, for whom harmony of both church and nature was preeminent. Interestingly, however, Kepler argued for a strong and explicit emphasis on confession and the physical truth of Copernicanism in his *Mysterium Cosmographicum* for precisely the reasons that Hafenreffer had argued for their exclusion: the goal of strengthening a divided church. When Kepler had earlier described his book to Michael Maestlin, he had asserted that he hoped it would serve to strengthen its readers' faith in God. As we saw earlier, this strengthening of the faith would be achieved by the book's emphasis on geometry as a means by which people might better understand both the world God had created and the nature of God himself. And in that same letter where Kepler had claimed that his theories in the *Mysterium Cosmographicum* would point the way toward a positive, rather than a negative, theology, Kepler asserted something more immediately relevant to the confessional disputes that dominated the church. Specifically, he believed that the geometrical arguments of the *Mysterium Cosmographicum* clarified an objection of the Calvinists against the Lutheran doctrine of "illocal presence," central to the Lutheran understanding of the Eucharist.[83] If he could use the *Mysterium Cosmographicum* to help settle one of the most pressing theological disputes of his day, would he not be *decreasing* the very churchly dissension that so worried Hafenreffer?

What Kepler had argued, in his earlier letter to Maestlin asserting the theological value of his *Mysterium Cosmographicum*, was that his book made the Calvinist objection to the doctrine of illocal presence understandable and, indeed, persuasive. Kepler had shown in his book, he noted, that everything in the physical world could be understood only through geometric quantities—that is, corporeally. He had shown that quantities were the foundation of reality, that they linked God, man, and nature, and that they could be grasped not only rationally but also intuitively, via the natural light. To attempt to describe the physical world in ways that turned the notion of quantity on its head was utter foolishness, according to Kepler. Therefore, Kepler maintained that the Lutheran claim of ubiquity—the claim that Christ's body was a physical body in any sense of the word and yet was not subject to local extension—was meaningless, much as the Calvinists insisted. To speak of a presence that was both physical and illocal was to speak incoherently and obscurely and "to assert nothing except mere negations." As Kepler explained: "From here comes that agitation of the Calvinists toward the phrase 'illocal presence.' For both the expression (presence) and the thing understood behind the expression were chosen from the creation of this world, which exists in space and time, and they indicate [the idea of] quantities, even to those who are most cautious. If anyone at all were to take the opportunity to carefully assess these and similar things selected from my little book, I think that the factions differing in religion would come one step closer together."[84] By forcefully demonstrating the centrality of geometry to the entire physical world in his *Mysterium Cosmographicum*, Kepler believed he had made clear to discerning readers that he agreed with the Calvinists against his fellow Lutherans when it came to the presence of Christ. He also believed that he might convince other Lutherans of this simple truth— after all, who could argue with the certainties of geometry?

After reading Kepler's letter to Maestlin, then, it is clear that Kepler saw the potential theological impact of his work to be far greater than simply the resolution of the seeming contradictions between scripture and Copernicanism. And while Matthias Hafenreffer had argued that Kepler should act as an abstract mathematician and *avoid* theological claims because of their potential to *divide* the church, Kepler seemed to believe that he should, on the contrary, *embrace* the linkages between mathematical and theological claims precisely because of their ability to *unite* the church. Kepler believed, that is, that in proving that the Calvinists were right about ubiquity and the Lutherans were wrong, he was clarifying and helping to eliminate a point of tension between the Lutherans and the Calvinists. In so doing, he believed he was helping to repair some of the breaches in the church and uniting the factions that were at war. Of course, some might dig in their heels, ignore Kepler's geometric claims, and insist still more strongly on the superiority of their own doctrinal positions; arguing theology, that is, might obviously backfire and make things worse. Yet Kepler, persuaded by what he saw as the clear and obvious proofs of his *Mysterium Cosmographicum*, felt that if people only opened their eyes and minds and considered his words, the potential for benefit was greater than the potential for harm. Unity might not be as elusive as it was starting to seem to so many others. Should not the natural light that had made these truths so obvious to him be clear to others as well, with the proper guidance? Even the slave boy in Plato's *Meno* had ultimately recognized the geometric truths that were unknown to him before Socrates's intervention.

Kepler and Hafenreffer, then, each made their cases about the role of mathematics and natural philosophy in theological disputes with the same goal in mind. Kepler believed that emphasizing the physical and religious aspects of his book would bolster the unity of the church by helping the hostile confessions better understand one another and perhaps resolve their differences. Hafenreffer argued that those emphases would further divide the church by creating more disagreement and that eliminating all mention of theology from the *Mysterium Cosmographicum* would far better preserve the unity of the church. It is clear, however, that Kepler and Hafenreffer had two separate notions of "the church" in mind. For Hafenreffer, the church whose unity he hoped to preserve was the Lutheran Church. As a theologian at Tübingen, a mainstay of Lutheran orthodoxy, Hafenreffer saw Lutheran unity as preeminent. The interconfessional doctrinal debates among Lutherans, Catholics, and Calvinists needed to occupy the full energy of the church. The church could not afford debates within its ranks if ultimately it hoped to maintain its integrity in the face of external opposition.

For Kepler, however, the "divided" church that needed repair was the whole of Christendom, not just the Lutheran confession. The struggles within the Lutheran Church were real, but they paled in comparison to the debates dividing the church understood in a more universal sense. While some believed that those interconfessional debates were irreconcilable, and indeed that such reconciliation was undesirable (for only one particular confession represented the *true* church), Kepler's ultimate goal was a united Christendom, and he believed that reconciliation of the confessions was indeed possible. Moreover, Kepler felt not only that reconciliation of the church was a priority but also that his cosmological work was an important tool in such an enterprise. He argued that the truths of astronomy, demonstrated a priori in his book, showed that some of the debates dividing the confessions, like the nature of the Eucharist, could be definitively decided. His mathematical work, that is, could be mobilized to settle some of the sharpest confessional disputes in ways that would brook no dissent, for they offered certain knowledge, demonstrated a priori, of the very fabric of the heavens. Of course, Kepler had demonstrated a truth that contradicted the beliefs of his own Lutheran Church—one that, as we will soon see, caused him a great deal of trouble. Yet though Kepler considered himself a devout Lutheran, the unified church he envisioned contained elements of all the confessions—for, as he later wrote, "Christ the Lord who spoke this word . . . neither was nor is Lutheran, nor Calvinist, nor Papist."[85]

Despite Kepler's belief in the importance of his discovery, Kepler evidently found at least some of Hafenreffer's arguments persuasive—or, perhaps, still felt too closely bound to his Tübingen roots to fully defy the advice of his mentor, particularly on so contentious an issue. In a letter to Maestlin, Kepler reiterated Hafenreffer's request "that I refrain from mention of the sacred scriptures in public. It would give offense to many good men. . . . Meanwhile, he bids me to proceed actively with these hypotheses, to the extent that they are helpful to astronomy."[86] Kepler confided to Maestlin that he would follow Hafenreffer's advice and omit the chapter on Copernicanism and scripture from the *Mysterium Cosmographicum* because of the possibility of creating unnecessary conflict. "What are we to do?" he asked. "The whole of astronomy is not worth one of Christ's little ones being offended."[87] Yet Kepler took pains to note that he did this out of respect for unity and not because he felt that there was anything objectionable about the material he wanted to include. Moreover, he argued to Maestlin that the same was true for Hafenreffer himself. Hafenreffer, he noted, had "eloquently praised the discovery," understanding full well its Copernican import. And though Hafenreffer pretended to find the idea of heliocentrism problematic, Kepler wrote that "I truly cannot believe that he is averse to this opinion. He pretends, in order that he may reconcile his colleagues, whom perhaps he offends with the promotion of my book. And this must be conceded to him. For peace with his colleagues is more important to him than with me."[88] Kepler could not accept the possibility that a close mentor and friend, one whom he so respected, could have read his book and not been persuaded by the Copernican arguments he had so clearly outlined.[89] He reiterated this point in a later letter, claiming that Hafenreffer was "not opposed to Copernicus but must necessarily stand among the other theologians for the authority (as they think) of scripture. Therefore, he does not explain to me his genuine opinion."[90] Yet he accepted Hafenreffer's seeming opposition to addressing the supposed clash between Copernicanism and scripture directly, for peace and unity, he believed, should be the ultimate guides for the behavior of all those who cared for the church, from the followers of established church doctrine all the way up to those who established it.

This did not mean, however, that Kepler was prepared to adopt the explicit position that Copernicanism was hypothetical or to deny what he saw as the unavoidable theological implications of his arguments—implications that were far broader than the question of Copernicanism alone. Rather, he explained to Maestlin, he would merely allude to his true beliefs obscurely, following the tradition of the Pythagoreans. The claims about illocal presence were never made directly in the text, after all—they were merely what Kepler saw as the inevitable consequences of his larger geometrical argument. He would leave things as they were, and those who read carefully would draw the appropriate conclusions, both about Copernicanism and scripture and about geometry and ubiquity. Those who were schooled and interested enough could ascertain his true intentions; others would be happy in their ignorance. He would recommend "the silence of Pythagoras and the riddles of Plato."[91] "Privately if someone approaches us, let us communicate to him candidly our opinion," he wrote. "Publicly let us be silent."[92] He conveyed to Hafenreffer his willingness to eliminate the offending chapter on scripture from the manuscript, and Hafenreffer appeared satisfied. Kepler's letter had filled him with joy, replied Hafenreffer, "for from it I understood that you esteem the tranquillity of the church more than any noble and beloved products of your own mental prowess. . . . [Therefore,] let the pious and holy tranquillity of the church live and flourish."[93]

Although Kepler omitted the chapter on Copernicanism and scripture, as he promised, he did include a brief discussion of the issue at the very opening of the *Mysterium Cosmographicum* (which Hafenreffer had indicated would be acceptable). He opened by sharply distinguishing between Copernican theory and scriptural truth, coming down very strongly on the side of scripture:

Although it is pious, immediately at the beginning of this discussion of nature, to consider whether anything is said contrary to sacred scripture, nevertheless I think it is untimely to raise this controversy here, before it is stirred up [by another]. I promise this in general: that I will say nothing which might harm the sacred scripture, and that if Copernicus is found guilty of this with me, I will consider him to have no use. And this was always my plan, from when I first became aware of the books of Copernicus's *Revolutions*.[94]

Kepler here left open the possibility that nature and scripture could, in theory, contradict one another and that in this case the faithful Christian would have to come down on the side of scripture—a possibility that he had denied earlier in private letters and would explicitly deny years later.[95]

Though Kepler hewed close to orthodoxy when it came to the relationship between scripture and natural science, he was more unequivocal in his challenge to the common astronomical approach to Copernican theory. Hafenreffer had urged Kepler to speak only as an abstract mathematician, without regard to the physical truth of Copernicanism. His discovery would then be widely hailed as useful to astronomy, Hafenreffer claimed, even by those who denied that Copernicanism was true. With this Kepler clearly did not comply, as would have been evident to anybody reading the text, which contained many arguments for the Copernican system as a physically true description of the cosmos. In the end, though Kepler eliminated the attempt to reconcile scripture and Copernican theory on the grounds that it was not worth damaging the integrity of the church, this amounted at best to an evasion of the issue, not exactly the approach Hafenreffer had recommended. If Copernicanism were true, and so incontrovertibly true that even Aristotle would agree, then it must, by extension, agree with scripture—or scripture itself was wrong, a position that was unthinkable, even to Kepler. Kepler had claimed that he would dismiss Copernicanism if it were shown to disagree with scripture, yet it was clear to the discerning reader that this was not a position he viewed as possible.

Maestlin also clearly stood on Kepler's side of the fence. Kepler had entrusted him with the coordination of the printing, and—against Hafenreffer's advice to Kepler, and without consulting Kepler first—Maestlin prefaced the manuscript with the full text of Rheticus's *Narratio*, which opposed the traditional Lutheran approach to Copernican theory and argued for its physical truth. Maestlin added a brief letter to the beginning of his addition, where he urged readers to approach the texts of both Rheticus and Kepler without preconceptions and to keep in mind that even in the works of the ancients "the question of the place and lasting rest of the earth is not settled." He further noted that Kepler had demonstrated Copernicus's claims "as drawn by genuine and appropriate arguments both from the nature of things and from geometry, due to which they cannot be contradicted."[96] Rheticus's *Narratio*, Maestlin's own preface, and Kepler's comments all made it clear that Kepler did not speak as an abstract mathematician. Even though Kepler removed the section on scripture and Copernicanism, Hafenreffer could not have been thrilled with the final product—and this even without recognizing (as it seems he did not) the deeper theological import Kepler saw in the text. Still, Maestlin noted that Hafenreffer's criticisms were all veiled and that he tended to address the issue with jokes. And Kepler remained, at least temporarily, in Hafenreffer's good graces, still a beloved friend and disciple despite their disagreements.

Conclusion: Kepler, Descartes, and Leibniz

The encounter between theology and metaphysics in the seventeenth century was often highly charged and deeply confessional. For some Lutherans, as I noted at the start of this chapter, this encounter was simply a rejection—metaphysics was not to be allowed into the theological arena. Other Lutherans and many Calvinists, as both Walter Sparn and Ian Hunter have shown, adopted metaphysical approaches that aligned with preexisting confessional commitments; by and large, members of each confession embraced a metaphysics that would allow them to uphold their own conceptions of the various Christian doctrines. Rather than simply relegating divine mysteries to the realm of faith, that is, many theologians on all sides of the confessional disputes endeavored to explain those mysteries metaphysically to bolster their own confessional stances.

How does Kepler fit into this story of confession and metaphysics? At the start of his career, as I've shown here, Kepler seems almost a reversal of the paradigm described above. Though a committed Lutheran, Kepler's independent metaphysical commitments—commitments that sprang from his belief in the centrality of geometry to the nature of God, man, and the physical world—led him to reject the Lutheran idea of ubiquity and adopt the Calvinist position on the Eucharist. His metaphysics, that is, dictated his theology, rather than the reverse. Moreover, this led him, not to abandon Lutheranism, but rather to try to convince his fellow Lutherans to modify their own stance on the doctrine of the Eucharist. As a young and naïve mathematician at the start of his career, Kepler seemed to genuinely believe that if he made his case strongly enough, and grounded it in the unshakable claims of geometry, his confessional allies would have no choice but to agree. He believed that this would be an important step in bringing the rival confessions together in a reconfigured but reunited church. As we will see in the next chapter, he was quickly proven wrong; his views were rejected and he was expelled from his own confession, while he saw all the rival confessions become more and more entrenched in their own positions.

Kepler's confessional commitments did not dictate his metaphysics, but this may be, in part, because he was not actually a theologian, much as he wished he were. His "secular theology" fits somewhat more neatly into the story told by Amos Funkenstein, whose *Theology and the Scientific Imagination* probed the ways that the new approaches to metaphysics and natural philosophy in the seventeenth century created unprecedented problems for traditional theological doctrines, chief among them the idea of God's omnipresence. As philosophers of the seventeenth century emphasized both the homogeneity of nature and the need for a clear and distinct language to describe it, the question of God's relationship to the natural world became a particularly pressing one; "now and only now," Funkenstein writes, "a clear-cut decision had to be made as to how God's ubiquity—to which the Lutherans added the ubiquity of Christ's body—had to be understood."[97] Descartes and Leibniz, in particular, both struggled with how the Eucharist—a focal point for the larger debate about God's presence—should be understood metaphysically. In closing, let us briefly consider each of their stances to more fully understand Kepler's position in this larger story.

Descartes famously argued that all that existed in the physical world was extended matter and motion. All other sensible qualities—color, taste, etc.—were only mental perceptions, not real properties but simply effects of matter in motion. This philosophy caused some very obvious problems for the traditional Catholic doctrine of transubstantiation, in which Christ's presence in the bread was explained via the Aristotelian idea of the accidents of the bread (appearance, taste, etc.) continuing to exist independently of the material substance of the bread, which was transformed into the body of Christ. Since Descartes had argued that accidents could not exist independently of substance, it was unclear precisely how his theory could be reconciled with the traditional doctrine of the Eucharist. Descartes came up with some creative—though incomplete and not always mutually compatible—solutions to this problem in his efforts to prove that his theory was consistent with traditional Catholic doctrine.[98]

The specifics of those solutions are less relevant to my analysis than Descartes's larger attitude toward the relationship between his metaphysics and his theology. Here, there are two possible interpretations. The predominant scholarly position on this question is that Descartes wanted to avoid questions of theology as much as possible and was drawn into theological debates against his better wishes and judgment, forced by the objections of others to his metaphysical approach. As Roger Ariew notes, this view has a long history—John Aubrey attributed it to Hobbes himself, who "was wont to say that had M[ieur] Des Cartes (for whom he had great respect) kept himselfe to geometrie, he had been the best geometer in the world; but he could not pardon him for his writing in defence of transubstantiation, which he knew was absolutely against his opinion (conscience) and donne meerely to putt a compliment [on] (flatter) the Jesuites."[99] By contrast, Ariew himself argues that Descartes willingly and of his own initiative addressed questions of theology in the hopes that the Jesuits might better teach and spread his system if they believed it to be theologically acceptable. Thus, he was merely doing what any other good seventeenth-century Catholic natural philosopher would have done, Ariew claims—discussing "the compatibility of their physical theories with such mysteries of the Catholic faith as the sacrament of the Eucharist."[100]

Whether Descartes discussed questions of theology willingly or unwillingly, two things are clear according to either interpretation: Descartes worried about the reception of his larger philosophical system because of its theological implications, and he was less interested in determining the truth of various doctrinal positions than in developing some possible ways to avoid the charge of heresy. He consequently wrote about transubstantiation to the Jesuit priest Denis Mesland only with the stipulation that if Mesland communicated his view to others, "it would be without attributing its authorship to me, and even that you would communicate it to no one, if you judged that it is not completely in conformity with what has been determined by the Church."[101] He likewise argued to Mesland that his interpretation of the Eucharist was just a possible way to adhere to orthodox doctrine and that he did not think it was the only conceivable solution: "there is certainly," he wrote, "no need to follow [that account] which I have communicated to you, in order to reconcile it with my principles. I only proposed it on that occasion because it seemed sufficiently convenient for avoiding the objection of Heretics, who claim that what the Church teaches is both impossible and contradictory."[102] For Descartes, as for

Kepler, metaphysical commitments came first; unlike Kepler, however, they did not lead him to change his doctrinal stance, even when they seemed to necessitate such a change. Instead, Descartes simply tried to find creative ways to uphold the old orthodoxy under his new system.

Leibniz, like Kepler, was eager to tackle theological problems directly and definitively, and not just as a defense against the possibility of heresy. He specifically faulted Descartes for relegating questions of doctrine to the sidelines, arguing that Descartes had "artfully evaded the mysteries of faith by claiming to pursue philosophy rather than theology, as though philosophy were incompatible with religion."[103] He cited Descartes's discussion of the Eucharist as an example of this lack of attention to questions of theology, for in his brief attempt to justify transubstantiation from the standpoint of his own theories, Descartes had "revived a doctrine rejected by the universal consensus of theologians."[104] This was no mere quibble, for Leibniz; as Christia Mercer argues, a primary reason that Leibniz rejected the mechanical philosophy of Descartes was that it was not compatible with the metaphysics of the Eucharist.[105] Metaphysics mattered to Leibniz, but so did theology, and he could not accept either a metaphysics that was incompatible with the Christian mysteries or a doctrinal position on the mysteries that did not withstand metaphysical scrutiny.

In his analysis of the Lutheran doctrine of *communicatio idiomatum* and the related question of ubiquity—the focus of Kepler's own disagreement with the Lutheran Church—Leibniz found the Lutheran stance irrational, much as Kepler did.[106] When one assumed an abstract transfer of properties from one nature of Christ to the other, Leibniz believed that one was forced to uphold logical impossibilities, like the prospect of divinity dying or of humanity being omnipresent. Instead, Leibniz argued that it made sense to conceive of the communication of properties only when referring specifically to the concrete person of Christ. Just as one could say that a poet treated disease if that poet happened also to be a doctor and treated disease in his medical guise, so one could say that a human was omnipresent if that human was understood to refer specifically to Christ and if his omnipresence was understood to refer specifically to his divine guise: "that he who is a man, though not *qua* a man, but *qua* God, is omnipresent."[107] Like Kepler, then, Leibniz seemed to tend clearly toward the Calvinist understanding of *communicatio idiomatum* rather than the Lutheran one.

Instead of doing as the young Kepler tried to do, however—seeking to use his metaphysics to convince others of the incorrectness of their own theological views and the superiority of a rival theological position—Leibniz chose to evaluate the question of the Eucharist anew. While he disagreed with the doctrine of ubiquity itself, he tried not to dismiss the Lutheran approach to the Eucharist more generally but rather to develop a new, broader kind of metaphysics that would embrace it—along with a variety of other confessional stances on the Eucharist. In particular, he developed a new metaphysical theory of substance that was broad enough, he argued, to support all views. He even suggested a variety of different possible definitions of substance—too complicated to analyze directly here—that would be compatible with his larger metaphysics and that might be adopted by different audiences to suit their unique doctrinal stances.[108] As he wrote to Antoine Arnauld, "I hope to show what no one has previously thought, [namely] that in the ultimate analysis Transubstantiation and real multipresence do not differ. . . . And consequently Transubstantiation, as most cautiously expressed in the phrase by the Council of Trent and [which] has been illustrated by me based on Saint Thomas, does not contradict the Augsburg Confession; indeed, it follows from it."[109] Leibniz hoped that his new and more expansive theory of substance might allow for real churchly reunification, for it would show the various confessions where their doctrines already agreed and where it was acceptable for them to disagree.

Like Descartes, then, the young Kepler's metaphysical commitments preceded his understanding of church doctrine. Yet unlike Descartes, who sought to retroactively reconcile those commitments with traditional doctrine, Kepler simply decided that Lutheran doctrine was wrong. He did not do so to further fan the flames of confessional conflict but because he believed that in making his metaphysical stance clear, and in highlighting its inevitable doctrinal implications, he could reunite the warring confessions around the one clear truth. As we will soon see, ultimately he came to realize the futility of this stance—no confession, it was clear, would willingly surrender its own position and adopt the stance of another on so central a question as the Eucharist. Leibniz's more expansive approach—to show the ways that the confessions already agreed and to leave room for the places where they disagreed—would eventually align more with the mature Kepler's hopes for a reunified Christendom, as we will see in later chapters. Kepler ultimately sought to develop just such a paradigm for broad agreement that allowed for individual disagreement under its expansive umbrella. Of course, even Leibniz's later attempts, offered to a world far less bent on mutual destruction than Kepler's own, fell flat—as Christia

Mercer notes, they didn't even convince Arnauld, "who was not impressed enough even to respond."[110] This is perhaps why, when it came to his own personal quest for readmission to the Lutheran Church after his excommunication, Kepler came, like Descartes, to try and avoid metaphysical arguments entirely, when possible, and to argue on other grounds: the grounds of conscience.

TWO

"Matters of Conscience": Kepler and the Lutheran Church

Kepler was a sincerely devout but stubbornly independent child when it came to questions of theology. Though he longed to be a priest himself, he found himself disagreeing with many of the Lutheran priests he encountered on matters of both doctrine and demeanor. Once, when he was twelve years old, he listened to a local deacon lambast the Calvinists and could not comprehend why the leaders of the church seemed so set on dividing Christian from Christian. At thirteen, he entered the Lutheran seminary in Adelberg, and there he listened to his teachers forcefully denouncing the Calvinist doctrine of the Eucharist. After intense reflection, the young Kepler realized that despite the arguments of his teachers, the position of the Calvinists seemed to him more sensible—and more reflective of the biblical text—than the one he had been schooled to accept.

Still, Kepler averred, he was a good Lutheran. "I imbibed the Lutheran creed from my parents' teachings," he would later insist, "from frequent reasoning about its foundations, and from daily exercises, and I embrace it."[1] His doubts and disagreements were merely sincere attempts to understand the confession with which he identified and to decide what his conscience supported and what it did not. Hadn't Luther, after all, insisted that "we are all equally priests"? Hadn't he stood his ground at the

Diet of Worms, announcing, "here I stand; I can do no other," and affirming the importance of holding fast to sincere belief? Ought Kepler then not rightfully use his intellect to understand the scriptures and their implications as best he could, and shouldn't his conclusions be respected if they were sincerely held?

His theological mentors believed the answer to be a resounding no. Not all interpretations were correct interpretations, and Kepler had crossed the line in his theological speculations. Heresy could not be tolerated, for it wounded the body of the church, a body that—with disputes raging between Lutherans, Catholics, and Calvinists—was already grievously ill. In 1619 Matthias Hafenreffer, Kepler's old friend and theological mentor, sent Kepler a letter both stern and desolate. Kepler had failed to subordinate his mental prowess to the sacred mysteries of Christian worship, wrote Hafenreffer, and as a consequence, his attempts to reason about theology had only thrown him "into a pitiable sort of confusion." For this reason, Hafenreffer asserted that neither he nor his colleagues could approve Kepler's "absurd and blasphemous fantasies."[2] Along with this message, Hafenreffer conveyed to Kepler the news that his exclusion from the Lutheran communion, first decreed seven years earlier, could not and would not be revoked. The young boy who had hoped to become a Lutheran priest was now cast out of the very community that had given his life its earliest definition.

Though early in his career Kepler had argued for the truth of his doctrinal position on the grounds of reason and metaphysics, as we saw in chapter 1, in this later dispute with Hafenreffer Kepler chose to emphasize not truth but freedom, and not reason but conscience. Because his beliefs were sincerely and conscientiously held, Kepler insisted, they should be tolerated. For this reason, he similarly wrote that he would tolerate the beliefs of his fellow Christians, right or wrong. Indeed, he linked his refusal to sign the Formula of Concord not only to his specific disagreement with one of its doctrines but also to its larger antagonistic stance toward the beliefs of Christians from other confessions. "It is better," he argued, "that I sin in justifying, in speaking well, in interpreting for the better, although I am not a doctor of the church, than in accusing, excluding, and destroying."[3]

When Kepler invoked the idea of freedom of conscience, however, he did not simply mean what we moderns often understand the phrase to imply: the right to believe (or disbelieve) anything and everything one chooses. Instead, he relied upon a long history of debates about conscience, debates that intensified with the Reformation. According to the vast majority of church fathers who invoked the term, like

Thomas Aquinas, conscience was somehow connected with a divine—and therefore objective—moral truth, and conscience was far from free; when properly understood, it corresponded directly to the voice of God and demanded absolute obedience.[4] And when misused, ignored, or twisted, as it often was, the false conclusions of conscience needed to be harshly condemned. Aquinas thus claimed that "heresy is a sin which merits not only excommunication but also death."[5] Conscience, in other words, did not lead to toleration and individual freedom for Aquinas but to their opposite.

When Luther stood up at Worms and cited his conscience as the grounds for his dissent, he was following in this scholastic tradition. He certainly argued, as had many before him, that faith was not something that could be compelled, and that heresy, too, could not simply be transformed into orthodoxy by force; "heresy is a spiritual thing," he wrote, "which cannot be cut with steel nor burned with fire nor drowned with water."[6] Yet like Aquinas before him, Luther believed that there was *one* truth. And like Aquinas, he did not believe that it was immediately obvious to everybody what that truth was. In fact—and here he differed from the opinion of Aquinas—Luther believed that conscience was not linked to reason at all, nor was it something an individual could choose to wield for good or for evil; it was, rather, a reflection of the total moral state of an individual, which needed to be liberated by God via grace and freed from the devil's grasp.[7]

When Luther spoke of Christian liberty and conscience, the liberty that he proclaimed was not the freedom to believe anything one chose but rather the freedom to absolutely obey the words of scripture against the opinions of the authorities.[8] As to whether those words were open to multiple interpretations, Luther's answer was clearly negative. When, for example, the capitulars of Altenburg argued for their right to celebrate Mass on the grounds of conscience, Luther spoke of their *"erdichtetes Gewissen,"* their fictitious conscience.[9] Only the Bible grounded true conscience, and there was only one true interpretation of the Bible—the interpretation of Luther and the future leaders of his movement. By extension, though Luther began his movement by speaking against the eradication of heresy by force and suggesting a policy of toleration, with intensifying opposition by Catholics Luther came to argue that toleration only led to the spread of false doctrine and that heretics needed to be silenced. In his assent to Melanchthon's Memorandum of 1531, Luther noted that "though it seems cruel to punish [heretics] with the sword, it is more cruel that they damn the ministry of the Word, have no certain teaching, and suppress the true, and thus upset society."[10]

A belief in the importance of conscience, then, did not necessarily imply the right to freedom of conscience and the subsequent religious toleration that might result from such freedom. Indeed, if one believed in the idea of an absolute religious truth and if one truly cared for one's fellow Christian, the idea of religious liberty might seem a distinctly pernicious one. As Huguenot Theodore Beza would later proclaim, "it means that everyone should be left to go to hell his own way."[11] Of course, there were already those who argued *against* the idea of absolute moral or divine truth—or at least against the possibility that such absolute truth could ever be discerned by an individual conscience. Montaigne's skepticism led him to contend that that which men labeled their "conscience" was really just custom or public opinion, so ingrained that men believed it to be truth.[12] Hobbes similarly claimed that "men, vehemently in love with their own new opinions (though never so absurd), and obstinately bent to maintain them, gave those their opinions also that reverenced name of conscience."[13] And Locke would later argue that what men took to be their divinely implanted conscience was often nothing more than the arbitrary influences of their childhood, grown over time into firmly held convictions.[14]

Yet such skepticism about the possibility of ascertaining divine or absolute truth did not necessarily entail freedom of conscience and toleration either. Montaigne noted that if one's ultimate concern were not for truth but for practicality—that is, if one hoped to ensure the good of the community—then the merit of toleration was an open question. In his essay on freedom of conscience, he argued for both sides: "it may be said, on the one hand, that to give factions a loose rein to entertain their own opinions is to scatter and sow division. . . . But on the other hand, one could also say that to give factions a loose rein to entertain their own opinions is to soften and relax them through facility and ease, and to dull the point, which is sharpened by rarity, novelty, and difficulty."[15] Hobbes and Locke both took separate sides in this debate. Locke believed that toleration was essential in the state, for "no peace and security, no, not so much as common friendship, can ever be established or preserved amongst men so long as . . . religion is to be propagated by force of arms."[16] By contrast, Hobbes felt that unity and conformity, even of belief, were essential for the maintenance of peace, since "men that are once possessed of an opinion that their obedience to the sovereign power will be more hurtful to them than their disobedience will disobey the laws, and thereby overthrow the Common-wealth, and introduce confusion and civil war."[17]

Kepler's own stance on both conscience and toleration was a compli-cated one. On the one hand, Kepler followed the scholastic tradition in his belief that his conscience—at its best—corresponded to an absolute and objective truth. He followed Luther more specifically in arguing that that truth was directly rooted in the words of scripture rather than the statements of the churchly authorities. "I persist in the pursuit of truth and the detestation of error," he wrote in a 1617 letter to Hafen-reffer.[18] As he argued to Maestlin one year earlier, he was justified in his insistence on the truth of his position—and in following the dictates of his conscience—because of the clear backing of scripture: "through the grace of God, I submit to the simple and plain sense of scripture."[19] Kepler's belief that God had clearly written both his books—the Book of Nature and the Book of Scripture—in terms that were accessible to human reason likewise helped to assure him that it was possible for men to know the divine mind and follow its true guidelines.

Yet instead of the intolerance for other perspectives that such be-liefs might inspire, Kepler argued on behalf of conciliation, tolerance, and the acceptance of difference. In his dispute with Hafenreffer over his exclusion from Communion, he grounded his own stance on these foundations, arguing that he wrote "not as a judge over you in my dif-ference of opinion . . . but rather as a custodian of a purely personal conscience."[20] He argued for tolerance because he still hoped to help unify the church, much as he had as a young mathematician in Graz. In later years, however, Kepler came to recognize that unity could not be achieved by persuading or forcing the various confessions to con-form to one image of the church in all its particulars. Unlike Hobbes, who would assert that conformity yielded peaceful unity, Kepler took the stance later articulated by Locke in the political sphere, when he linked a peaceful and united community to a policy of toleration. For Kepler, this still did not imply toleration of anything and every-thing—as, indeed, it did not for Locke either—but rather toleration of a range of perspectives and opinions within the three major confes-sions, based on a sincere and careful attempt to read and understand the scriptures and ancient authorities.

Of course, Kepler was not alone in making such claims for religious toleration or conciliation in the early seventeenth century. Kepler's ar-guments for unity undergirded by tolerance have a great deal in com-mon with the adherents of religious irenicism, a tradition that flour-ished in the sixteenth and seventeenth centuries following the model of Erasmus. Irenicists hoped to create a middle ground that would allow Catholics, Calvinists, and Lutherans to live together peacefully and that would replace the divided confessions with one unified but not necessarily absolutely uniform church.[21] There were various suggestions as to how this might be accomplished, many of them put forth by Erasmus himself. One, which did aim for a more uniform unity, was to allow a new general council to decide upon the particulars of doctrine that might unite the confessions—this, in fact, was a view that Kepler himself considered over the years. Yet he leaned more toward Erasmus's other suggestions, some of which implied that there need not be absolute agreement on doctrine, not immediately and perhaps not ever. Certain things might be left to the individual conscience to decide; likewise, each side might choose to accommodate the other and compromise on specific points even if they ultimately disagreed.[22] Kepler came to argue for this kind of unity without uniformity, following a model of particular importance to him: the model of harmony. The musical metaphor of harmony, of *concordia discors*, made clear that differences and discordant elements were central to the achievement of some kind of larger concord. There could, in other words, be room for disagreement within the Lutheran confession and, especially, within the larger church.

At the time of his ultimate excommunication from the Lutheran Church, Kepler had come to believe not only that faith need not imply absolute compliance with all the doctrinal positions of one confession but also that it wasn't necessary for those guiding the church to pursue a single, unitary truth. The larger body of the church, he came to argue, would not be harmed by small splinters in its midst. Still further, Kepler had come to believe that geometry was too forceful a tool to fully heal the body of the church when it came to individual points of doctrine; tolerance, rather than intellectual rigidity, would allow the church to thrive best. Though Kepler's own theological views were likely still guided by his personal metaphysics, in making the argument for religious harmony Kepler marshaled his metaphysics more loosely and broadly. The metaphysics of harmony became, for Kepler, a larger model of concord rather than the metric by which specific points of doctrine might be accepted or rejected. Of course, as Erasmus had realized years earlier, neither side appreciates a peacemaker. "Far be it from me," Erasmus had demurred, "to play the arbiter in this inextricable tragedy, for I will get no thanks from either party."[23] Unfortunately, this is a lesson Kepler learned all too well.

Kepler's Exclusion from the Lutheran Communion

Kepler's ultimate conflict with the Lutheran Church—the conflict that led to his excommunication—did not rest on his astronomical claims, controversial though they were. Though Kepler's realist Copernican claims did alarm Hafenreffer, as we saw in the previous chapter, Kepler ultimately acceded to Hafenreffer's request to avoid the issue of Copernicanism and scripture directly in the *Mysterium Cosmographicum*. And even after the publication of his *New Astronomy* in 1609, which did explicitly address some of the scriptural interpretations he had agreed to leave out of the *Mysterium Cosmographicum*, Kepler was not subjected to any censure or reprimand by the Tübingen theologians.[24] Instead, Kepler's confrontation with the Lutheran Church stemmed specifically from Kepler's unorthodox views about the theological doctrine of ubiquity. As we saw in the previous chapter, Kepler had made clear privately to Maestlin as early as 1597 that he believed that Christ's body was not omnipresent in a physical sense and that even the presence of Christ in the Mass should not be understood physically—a position that had far more in common with the Calvinist Church than with Kepler's own.

It is important to recognize that though Kepler's position certainly contradicted the official Lutheran stance on the presence of Christ in the Eucharist, his argument about the doctrine of ubiquity in particular was actually less radical than it may seem at the outset, for it was only the Tübingen theologians who so strongly insisted on the physical omnipresence of Christ's body. Indeed, the proper Lutheran approach to the doctrine of ubiquity was hotly debated within the ranks of the confession itself, and not even the drafting of the Formula of Concord conclusively settled this debate.[25] Even at the start of the movement, Melanchthon had disagreed with Luther and adhered to a more Calvinist understanding of the presence of Christ's body, and his views had at first dominated the movement; indeed, in the Consensus Dresdensis of 1571 the faculty at Wittenberg and Leipzig had openly rejected the strict form of the ubiquity doctrine. Martin Chemnitz, a follower of Melanchthon and one of the foremost Lutheran theologians of his day, advocated a milder form of "relative ubiquity," which stressed the omnipresence of God's will rather than his body. It was primarily in Württemberg that Luther's strong notion of absolute physical ubiquity held sway, mostly through the influence of Johannes Brenz, another leading Lutheran theologian—born, by chance, in the same town as Kepler— who feared that "the devil intended through Calvinism to smuggle heathenism, Talmudism, and Mohammedanism into the church."[26] The Württemberg Confession of 1559, drawn up by Brenz, was the first document to officially endorse this understanding of ubiquity.

With the Formula of Concord, Lutherans attempted to solidify some kind of confessional unity, as noted earlier. The notion of ubiquity proved a particularly tricky one to resolve; in the end, the document tried to effect a compromise between the absolute and relative approaches but leaned toward the absolute one, given the prominent role the Württemberg theologians played in drafting it. Martin Chemnitz, also one of the document's drafters, reluctantly signed the Formula for the sake of peace, despite his reservations about the doctrine of ubiquity. Even so, the controversy continued; it was carried on in particular between the theologians of Giessen, who adhered to Chemnitz's relative view, and the theologians of Tübingen, who followed the absolute view of Brenz. Matthias Hafenreffer himself was one of the key Tübingen figures in this controversy.[27] There were, in fact, many Lutherans who shared Kepler's reservations about ubiquity, just not in Tübingen. Kepler could not have picked a worse place to voice his concerns and still hope for acceptance.

Kepler found himself opposed to the doctrine of ubiquity for several reasons. First and foremost, as we saw in the last chapter, he believed the doctrine to be rationally unintelligible, particularly given the geometrical metaphysics that underpinned his larger worldview. In 1597, as the *Mysterium Cosmographicum* was nearing completion, he had written to Michael Maestlin and linked his disagreements with the doctrine of illocal presence closely to the arguments of that book. Since he had demonstrated that geometry was the central archetype underpinning the cosmos and all things in it, and since geometry implied quantity—physical extension—there was no way to make sense of the idea of a physical presence that was real and not local. Kepler made similar arguments against the doctrine of ubiquity, citing reasons of geometry and intelligibility, in private correspondence to friends at the same time and in later years.

One friend with whom he appears to have corresponded on this topic in great depth was Colmann Zehentmair; unfortunately, none of Kepler's letters to Zehentmair have survived. However, Zehentmair's replies to Kepler are extant, and in them he summarized what Kepler had written to him before responding. In October 1599 Zehentmair wrote to Kepler that he had received Kepler's letter about the Eucharist, in which Kepler had apparently argued that in the Mass one could find "the fruit and merit of the Lord but not the bodily substance." As Kepler had apparently made this argument on the basis of reason, Zehentmair disagreed by questioning Kepler's entire approach. "Thus far," he wrote,

I suspend my assent, not because I am held back by the prejudgment and authority of our great theologians, but rather because I believe that it is far preferable and safer that simple people accept mysteries of this sort rather than scrutinizing those same mysteries with the sharpness of our reason. If something absurd emerges (for all the Christian religion is absurd according to reason), we should either deflect it or bend it to our understanding. For I plainly think that the more someone occupies himself with examining this or that in the sacred mysteries according to reason and tries to accommodate it agreeably to everybody's intellect, the farther he recedes . . . from the true sense.

In fact, Zehentmair contended, Luther and the Lutheran theologians took care to label the Mass a mystery specifically to combat the attacks of those, like the Calvinists, who attempted to argue with them on rational grounds. Unlike the Calvinists, who tried to understand the Eucharist rationally, "we observe these things with eyes of faith and commit to the omnipotent Lord that which we cannot understand through reason."[28] Though Zehentmair clearly disagreed, Kepler, it seems, had tried to convince him in his letter that reason was a legitimate tool to use when considering contentious issues of confessional doctrine.

Likewise, Kepler based his disagreements with the doctrine of ubiquity on similarly rational grounds in a poem about the Eucharist that he wrote sometime around 1610 or 1611 and sent to his friend Christoph Donauer, a pastor in Regensburg. Donauer apparently enjoyed expressing his theological views in verse and had sent Kepler some poems criticizing the views of Thomas Wegelin, a Lutheran theologian then teaching at Tübingen. Donauer asked Kepler for his opinion, and Kepler responded with his own poem about the Eucharist, which also disagreed with Wegelin and with the Lutheran position more generally. Kepler opened his poem directly by citing Calvin and his disagreement with the doctrine of ubiquity, noting that "Calvin denies that the flesh of Christ is everywhere."[29] He then expressed his personal agreement with Calvin, on the grounds that Christ's human nature must be bound by spatial location in virtue of the fact that it is human—it simply could not logically exist everywhere. Human laws and divine laws could not be substituted for one another—the finite, in other words, could not be made infinite. Kepler argued in the poem that while Christ's body could not be physically present everywhere, Christ as God was still present everywhere via his will: "the place of the flesh is in the heavens, but the will is everywhere."[30] God's actions were omnipresent, according to Kepler, but his body was not.

Yet over the years, Kepler began to offer other reasons for his disagreement with the doctrine of ubiquity and to de-emphasize or omit mention of the rational and geometrical objection that had motivated him at the start. One of the primary grounds on which he anchored his disagreement was holy scripture itself, which included no mention of the doctrine of ubiquity as the Lutherans understood it, and which Kepler believed seemed to suggest the opposite. Scripture itself and the ancient church fathers, he emphasized in correspondence time and again, were quite clearly opposed to the notion that held current Lutherans in thrall; Kepler insisted that he was merely embracing the beliefs of the traditional church and that he necessarily opposed any doctrine that was so obviously a modern innovation. "I feel the force of antiquity deep within my heart," he wrote to Matthias Hafenreffer in 1610, attempting to justify his reservations about ubiquity. "I am wrongly suspected by you of Calvinist belief when it comes to the person of Christ. There is nothing in Calvinism that is agreeable to me that is new—[I believe] only that of which antiquity convinces me."[31] His disagreements with his Lutheran brethren, Kepler claimed, revolved only around "that which Luther first invented and Jakob Andreae and others elaborated concerning the ubiquity of God and his union with the flesh, [leading toward] the omnipresence of the flesh."[32] Likewise, in 1616 Kepler asserted to Michael Maestlin not that geometry proved him right, as he had argued in 1597, but rather that scripture did. "I submit to the simple and plain sense of scripture," he declared.[33] Though he agreed with the Lutheran Augsburg Confession, he could not embrace a doctrine that so clearly seemed to go against the traditional understanding of the Mass. "Those who accuse me even minimally of novelty," he insisted, "do me injury."[34]

Though Kepler did express his disagreements with the doctrine of ubiquity from time to time over the years, mostly in private correspondence, until 1612 he encountered no strong opposition from the guardians of orthodoxy in Tübingen. After the expulsion of Lutherans from Catholic Graz in 1600, Kepler had made his way to Prague, a cosmopolitan city of many beliefs and confessions. This diversity, along with Kepler's own position as imperial mathematician to Rudolf II, likely shielded him from the censure of the theologians in the more conservative Württemberg. Yet as the political situation in Prague became increasingly fraught, Kepler came to feel that he would be better off in

Württemberg, his *"vaterland,"* as he often referred to it. He still saw his old university, Tübingen, as a place of refuge and tranquillity, despite any disagreements he may have had with some of the faculty there. He wrote to Duke Johann Frederick of Württemberg in the spring of 1609 requesting a position for himself at Tübingen, as a professor either of mathematics or of some other discipline that the duke thought worthy of him. The duke referred the request to his council, who noted that Michael Maestlin, the current professor of mathematics at Tübingen, was getting older and that Kepler, a famous mathematician in his own right, would potentially make a good replacement. However, they directed him to turn down the request at present; Tübingen did not yet need to replace Maestlin. As a token of goodwill, however, the duke sent Kepler a goblet worth twelve to fifteen florins.

Encouraged by what he took as a sign of the duke's favor, Kepler quickly replied to the duke, thanking him and reiterating his interest in the position. Kepler then devoted the rest of his letter to a summary of his position on the issue of the Eucharist and on confessional disputes more generally. Recognizing, perhaps, that his theological objections to the Formula of Concord might prevent him from obtaining a job at Tübingen, Kepler hastened to explain those objections to the duke and convince him that they presented no impediment either to his own status as a true Lutheran or to his potential employment at the university. In this letter, Kepler argued that his opposition to the Lutheran stance on ubiquity—and his consequent refusal to sign the Formula of Concord—rested primarily on the attitude of the Lutheran Church toward doctrinal conformity and its consequent strident opposition to all other opinions and confessions. Christian peace and unity were more important than anything else, Kepler insisted, and the intolerance of his Lutheran brethren made those goals less and less possible to achieve.

In his opinion, Kepler wrote, when it came to contentious issues like the Lord's Supper or the question of predestination, each side ought to "reach out a hand" to the other and try to find common ground. The Formula of Concord, however, took the opposite approach, by pointing an accusatory finger at the Calvinists and only increasing the divide in the church. For this reason, he would sign the Formula "only conditionally," and he would continue to believe that even someone who adhered to the Calvinist doctrines should be deemed "our brother in Christ."[35] He hoped that this conditional subscription could still allow him to serve at the university (which did require a subscription from all its faculty). He added that his appointment could even lead to increased peace in the church, by demonstrating to others with specific objections that there was still a place for them within the Lutheran confession. [36]

The argument against the Lutheran doctrine of ubiquity on the basis of peace and brotherly love was one that, like the argument on the basis of scripture, Kepler repeated numerous times in his personal correspondence. In 1616, at the same time that he insisted to Maestlin that the Lutheran position was not to be found in the scriptures, he argued that what motivated him most on this issue was that "I desire peace among the reformers, nor do I approve all the deforming and exaggerations of the doctrines of the Calvinists."[37] In fact, he wrote, when it came down to it, there was much on which all the confessions agreed with respect to the nature of Christ, and only little on which they disagreed: "I cannot see, at root, that there is that much of a distinction concerning the person of Christ" among the disputing parties, he wrote.[38] In fact, it was only the Lutherans who had chosen to distance themselves from tradition, scripture, and reason so dramatically on this issue. As Kepler would later explain, "Gradually I had learned that the Jesuits and Calvinists agree about the issue of the person of Christ, and they both rely on the church fathers . . . [while] the disagreement of ours is new."[39] If the Lutherans could but recognize the accord between the other confessions on the question of *communicatio idiomatum*, they might better align their own position with the traditional approach. So much that was broken could be fixed, Kepler believed, if people simply followed Paul's exhortation in Romans 14:19 to "follow after the things which make for peace."

The Duke of Württemberg did not reply to Kepler's second letter pleading for a position and arguing for tolerance, and two years later, with no offer of a position at Tübingen yet forthcoming, Kepler put pen to paper once again and asked the duke if a post might now be found for him in the province to which he so longed to return. This time, the duke's advisers seemed inclined to grant the request; they once again reported to the duke that Kepler was a reputable mathematician who himself had studied under Maestlin, and who could clearly fill the job of mathematics professor with great skill. The final word, however, went to the theological consistory in Stuttgart, to whom Kepler's request was also forwarded. The consistory responded a few days later with a resounding negative—Kepler should certainly not be offered a post at Tübingen. Reviewing Kepler's correspondence, and in particular the letter he had sent to the duke two years earlier explaining his opposition to the Formula of Concord, the consistory deemed

him a "sly Calvinist" who would taint the students with his unorthodox theological views. Likewise, he had proven himself to be an "opin-
ionist in philosophy," one who would cause a great deal of unrest at
the university.[40] Since the statutes of the university clearly required un-
conditional acceptance of the Formula, the duke should deny Kepler's
request on these grounds alone and avoid bringing a troublemaker into
the halls of Tübingen.

This is the first clear instance where Kepler's unorthodox theological beliefs brought him into conflict with his own church, though it
would not be the last. While his refusal to sign the Formula of Concord prevented his employment at Tübingen, at first it did not affect
his personal relationship with his local church in Prague. This changed
when Kepler moved to Linz in 1612, shortly after the death of Rudolf II
(though he retained his position as imperial mathematician to Rudolf's
successor, Matthias). In Linz, Kepler met Daniel Hitzler, who had assumed the position of the head of the Lutheran Church there in 1611,
shortly before Kepler's arrival. Kepler may have heard of Hitzler even
before arriving in Linz; both had studied at Tübingen, though Kepler
preceded Hitzler by five years. Hitzler had a reputation as an orthodox
Lutheran, in keeping with his training in Württemberg, and as a "wel-
coming and loving [man and] a theologian of eloquence and a sincere
heart."[41] Moreover, he had wide-ranging interests that would have appealed to Kepler. A pamphlet of 1660 memorializing Hitzler spoke of
his passion for the study of both music and astronomy; according to
Tobias Wagner, the Tübingen theologian who wrote the memorial, Hitzler not only had studied both fields but had also composed his own
musical arrangements and invented his own mathematical and mag-
netic instruments.[42]

Given their shared backgrounds and interests, Kepler likely felt fairly
comfortable when he approached Hitzler to discuss his theological
views, despite Hitzler's own reputation as an orthodox thinker. After
all, Kepler continued to self-identify as a Lutheran and wanted only
to make clear that he objected to one—in his view unsubstantiated—
aspect of the Lutheran approach to the Eucharist. He hoped to continue to partake in all the church's activities and would even sign the
Formula of Concord so long as his objection on this one point was duly
noted. Hitzler's response, however, was far more extreme than Kepler
anticipated. Not only did Hitzler emphatically disagree with Kepler's
approach, but he insisted that with this approach Kepler had firmly
placed himself outside the acceptable bounds of Lutheran belief—and
consequently he refused to allow Kepler to participate in Communion.

Stunned and grieving, Kepler immediately appealed Hitzler's decision to the consistory in Stuttgart—he did not know, of course, about the negative assessment of his theological views that they had conveyed to the duke the previous year. Though we don't have an extant copy of Kepler's letter to the consistory, we do have a copy of their response, which counters, in some detail, the various claims he must have made in his letter. If we judge by the reply of the consistory, Kepler argued to them that his disagreement with Hitzler was a matter of personal conscience, and that Hitzler had no authority to exclude him from Communion on that basis. He remained a committed Lutheran, aside from the one disputed question of ubiquity; surely, it was wrong for the church to cast him out simply because he tried to remain true to his own inner beliefs. Was it better, asked Kepler, for him to remain a member of the church with some small disagreements or to be forced to leave entirely?

In an answer drafted in September 1612, the members of the consistory countered that Kepler's problems were his own fault and that Hitzler had acted rightly. "If someone," they wrote,

openly expresses in speech that he considers himself a true member of the evangelical religion but says that not all of its religious articles are correct . . . and deludes himself with his own thoughts in matters of faith and divine mysteries and does not base himself on any certain form of pure doctrine by subscribing to the Formula of Concord, which is the symbol of public orthodoxy of our times and which is grounded in the divine holy scriptures, but rather contradicts it himself, in one or more articles: [in such a case] a minister of the church, who is a true guardian of the mysteries of God, does rightly when he does not admit such a one to Communion.

Kepler's problem, they maintained, was that "in contentious religious matters he bases his faith and opinion not on the holy, prophetic, and apostolic writings but rather on his own mental prowess." But the mysteries of Christ were not supposed to be probed via reason; they were to be accepted as secrets, or mysteries, via faith alone. Kepler's mental prowess had only led him astray. He ought to recognize that "the mysteries revealed in the scriptures are immeasurably greater and superior to the cleverness of Plato and Aristotle, Ptolemy and Copernicus."[43] The members of the consistory reminded Kepler of Paul's words in 1 Corinthians 2:1: "When I came to you, I did not come with eloquence or human wisdom as I proclaimed to you the testimony about God." God's cause was not served, they insisted, by new subtleties and questions.

If Kepler recognized this and embraced true doctrine, he could be re-admitted to the church.

As framed by the Stuttgart consistory, the trouble in which Kepler found himself embroiled revolved around a larger question that the Lutheran Church struggled to settle: what was the proper role of reason vis-à-vis faith? As we saw in the previous chapter, the discipline of metaphysics gradually crept back into Lutheran universities in response to the increasingly heated polemics with the Calvinists. And though some felt that it was a necessary weapon when used against those who would wield it in the name of error, by and large the seventeenth-century Lutheran Church positioned itself strongly against the use of reason when it came to the *resolution* of questions of theology. That is, though reason might be wielded to defend established doctrine, it could not be used to determine which doctrine should be embraced in the first place. Luther had, after all, begun the movement by sharply attacking medieval scholasticism and its inappropriate emphasis on reason in theology, and even Melanchthon had argued that "ignorance of God" was one of the irreparable effects of original sin.[44]

The Lutheran theologians of the sixteenth and seventeenth centuries nearly unanimously affirmed the need for a sharp separation between reason and doctrinal decisions: doctrine always came first, they held, and reason was more likely to lead one into error than into truth. Summing up a century of Lutheran discussion on the issue, Abraham Calov insisted that reason could not be used to understand matters of faith.[45] He cited 1 Corinthians 2:14 as proof, where Paul had stated unequivocally that "natural man receiveth not the things of the Spirit of God, for they are foolishness unto him; neither can he know them, because they are spiritually discerned." It was only by faith that God and true doctrine could be discerned. In his *Loci Theologici*, Johann Gerhard had likewise argued earlier that "reason within its own sphere contains nothing that is contrary to Scripture" but that it became a danger "when it wants to exceed its sphere and make judgments about the greatest mysteries of faith from its own principles." Reason was simply not strong enough to penetrate the divine mysteries, and when it tried, it revealed only error rather than truth. "He who denies or impugns the mysteries of faith revealed in the light of grace," continued Gerhard, "on the ground that they are incongruous with reason and the light of nature, does, at the same time, fail to make a proper use of the office and benefits of reason and the light of nature."[46]

As a consequence of this dismissal of reason, Lutheran theologians insisted that the doctrines agreed upon by the church and enshrined in the Formula of Concord must be accepted as true; individual disagreement on rational grounds simply held no sway for them, since reason could not penetrate doctrinal mysteries in the first place. They saw the imposition of doctrine from above and the necessity of doctrinal unity as consequences of the preeminence of faith over reason. They certainly wanted to resolve the disputes plaguing the confessions, but on their own terms, on the basis of the doctrines agreed upon by the theologians who spearheaded the Lutheran movement. Such a refusal to compromise on questions of doctrine might certainly lead to still further schism, but schisms based on doctrinal truth were to be embraced, for they would ultimately lead to a church unified by truth, with the heretics cast out and no longer plaguing the *corpus Christianum*. Gerhard could thus argue that "we grant that in a certain sense we are schismatics, because we have left the Roman Church and its head, the Roman Pope, but we have not thereby separated ourselves from the unity of the Church Catholic and its head, Jesus Christ. O blessed schism, through which we are united with Christ and the Catholic Church."[47]

The position of both the Stuttgart consistory, who maintained that Kepler had removed *himself* from the community, and of Gerhard, who argued that schism brought unity, rested on the notion that a community must have a solid and agreed-upon doctrine that united it. How else was a community to define itself if not by doctrinal accord? What was to prevent the Lutheran Church from dissolving or being overtaken by one of the rival confessions if it didn't hold firmly to its own positions, agreed upon by its most eminent theologians and undisputable on the grounds of reason, which had no place when it came to questions of doctrine? Moreover, if salvation was important for all in the community, it was the responsibility of those who guided the community to ensure that its members were on the path that led to it—and to excise those spiritually sick members who might lead other members of the communal body astray.

The Kepler that we saw in chapter 1 would have agreed with the idea of doctrinal unity but not with the dismissal of reason as the means by which it might be achieved. As a young man, Kepler had hoped, like his Lutheran mentors, to unite the community and help lead all Christians toward the one truth and the salvation that it guaranteed. This truth, he believed—in sharp contrast to his Lutheran mentors—was not always captured in the doctrines of a single confession, and the positions of the various confessions needed to be considered carefully to ascertain it. Still further, Kepler believed—also in contrast to the prevailing Lutheran approach—that it was possible to achieve rational knowledge of God and to use metaphysical or mathematical arguments to demonstrate specific theological claims; his *Mysterium Cosmographicum* made that quite clear. If, then, one applied one's reason appropriately to questions of theology and doctrine, relying on the scriptures as a guide, one might identify those components of orthodox doctrine that were valid and authentic and those that were irrational and not rooted in scripture or tradition. The young Kepler felt that using reason in this way would achieve the kind of doctrinal unity sought by the Lutheran theologians, though it would be a doctrine far broader than the orthodoxy they envisioned. It would require a willingness to admit error and to modify one's doctrinal positions when they were revealed to be wrong. This doctrinal modification would, Kepler hoped, lead to a greater unity, with Catholics, Lutherans, and Calvinists all adhering to the very same doctrinal positions, buttressed by both scripture and reason.

Yet by 1612 and in the years following, Kepler, as we have already begun to see, had begun to lessen his emphasis on reason as the grounds for his dissent and as the justification for others accepting his position. He continued to insist that he believed his own position to be the correct one, both on rational and on scriptural grounds. But instead of simply claiming that his disagreements should be tolerated because he was *right*, in any absolute sense, Kepler began to argue that his disagreements should be tolerated simply because he *believed* himself to be right. In other words, Kepler moved from an emphasis on truth as the legitimate grounds for dissent to an emphasis on something different, something that he called "personal freedom" or "conscience." It is to this notion that we now turn.

Personal Conscience: Kepler and Hafenreffer Revisited

In 1617, five years after Kepler had received his letter from the Stuttgart consistory, Kepler visited Tübingen, in the hope that his supporters there might intercede with the consistory on his behalf and get him readmitted to Communion. In particular, he met with Matthias Hafenreffer. As Hafenreffer was both a dear friend of Kepler's and a theologian who stood high in the ranks at Tübingen, Kepler felt that if anybody could help him, Hafenreffer could. When Kepler left Tübingen, he seemed convinced that Hafenreffer would take up his cause soon afterward, but he was disappointed when no word from Hafenreffer arrived. One year later, in November 1618, he took the initiative and wrote directly to Hafenreffer, reminding him of his situation and pleading once again for Hafenreffer's aid.

He assumed, he wrote, that Hafenreffer's silence stemmed from the fact that he had no good news and was worried about offending a close friend like Kepler with a letter of negative tidings. Yet Kepler implored Hafenreffer to revisit his case and to consider to whom he owed more allegiance—his colleagues who had drafted the Formula of Concord or the ancient church fathers with whose views Kepler insisted his own position better agreed. This was not just a matter of a lone individual disagreeing with the majority position, Kepler argued, but of scriptural authority disagreeing with modern innovation. Of course, this was easy to say and impossible to prove—after all, Hafenreffer and his fellow theologians believed that *their* interpretation of scripture and the church fathers was correct and Kepler's was erroneous. In this case, Kepler's situation might easily still seem to be a case of the individual pitting himself against the accepted churchly authorities. Thus, for the majority of his letter to Hafenreffer, Kepler did not try to prove that his own position was the correct one. Rather, he argued that since he personally *believed* his position to be the true one, on the basis of scripture and tradition, his beliefs should be tolerated because of their sincerity. "In matters of faith," he insisted, "full conviction is demanded." How, then, could they ask him to subscribe to an opinion with which he so strongly disagreed? He prayed that there would never come a time when, in despair, he would choose to do something "that would violate my conscience."[48]

He asked Hafenreffer what kind of precedent he wanted to set. What would the Lutheran Church do about other members with dissenting opinions? Would they expel everyone who disagreed on any point? It might, Kepler suggested, be better for the church in the long term if they tolerated his own disagreement on this one point and demonstrated to future generations that even those with qualms on specific points of doctrine would still be welcomed. This did not mean that the theologians could not still be the primary determinants of doctrine, nor that Kepler did not accept them as his superiors. "I embrace all of you as my spiritual fathers in all questions of faith," he insisted, "but as human beings."[49] Kepler insisted that while his superiors in the church certainly had more experience and training than he, they did not have any greater access to the divine word. While he was willing to abide by their decisions for the most part, he could not do so when they conflicted with what he saw to be the truth of scripture and his own personal conscience.

And it was personal conscience that Kepler held up as the primary reason why Hafenreffer and his colleagues should stop persecuting him. "With regard to my exclusion," he wrote, "I do not intend to eliminate it through a crafted confession."[50] However, he still lamented it and longed to rejoin the churchly community with which he still identified. Further, he would never insist that others accept his personal opinion about true doctrine, nor would he try to convince others that he was right: "I consider myself not your judge in the matter of my difference of opinion. I am rather a layman. But as a custodian of a purely personal conscience . . . I will not undertake to use the popular zeal and [persuade] the people to defend me."[51] Kepler made clear here that he had no plans to try to convert the public to his position on ubiquity, not even by means of the mathematical arguments that seemed to hold such promise years earlier. Conscience was a personal thing, a kind of inner knowing, and he could be secure in his sense of rightness without trying to force that sense on others.

At the same time, conscience was not *just* private. It was, quite literally, shared knowledge—*con-scientia*. Kepler rested his appeal to conscience on the fact that all Christians were brothers, who shared a place in the communal body of Christ. And as conscience had long had a legal resonance, for it was long understood as the kind of knowledge that one might share in court about the actions of another,[52] Kepler argued that it would go against his conscience to prosecute his fellow Christians simply because of their differing interpretations of an individual point of doctrine. In 1612 he had written to the Duke of Württemberg that it was not the errors in the Formula of Concord that most troubled him but its intolerant attitude toward other Christians. Kepler made this point still more forcefully when he wrote to Maestlin in 1616 and insisted on his own right to believe what his conscience dictated. He explained there that

I don't want to receive the wrath of the theologians, but I will not judge brothers, who, whether they stand firm or whether they fall, are of the Lord and are my brothers. It is better that I sin in justifying, in speaking well, in interpreting for the better, although I am not a doctor of the church, than in accusing, excluding, and destroying. . . . The Jesuits shout, the Calvinists shout, that injury is caused to them in the article concerning the person of Christ in the Book of Concord, and that much is imputed to them that they themselves do not teach. Which if I subscribe to without excluding those points, then I make myself a judge condemning those for intending a crime, or I become an accuser of those against my conscience, which defends them rather than accuses them in certain matters.[53]

While in his letter to the duke, Kepler may have avoided arguing on doctrinal grounds deliberately, to Maestlin Kepler was always open and up front about his doctrinal views—recall that he had told Maestlin explicitly in 1597 that he disagreed with the doctrine of ubiquity on geometric grounds. It is therefore noteworthy that by 1616 Kepler linked his refusal to sign not primarily to the content of the propositions in the Formula of Concord but rather to their hostile judgment of all other Christians. What mattered most, Kepler claimed here, was not truth or error, not whether other Christians "stand firm or whether they fall," but whether they might live together in peace and brotherly love. Kepler refused to sign because to do so, he argued, would be to judge both Catholics and Calvinists to be unredeemable heretics, and this he could not do. All Christians were brothers, and brothers did not accuse brothers.

Kepler would not condemn and focus on blame, he continued in his 1618 letter to Hafenreffer begging for readmittance to Communion, because his greatest wish was that the church might one day be reunited. "My family and I," he noted, "say daily prayers for the unification of the thrice-divided church."[54] He certainly recognized that Hafenreffer and his colleagues, too, wanted a united church—but they wanted the church united on their own terms, following their doctrinal positions. Kepler hoped for a different model of reunion, one that might reach out a hand to Catholics and Calvinists rather than pushing them away. As an example of such a model, in his letter Kepler cited someone whom many considered an apostate of the worst kind: Marcus Antonius De Dominis.

De Dominis was a Catholic archbishop of Spalato who left the Catholic Church and moved to England in 1616, where he wrote several anti-Catholic works. He ultimately left England and rejoined the Catholic Church before his death in 1624.[55] When Kepler cited De Dominis in his letter of 1618, De Dominis had already begun his *De Republica Ecclesiastica*, in which he argued for churchly reunification. Kepler cited that text rather than De Dominis personally, though it is possible that he met De Dominis at Rudolf II's court in Prague during one of De Dominis's visits there. He certainly kept abreast of De Dominis's work in general, both theological and scientific; Giovanni Francesco Sagredo, Galileo's friend, noted in a letter to Galileo that he had purchased a copy of De Dominis's *De Radiis* directly from Kepler.[56]

On the question of churchly unity, De Dominis insisted that there was much that united all the confessions already, and—like Kepler—he believed that that unity rested on the scriptures and the ancient church practices. De Dominis raised the possibility, cited by Kepler in his 1623 *Confession of Faith*, of a new church council, attended by all the major confessions, which would decide on mutual points of agreement that might form the basis of the newly unified church. This new council would allow the learned members of all the confessions to speak calmly and peacefully together and at the same time would put a halt to the increasing rages of the people and fanatical rants of the town preachers, who only increased strife and dissension in the name of an ostensible righteous zeal.

In his *De Republica Ecclesiastica*, De Dominis also suggested that Christians could all be united under the canopy of one church without settling each and every difference that divided them. He offered a model of loosely united regional or national churches, governed individually but bound together by the practices that united them and by one leader of all the various churches who encouraged unity while respecting autonomy at the same time.[57] Though he did not make the analogy explicitly, the model for the church described by De Dominis has much in common with the model of the Holy Roman Empire itself, which gave authority to its individual princes and electors while uniting its provinces under the larger authority of the emperor. And it was a model that appealed to Kepler, who cited De Dominis as a "prophet" and linked De Dominis's words to those of the prophecy that Kepler himself had made about the future unity of the church in the *De Stella Nova* of 1606.[58] In particular, Kepler, like De Dominis, saw room for diversity within the future unified church; like De Dominis, he believed that unity need not imply uniformity or absolute conformity with one doctrinal position. In his own descriptions of what such a unified diversity might look like, Kepler referenced not the political system of the empire but rather a model that was, on the surface, far more removed from the specifics of earthly life: the model of musical and cosmic harmony.

Kepler elaborated elsewhere the specific continuities that he saw between cosmic and churchly harmony, as we will see in later chapters—he did not discuss musical or heavenly harmony directly in his letter to Hafenreffer. Yet he referenced it clearly by appending to his letter to Hafenreffer two seemingly unrelated documents: a copy of the title page to *The Harmony of the World*, his latest major manuscript; and a copy of his *Unterricht von H. Sacrament des Leibs Bluts Jesu Christ unders Erlösers*, a short pamphlet he had written during the previous year summarizing his views on the Eucharist. The *Harmony of the World* was Kepler's masterpiece, to his mind, for it was the work in which he linked music, mathematics, astrology, and astronomy to the harmonic archetypes that underpinned all reality. The *Unterricht* was intended primarily for his children, and it was written simply and in question-and-answer format; Kepler hoped that through it his family might better understand both the theological basis for Kepler's own views on the body of Christ and the nature of the conflict in which he was embroiled.

By including both documents with the letter to Hafenreffer in which Kepler argued for personal conscience and churchly unity, Kepler made it clear that his position on the Eucharist, and on the nature of the church more specifically, had something important to do with the harmony that God had imprinted throughout the cosmos. The church and the cosmos were linked, according to Kepler, through the archetype of harmony, and harmony was a very different metaphor than simple and undifferentiated unity. As he elaborates in the *Harmony of the World*, harmony is pleasing only *because* of its diversity: "just as . . . individual consonances considered separately are pleasing on account of the fact that they are plainly not identical notes, but in a way figured and different notes, . . . in the same way . . . the harmonious singing of parts . . . without any variety in them ceases to be pleasing altogether."[59] He hoped that this message, and the larger plea for peace and acceptance that underpinned his letter, would be clear and compelling to Hafenreffer.

In this letter, however, Kepler did not expand on the implications of his *Harmony* for questions of theology; he simply made clear that he would restrict his public emphasis to cosmic harmony and keep his religious views private. He asked Hafenreffer to do his utmost to publicize Kepler's forthcoming book on harmony and to distribute copies of its title page to the booksellers in Tübingen. If Hafenreffer wanted Kepler to avoid contentious theological debates, let him help Kepler to restrict his focus to mathematics and astronomy. "I submit," he wrote, "to the distinction which your reverence has made between me the mathematician and me the theologian. I do not claim the latter title at all."[60] Kepler's insistence that this theological dispute was simply a question of conscience emphasized that he had no interest in debating doctrinal issues with Hafenreffer or with others; his personal qualms would remain personal. Yet he was an expert mathematician, and if he hoped to convince others of anything, it would be in this regard: in demonstrating the harmony of the cosmos, a harmony that could serve the church by providing a model for the church to emulate. And unlike in 1597, Kepler made no mention of any more direct ways that his mathematics might reflect on doctrinal issues like ubiquity; the implied linkage between his theological beliefs and his mathematical ones was now far more indirect.

Hafenreffer's reply to Kepler, written in February 1619, was supportive but troubled. While Kepler had guessed that Hafenreffer hesitated to write to him after their Tübingen meeting out of fear that he might offend a friend, Hafenreffer insisted at the start that this was not the case: "I have always been most certain," he wrote, "that it ought to be considered no offense, but rather the greatest kindness, to lead a friend—even if he suffers and resists—on the path of truth." If Hafenreffer had hesitated, it was because he had seen all his efforts on Kepler's behalf fail so far owing to Kepler's own obstinacy. But he would try yet again to lead Kepler toward the truth by reminding him of the proper outlet for his inquiring mind. He wrote: "I always esteem and make much of your remarkable and celebrated thoughts, knowledge, and work dealing not only with the sublunar world but also the entire upper world, all the way to the outer surface of Saturn. But with regard to that 'upper' world which is heavenly in a spiritual sense— here, hands off! Here all the sharpness of human intellect necessarily becomes foolish." He added that if Kepler could not help but apply his discerning mind to questions of theology, he should focus on one phrase alone: "et verbum caro factum est" (and the Word was made flesh). Hafenreffer implored, "If you love me, if you have ever loved me, forget the passions which you affected in your last letter and focus on this: There are three words: 1. *Verbum*, 2. *Caro*, 3. *Factum*. Either I am foolish or crazy, or you will understand how foolish, crazy, and absurd we are when we attempt to investigate the divine mysteries with our rational method of stupidity."[61] The clarity of this message from the Gospels, Hafenreffer hoped, would help Kepler realize that even his lofty intellect could not scale the divine mysteries; limited and fallible human minds could only hope to appreciate wonders like both the Incarnation and the Eucharist as God had intended—as mysteries.

As to the second harmony Kepler had mentioned—his astronomical *Harmony of the World*—Hafenreffer noted that he had distributed copies of the title page as Kepler had asked and that he wished the book

well. Though Kepler had linked—however indirectly—his astronomical harmony to his hopes for religious harmony, Hafenreffer took the
opportunity to exhort him yet again to steer clear of theological questions and to remind him of "my Christian distinction between you the
mathematician and you the theologian, a title suited only to the dis-
ciples of the heavenly [i.e., divine] Word."[62] Hafenreffer took Kepler's
Unterricht, the pamphlet that Kepler had written on the Eucharist and
had appended to his letter to Hafenreffer, to be further proof of the
problematic overlap between reason and theology, for he felt it contradicted the clear words of scripture on which he urged Kepler to reflect.

Theology, Mathematics, and Expert versus Lay Knowledge

In Hafenreffer's reply, Kepler sensed a door closing to him. If even his
close friend and mentor believed him guilty of heresy and refused to
consider things from his point of view, what hope was there for him?
Unwilling to fully give up hope, Kepler replied once more to Hafenreffer in a lengthy letter of April 1619 and attempted one last time to
convince Hafenreffer of the justice of his position. He began the letter
by noting that he had debated whether or not to write back at all after
his last letter but had decided that he could not leave "a man who had
done me the utmost good with the wrong opinion of my words and
actions; for to be condemned by [you] would truly mean my end."[63]
Kepler insisted that he felt no bitterness toward Hafenreffer himself but
merely sorrow over his exclusion from Communion. Moreover, despite
all that had been said, he still retained the hope that Hafenreffer and
his fellow theologians might alleviate his sorrow by granting him the
right to rejoin the church and participate again.

To that end, Kepler asked Hafenreffer the following. If the theologians insisted that he abandon his qualms, leave theological questions
aside, and act as a simple layman, then could they not at the very least
treat him as a layman when it came to questions of communal participation? Why, that is, should he be forced to sign the Formula of
Concord at all? He lived in Linz simply as a layman—he was neither a
member of the university nor an official in the Lutheran government.
The theologians, he wrote, should "give me the right which is enjoyed
by the majority of simple laymen, who are not commanded to subscribe. In what way am I more at fault than these men? . . . Why can
I not be accepted as a simple layman? If the fact that I refuse to subscribe makes the theologians suspicious (which a simple layman does

not have to do), then the fact that the theologians ordered me to sub-scribe made me suspicious first, since they would not order a simple layman thusly."[64] Here, Kepler tried to use Hafenreffer's own arguments against him and to move the blame from his own shoulders to those of Hafenreffer and his colleagues. Kepler was not the one who had tried to assume the role of theologian—the consistory and the faculty at Tübin-gen had done that for him.[65]

Kepler added that he was perfectly willing to act the simple layman and ignore his theological qualms, but only if he did not have to sub-scribe; the very requirement that he sign the Formula made it impos-sible to subsume his theological concerns in favor of "simple" faith. After all, he wrote, "a signature presupposes approval of each and every thing which is written in the book."[66] While the layman could simply overlook many of the theological details that were subject to debate, one who signed the Formula necessarily took a stand on everything within it—even the many things, Kepler pointed out, that were not strictly doctrinal. He had personal reservations about the idea of ubiq-uity, as he had made clear earlier; he also believed that much of what the Formula detailed should not require universal assent at all. With these doubts, he felt that he could not subscribe, for if he did, it might seem "that I subscribe to authority rather than to truth."[67]

The dispute between Kepler and Hafenreffer, then, involved not just the question of reason's proper use or simply the proper interpretation of the words of Christ about the Eucharist but also the role of laymen in the church more generally. This was a much-debated question even before Luther called for the supremacy of individual belief and faith against the entrenched hierarchies of the Catholic Church. In his in-sistence that he acted as a layman and not a theologian, Kepler seemed to be stressing, at times, that as a layman he could afford to ignore his reason and not worry about his disputes with established church doctrine. Yet Kepler's emphasis on his status as a layman also allowed him to stress something quite different: that as a layman, he had as much right to interpret scripture as anyone and that his interpretations might even be preferable to those of the theologians at Tübingen.

Here, Kepler was resting not merely on the Lutheran idea of per-sonal access to scripture captured by Luther's *sola scriptura* but also on the earlier idea, advanced most prominently by Nicholas of Cusa, that the layman somehow had *better* access to truth than those in positions of authority.[68] Cusa had devoted three treatises to the idea of learned ignorance and highlighted the figure of the *idiota*, or layman, whose knowledge ultimately surpassed that of the professionals with whom

he debated. Like Cusa's *idiota*, Kepler insisted that he needed to follow his layman's knowledge, obtained by studying God's two books
directly rather than by blindly acceding to the interpretations of the
professional theologians.

Luther and the other early reformers would likely have agreed with
Kepler's emphasis on the knowledge of the layman, though not necessarily with the specific positions that he espoused. According to legend,
Luther argued at the Diet of Worms that "a simple layman armed with
Scripture is greater than the mightiest pope without it," and Calvin had
written in one of his letters that he had "never been anything else than
an ordinary layman, as people call it."[69] Yet those arguments could not
mean that every individual had the right to interpret doctrine as he
chose; Luther had argued for one interpretation as the correct one, and
it soon became clear that to sustain itself as a movement, Lutheranism needed to develop some kind of doctrinal accord. Consequently,
a hierarchy and an orthodoxy quickly developed in the post-Luther
Lutheran Church.[70] Kepler, in many ways, seems to hearken back to
the early days of the movement and to a late medieval world that tolerated heterodoxy more readily than the confessional world of the seventeenth century, where each side felt the need to stick rigidly to its own
orthodoxies.

It was to this latter world that Hafenreffer squarely belonged, and
he focused on the figure of the theologian as authority figure: as Kepler
was no theologian, he ought to avoid all questions of faith and interpretation and leave those questions to those experts who, ensconced
in the halls of Tübingen, could answer them best. The distinction between layman and theologian for Kepler meant that he should not be
compelled to sign anything; for Hafenreffer, it meant that he *needed*
to be compelled to sign, for he had no right to decide these issues for
himself.

Moreover, when Hafenreffer urged Kepler to steer clear of theological questions, he emphasized Kepler's astronomical expertise. Kepler
should stick to those areas he knew well, Hafenreffer insisted and not
venture into those areas in which he was only a novice. Kepler himself
agreed that he was no expert in theology and—in reference to Hafenreffer's seemingly oft-remarked distinction between Kepler as mathematician and Kepler as theologian—agreed to name himself only the
former and not the latter. Yet in his most recent letter, when Hafenreffer had asked Kepler to remember his distinction between "you
the mathematician and you the theologian," he had seemed to imply
something further, beyond the fact that mathematics was a discipline

Kepler knew well, and theology one he did not. He had implied that mathematics and theology were not only different areas of expertise but fundamentally different ways of looking at the world, and he had warned Kepler to avoid thinking "mathematically" about theological questions.

Rather than address this distinction immediately, Kepler first insisted that Hafenreffer's suggestion about his motivations was wrong. He refused to sign the Formula of Concord not because of any "sharpness of intellect but rather [because of] the esteem of brotherly charity." As he had argued earlier, Kepler again emphasized that aside from any specific theological qualms, his refusal to sign the Formula was linked on a fundamental level to the antagonistic attitude that the text and its formulators took toward other confessions, or those with differing opinions. Even if the Calvinist and Catholic opponents of the Lutheran Church did not themselves seem interested in churchly harmony, it was still incumbent on Kepler and his Lutheran compatriots to take the initiative; citing Matthew 5:44, Kepler wrote that "we have to be good to our enemy and love those who hate us."[71]

Kepler then took up the challenge from Hafenreffer's letter and tried to explain the words "verbum caro factum est." In the context of this passage, Kepler explained, Word and flesh were diametrically opposed concepts—the one was so all-encompassing that the other was minuscule by comparison. Yet somehow, the infinite had entered the finite via the mystery of the Incarnation; the words of scripture cited by Hafenreffer and all the churchly authorities clearly agreed on this point, and thus, Kepler must agree with it too. "There happened here," Kepler made clear, "something beyond my geometrical grasp, which is to be seen by the eye of faith, since a finite womb was made for this task able to hold the infinite Word." Kepler insisted that he did not apply mathematical standards to *all* theological claims, for if he did, he would have to reject far more than simply the doctrine of ubiquity:

If I consider geometrical reason, it seems to me that there are many contradictions: not to give up omnipresence but nevertheless to descend whole into the womb, to enter whole into circumscribed flesh, for the whole Word to dwell in us, and nevertheless to be in the heavens, and also to be beyond worldly place, and yet to hang on a local cross for the salvation of men, or for the cross, to which all of our sins are attached, to be affixed in a specific place in Judea. Nevertheless, I believe all these things because they are the clear prescribed word of God and are interpreted by the church of God and all the fathers, not excluding Luther or Hafenreffer.[72]

He accepted much that seemed to contradict his geometrical understanding of the world, Kepler argued, because he had to—one could not reject the clear word of God and the combined assent of all authority and tradition.

Kepler insisted to Hafenreffer that it was precisely for this reason that one needed to be very careful about the implications of the scriptural claims on which doctrinal positions were based, particularly given that they were so difficult to understand. True, the divine Christ was made flesh in the finite womb of Mary. Yet because the Word was omnipresent, did this mean that Mary's womb, too, was omnipresent? Surely not. Here Hafenreffer, displeased with this comparison, wrote in the margins of his copy of Kepler's letter: "Mathematician, you become stupid."[73] Kepler, as though sensing Hafenreffer's criticism and its expression—the derogatory "mathematician"—finally took up the distinction Hafenreffer had insisted on earlier, between mathematical and theological ways of thinking. And he turned the distinction on its head by arguing that it was, in fact, Hafenreffer himself who was confusing the issues by trying to think mathematically and physically about a question that was fundamentally a spiritual one. The focus of scripture was on the Word in a spiritual sense. Why, then, were the Lutheran theologians obsessed with the *physical* implications of Christ's body, when nothing in the scriptures indicated that the divine aspects of Christ were supposed to be conceptualized physically? Ubiquity was relevant only if one thought in terms of bodies, and it made no sense to do so when speaking of divinity. Scripture insisted merely that "the Word is everywhere, but the flesh is not everywhere."[74] Christ's ubiquitous presence could be understood in a spiritual and voluntary sense without contradicting either the words of scripture or the majority of its authorities.

Kepler insisted that Hafenreffer and his colleagues needed to reassess their interpretation by carefully considering the actual words of Christ in the Gospels. Consider, he wrote, the words of Christ in Matthew 18:20: "For where two or three gather in my name, there am I with them." Christ's presence typically had a spiritual meaning, as in this instance, Kepler insisted. When it came to the bread of the Last Supper, Christ had not said "eat my body," nor had he said "the bread is my body." In saying far more generally "this is my body," he was making a spiritual, not a physical, claim—something that Kepler insisted earlier Lutherans like Martin Chemnitz and Aegidius Hunnius had recognized in their arguments against the Catholic doctrine of transubstantiation. Kepler asserted that in focusing on the physical body of Christ as the basis for their Eucharistic doctrine rather than on his power and will, it was theologians like Hafenreffer who were relying on mathematical and physical reasoning, rather than himself. And if the theologians were to simply insert a spiritual interpretation into the Formula of Concord alongside their own, Kepler would have no qualms about signing—for this would allow him to allay his personal conscience and prevent him from unjustly accusing all others who subscribed to different views on this subject.

Kepler's willingness to dismiss the relevance of mathematical thinking to theological problems stands in sharp contrast to his claims in 1597 on this very issue of Christ's presence in the Eucharist. Then, he had emphasized that nothing could be more relevant to theology than geometrical reasoning, for geometry mirrored the mind of God. Kepler had applied these ideas to the contentious issue of the Eucharist and had dismissed the Lutheran doctrine of ubiquity precisely on the grounds that it was geometrically incomprehensible. He had, in other words, been guilty of precisely what Hafenreffer had accused him—deciding his theological positions on the basis of his mathematical expertise and allowing mathematical modes of thinking to determine his approach to theological mysteries. In his letter of 1619, he still argued for the incomprehensibility of the doctrine of ubiquity but endeavored to do so primarily on the grounds of scripture, tradition, and personal conscience. He went so far as to readily agree with Hafenreffer that geometry and theology were incompatible modes of thought and to say that he considered questions of theology that were bolstered by authority and tradition only "with eyes closed to geometry."[75] What are we to make of this seeming change of mind?

On the one hand, there is always the possibility that Kepler simply told Hafenreffer what he thought he wanted to hear. It would have done him no good to insist on the relevance of mathematics to this theological dispute; if he had any hope of convincing Hafenreffer of his point of view, it likely lay in scripture and tradition rather than geometry. This may be part of the story, yet it cannot be the full story, for Kepler was never one to shy from or alter the truth when it came to his faith. He had stated earlier that "it is not in me to dissimulate in matters of conscience,"[76] and this is precisely what had gotten him into trouble in the first place. On the other hand, Kepler's change of heart may have simply stemmed from a loss of the starry-eyed idealism that characterized him in his early days. In 1597 Kepler was a thinker deeply committed to the absolute truth of his own ideas, the relevance of those ideas across all domains of knowledge, and the power of those ideas to effect real change in the world around him. As is often the case, this kind of idealism may have given way, over time, to more modest goals, and more modest beliefs about the power of his intellect and the reach of his theories.

Yet this, too, cannot be the full story. For in most cases the Kepler of 1619 did not sound all that different from the Kepler of 1597—at least when it came to the certainty of mathematics and the broad reach of its relevance. Kepler continued to emphasize that geometry originated in the mind of God, that it underpinned all aspects of creation, and that God had given man the power to use it to interpret the Book of Nature. And because geometry linked God, man, and nature, Kepler continued to emphasize that all three were directly relevant to one another—thinking geometrically meant thinking about God as much as about his creation. He completed his *Harmony of the World* the very same year as this letter to Hafenreffer, and it was there that he wrote that "geometry, which before the origin of things was coeternal with the divine mind and is God himself (for what could there be in God which would not be God himself?), supplied God with patterns for the creation of the world, and passed over to Man along with the image of God."[77]

But if Kepler's geometrical views had not changed substantially between the publication of the *Mysterium Cosmographicum* and the *Harmony of the World*, his religious views had—in particular his understanding of the ability of man to comprehend, model, and spread God's will and message. And central to this change was Kepler's changing conception of the role of the theologian. In 1597, having only recently left the halls of Tübingen and with little experience of the real world, Kepler strongly believed not only in the ideas and ideals of theology as a discipline but also in the person of the theologian. He himself had wanted to be a theologian, he wrote in 1597; he reconciled himself to the discipline of astronomy because there, too, he could conceive of himself as a priest, though with respect to the Book of Nature rather than the Book of Scripture. And though he continued to speak of his astronomical work in these terms, he ultimately came to see theologians not as a solution to the problems plaguing the church but as a primary source of those problems. In a letter to Maestlin in 1616, Kepler insisted that he did not want "to join the fury of the theologians."[78] He linked his own more forgiving behavior to his status as a layman, and the negativity of the theologians to their official positions. His insistence to Hafenreffer in the letters of 1618 and 1619 that he considered himself a layman, and not a theologian, may have stemmed not only from his concerns about signing the Formula of Concord but also from his growing distaste for the attitudes and behaviors of the theologians he saw around him.

Similarly, he attributed "the beginning of all evil" in Styria, the province in which Linz was located, to the behavior of the Lutheran theologians there. In the same letter to Hafenreffer of 1619 in which he argued for his status as a layman and his right to be readmitted to Communion, Kepler insisted that the religious crisis in Styria stemmed from the harsh pulpit sermons of two pastors, Fischer and Kelling. In particular, in arguing against the Catholic veneration of the Virgin Mary, Fischer had "extended his cloak at the pulpit and asked whether it would be decent for women to crawl under it"; it was still more indecent, he had insisted, to paint monks under the cloak of the Virgin as was the common Catholic practice.[79] Such a speech could only have been intended to shame and offend the Catholics in Styria, Kepler insisted, and therefore could be blamed for much of the trouble there. He had earlier complained of just the same thing to Georg Friedrich von Baden, noting that the Lutheran theologians of late tended to "confuse teaching and ruling" and to rely on the power of their princes to strike at anyone they perceived as an enemy.[80] And in yet another letter, whose recipient is unknown, he wrote that the "preachers are being too arrogant in their pulpits and arouse much dispute; they raise new issues through which devotion is hindered and often accuse one another falsely."[81] If this was what it meant to be a theologian, then Kepler wanted none of it. Still more, he no longer felt that he could trust such people to repair the divided church, particularly if that repair required an absolutely unified opinion on contentious issues like the Eucharist—regardless of what geometry had to say about the matter. Even if Kepler still believed that his geometry presented some clear solutions to those thorny theological issues that divided Christian from Christian, he no longer believed that it would be simple to implement those solutions. The possibility of a general church council that De Dominis had argued for—and that Kepler still hoped for, as one option among many—seemed more and more remote.

This accounts both for some of the change in Kepler's articulation of the relationship between theology and geometry between his 1597 letter to Maestlin and his 1619 letter to Hafenreffer and also for the change in his perception of the relevance of his two books—the *Mysterium Cosmographicum* and the *Harmony of the World*—to theology. In 1597 Kepler hoped that the *Mysterium Cosmographicum* would have direct effects on the confessional disputes of his day. Geometry would unambiguously decide the question of the Eucharist, such that Lutherans would no longer have grounds to dispute with Calvinists. In 1619 he still argued that the *Harmony of the World* had clear theological relevance, yet this relevance was described in much more general terms— terms that deliberately avoided the specifics of confessional disputes. There Kepler claimed (quoting Proclus's commentary on Euclid) that mathematics was "the preparation for theology" insofar as it "it perfects us in moral philosophy, implanting in our behavior order, propriety and harmony in social relations."[82] The *Harmony of the World* was theologically instructive only as a general model for proper personal and communal behavior, and not as a determinant of appropriate confessional stances on specific issues. Kepler's agreement with Hafenreffer that geometry could not speak to theology may best be understood on this level; Kepler had ceased to believe that geometry could aid in specific theological disputes. And this was primarily due not to any change in his understanding of geometry but rather to his growing disillusionment with theology and theologians.

Kepler closed his 1619 letter to Hafenreffer by pleading yet again for Hafenreffer to take up his cause with the consistory in Stuttgart. "My exclusion has now lasted for seven full years," he lamented, "and up to now I have never received a categorical decision from the theologians that would give me to understand whether or not I am to be excluded forever."[83] Even the 1612 letter from the consistory, in Kepler's view, had not fully decided the matter—perhaps because Hafenreffer himself had not taken a definitive stand. For this reason, Kepler begged Hafenreffer to stand by him and help resolve things in his favor; he hoped that because of their years of close friendship, Hafenreffer would "take this one step forward, although it is somewhat unpleasant, so that I know that I am not lost to you."[84]

Though Hafenreffer was clearly unconvinced by Kepler's arguments, as his many marginal notations to Kepler's letter make clear, he did forward Kepler's letter to both the theological faculty at Tübingen and the Stuttgart consistory. Perhaps he really did try to make one final attempt on behalf of his old friend, as Kepler had asked. Yet the consistory's final reply to the question of Kepler's fate was uncompromisingly firm. Erasmus Grüniger wrote on behalf of the consistory that "with regard to Kepler, we have long dealt with his craziness, but unsuccessfully. . . . [Therefore,] we did not want to refrain from communicating with the theologians at Tübingen what the consistory wrote to him several years ago about this matter . . . for there can be no other perspective on his foolishness."[85] Under no circumstances, wrote Grüniger, would Kepler be readmitted to Communion or the Lutheran Church. With this final reply on the part of the consistory, Hafenreffer wrote back to Kepler and placed himself fully alongside the theologians of Tübingen and Stuttgart and against his former friend and student.

Hafenreffer also took particular issue with Kepler's approach to the verse he had suggested: "et verbum caro factum est." Kepler, he claimed, had approached the verse exactly as Hafenreffer had urged him not to: he had analyzed it, wrote Hafenreffer, not as a theologian would, but rather as a geometrician, imagining "physical and geometrical things." His mental prowess had once again only confused him when it came to the "sacred mysteries of worship." Hafenreffer concluded that neither he nor the members of the consistory or the theological faculty could "approve your absurd and blasphemous imaginings." Rather, they could only offer him the following advice: "that you abandon the imaginings of foolish reason, cling to the heavenly truth in true faith and the divine mysteries in simple faith, which all true Christians do, and worship and honor [God] in pious obedience."[86]

If Kepler refused to do this, wrote Hafenreffer, "we see no remedy for that unfortunate wound inflicted on you by the sword of foolish human reason." Nor could they see how Kepler could possibly be readmitted to Communion: "how can he who does not cultivate or professes the same faith as the orthodox church enjoy the sacraments with that same church from whose faith he departs?" Kepler needed to reject his erroneous beliefs and embrace the truth, "or shun the community of our church and confession. For Christ must not be ridiculed, nor does the purest bridegroom of his church share his love with vain and blasphemous opinions." With a final exhortation that reason was "blind and foolish in matters divine," Hafenreffer urged him to instead embrace the ideals of simplicity and humility. He prayed that Kepler would take his letter in the brotherly spirit in which it was intended and recommended Kepler's soul "to the omnipresent Christ, your savior."[87] The letter concluded with the signatures of not only Hafenreffer but his colleagues in the theological faculty as well, along with a copy of the 1612 letter from the Stuttgart consistory, testifying that the Tübingen theologians agreed with everything it said.

Even as Kepler insisted that he no longer sought to link geometry and theology, then, Hafenreffer insisted that this was precisely what he continued to do. Though in all his writings to Hafenreffer Kepler had avoided the explicit linkage between geometry and confession that he had emphasized in 1597 and had claimed to approach the Eucharist issue strictly theologically, Hafenreffer still saw his primary flaw to be his mathematical thinking, his mental prowess leading him astray. Once a geometrician, always a geometrician, to Hafenreffer—and perhaps he was right. Kepler's firm conviction that God was a geometer who had used the model of geometry in all he had created made it difficult for him to ever firmly separate the geometrical from the theological, despite his protestations to Hafenreffer. The Book of Nature and the Book of Scripture were too closely linked in Kepler's mind for him, "with eyes closed to geometry, [to] adhere faithfully to scripture" as he had tried to argue to Hafenreffer. Even the inscription that was placed on his tomb, at his request, betrays this. "I measured the heavens, now I measure the shadow of earth," Kepler had written. "My mind was from heaven; the shadow of my body lies dead." The two heavens, astronomical and spiritual, were always necessarily connected for Kepler; his disagreements with theologians and his despair at the state of his church could never truly shake his belief that geometry, at a fundamental level, could ultimately point the way to an omnipresent truth.

Conclusion: Deciphering Mysteries

The story of Kepler and Hafenreffer is in many ways a story of two men at odds despite themselves. Over the course of their dispute, they used many of the same words, yet those words meant dramatically different things to each of them. Perhaps the most significant example of this divide is their use of the word "church." When Hafenreffer spoke of the need to preserve the strength and unity of the church, he meant the Lutheran Church in particular—for only the Lutheran Church, in his mind, represented the "true" church. Kepler, by contrast, consistently articulated a conception of the church that was far more expansive than that of many of his contemporaries. The church that he hoped to strengthen was far broader than Hafenreffer imagined, and the union he sought was not limited to the Lutheran confession alone but rather encompassed all three major confessions. In his mind, geometry was central to this potential union, since it represented common ground, and provided a tool for understanding God's creation and, indeed, for understanding the mind of God itself. While at first Kepler saw the implications of geometry in specifically doctrinal terms, he came to ultimately emphasize geometry as the language in which God's harmonies were written—harmonies that might be modeled by men on earth, seeking to create a way for those with different beliefs to live together.

This leads us to yet another instance in which Kepler and Hafenreffer used the same word with very different implications—in this case, the word "mystery." In titling his first book the *Mysterium Cosmographicum*, Kepler had deliberately invoked the theological overtones of the word *mysterium*, a word that had never before been used with astronomical or natural philosophical implications.[88] The sacraments, in particular, were often referred to as "divine mysteries," which typically implied both that they conveyed some spiritual message in a physical way and that the means by which this was accomplished was inscrutable to man. Hafenreffer used the word again and again in this way in his letters to Kepler; divine mysteries, he insisted, could not possibly be explained by recourse to reason or intellect. Following Luther, Hafenreffer emphasized that man, in his fallen state, could not possibly hope to access the full heavenly truth.[89] Mysteries were those truths that fell beyond the grasp of mere mortals; simple faith was the only way that man could access the message of salvation they conveyed. Kepler, however, intended something entirely different in his use of the word "mystery." Mysteries were manifestations of the divine, to be sure, but his book argued that God's ways were fundamentally geometrical. In using geometry to explain the workings of the universe, then, Kepler was explaining precisely that which many believed to be inexplicable—he was demonstrating the way God's mind worked. The theological connotations of the word "mystery" in the title of his first book were likely intended to point to just this—to the ways that his mathematical and astronomical text revealed the true divine mysteries for all to understand. And it was ultimately this divergence that led Hafenreffer to confirm Kepler's exclusion from Communion.

Kepler's arguments here place him alongside those medieval exegetes who, influenced by the work of Plato, believed that the secrets of God might be revealed through a study of the natural world. Though it was understood by all that the Fall had alienated man from God's truth, the study of nature allowed man to reunite with the divine once again; "to know the world," as Peter Harrison has written, "[was] not merely to come to know God . . . [but] to become *like* God: . . . to restore a likeness which had been lost."[90] Men like Hugh of St. Victor maintained that the study of nature allowed man to rediscover those divine mysteries that had been lost in the Fall and to reunite that which the Fall had divided. The study of nature was a religious enterprise, in this view, because it allowed man to regain the godlike Adamic view in which all of nature was open and explicable to his gaze. By the seventeenth century, this viewpoint had begun to change. Francis Bacon famously argued that man could never truly know the mind of God—Adam's very fault in Eden, according to Bacon, was his "approaching and intruding into God's secrets and mysteries."[91] Only if men recognized the limits of their reasoning capacities and the proper boundaries of their inquiries, wrote Bacon, could they begin to build up a true foundation of knowledge.[92] Kepler positioned himself alongside the earlier, more optimistic approach; much as he denied the view of Hafenreffer and of Luther before him that man could never truly penetrate the divine mysteries, he denied too the possibility that man's flawed intellect needed to be reined in. Like the medieval Platonists, Kepler continued to believe that man, a true microcosm of the heavenly macrocosm, was capable of being like God and that, as Hugh of St. Victor had written, "every nature tells of God."[93] And the specific mystery—the "wonderful secret" that he believed he had uncovered—was that "nature is God's image and geometry is the archetype of the beauty of the world."[94]

The fact that Kepler and Hafenreffer believed that the words they used—"church" and "mystery" in particular—signified completely different things leads us to another issue central to our larger discussion: the ways that language is implicated in the pursuit of knowledge. Language has long been used as a metaphor for philosophical inquiry; the ancients often used the alphabet, in particular, as a metaphor for the basic elements of any object of study. Sextus Empiricus, for example, tells us that the Pythagoreans claimed to do as linguists did. Just as linguists "first examine the words . . . and since the words are formed from syllables, they first investigate the syllables; and as the syllables are resolved into the elements of the written speech, they investigate these first," so too natural philosophers, "ought in the first place to establish the elements that all can be resolved into."[95] In his *Timaeus*, Plato likewise relied upon the idea of nature as a written book and argued that the letters and syllables of the alphabet in which this book was written were not the four elements—earth, air, fire, and water— but rather something far more basic, the elementary triangles that gave each element its unique essence.

Yet even as Plato and the Pythagoreans suggested that nature might be conceived via the metaphor of language and alphabet, Plato raised a question about the nature of language itself in his *Cratylus*. Was language something real, something that actually corresponded to the essences of the things it described? Or was it, by contrast, simply convention, with no connection to the things themselves? This question formed the crux of the debate between realists and nominalists that began to rage in the twelfth century and culminated in the seventeenth.[96] Nominalists typically aligned with the skeptical approach to knowledge more generally, the one that we saw characterized Hobbes and Locke in their attitude toward innate conscience at the start of this chapter. According to this approach, since human conceptions did not have any direct relationship to absolute or objective truth, neither did the signs—words—that humans assigned to those conceptions. Words were merely arbitrary conventions and could not themselves supply true knowledge. Hobbes for this reason framed his objection to the truths revealed by conscience in linguistic terms: people, he noted, try to pretend that they have access to truth by giving "their opinions . . . that reverenced name of Conscience . . . and so pretend they know that they are true."[97] Locke similarly argued against the kind of knowledge supplied by words: let someone, he wrote, "try if any words can give him the taste of a pine apple."[98]

Realists, by contrast, argued that language, like the world itself, originated with God, who had established a direct linkage between the two. It was therefore possible for a language to go beyond convention and speak to real essences, to a priori truth, though clearly most languages did not come close to this ideal. In the sixteenth and seventeenth centuries, some realists became obsessed with the search for this natural language, a language that eliminated the gap between words and things. This language would be unequivocal and universal, and it would capture the basic elements of the world and their unique combinations. And while for many this was strictly an "alphabet of nature," as Francis Bacon described it, for others it was an alphabet that would speak to all truths everywhere. Comenius hoped to create a universal language explicitly for the purpose of uniting all Christians and removing religious dispute; John Wilkins likewise hoped that a universal language might end "modern difference in religion."[99] Leibniz similarly searched for "l'alphabet des pensees humaines," the alphabet of human thoughts. This universal characteristic, in which every simple idea could be expressed via a corresponding simple sign, would allow men to calculate the answers to complex questions about the natural, economic, social, and political orders.[100] Leibniz believed that such an alphabet could eliminate conflict by eliminating the misunderstandings or baseless opinions that divided men.

Kepler, like the Pythagoreans and Plato before him, understood nature as a book that could be read and believed that his geometrical cosmology offered the code that would allow others to do so: "I truly believe," he had written, "that as astronomers we are priests of the Lord

Most High with respect to the Book of Nature."[101] In this way, his views aligned with those of Galileo, who famously argued that the Book of Nature was "written in the language of mathematics, and its characters are triangles, circles, and other geometric figures."[102] At the same time, Kepler referenced language more broadly, as did Leibniz after him. Beyond the idea of nature as a book, he referred, in his 1623 *Confession of Faith*, to an "alphabet of Christendom." He did so specifically when speaking of Marcus Antonius De Dominis, the "prophet" whom Kepler believed was on the right path in his desire to reunite the church. In writing of De Dominis's quest for churchly reunification, Kepler wrote that De Dominis "teaches the correct alphabet of Christendom . . . [and] takes syllables from this alphabet."[103] This is a striking metaphor, though its meaning is not immediately clear. An "alphabet of Christendom" might suggest that there was a single correct answer to theological questions. For if there was a clear correspondence between words and things, then the correct alphabet might point directly at the proper theological truth and might reveal the singular solutions to the doctrinal disputes plaguing Christendom.

This chapter suggests that Kepler's reference to an alphabet of Christendom would likely not have implied the possibility—or, indeed, the desirability—of a singular solution to doctrinal problems. In contrast to his hopes in 1597, he ultimately suggested in his dispute with Hafenreffer that the model of harmony ought to be applied to religious disputes. We might, therefore, apply the model of harmony to the idea of an alphabet that might pertain to questions of confession. For while many conceived of an alphabet in terms of symbols that formed words—in terms of *written* language, that is—this is likely *not* how Kepler used the metaphor. In fact, Plato, too, probably used letters and syllables as phonetic, rather than written, terms—in the majority of cases, he seems concerned with the sounds of language rather than its visual appearance.[104] Kepler himself specifically used the term "syllable" to refer to a unit that composed an audible harmony; he wrote about "the six syllables ut, re, mi, fa, sol, la."[105] When Kepler invoked the alphabet and syllables of Christendom, then, he simply invoked in slightly different terms the notes and chords that combined to produce the harmonious unity for which he yearned. There was no single correct formula for crafting a harmony, but many; rather than a one-to-one correspondence between words and things, Kepler hoped to find the various different combinations that might ultimately yield pleasing harmonies for Christendom. And though individual symbols and written words might stand alone and still have meaning, the syllables that made up a harmony could not—they became meaningful only in combination with one another.

To Kepler, in sum, the mysteries of God—be they theological or cosmological—were not secrets to be hoarded by a select few, nor were they enigmas impervious to rational inquiry. Rather, God had provided a blueprint by which they could be interpreted—geometry—and a model by which that interpretation might be judged—the model of harmony. That model demonstrated that the church was not simply the specific confession that most closely adhered to one narrow perspective on theological truth but rather all the confessions, united by common beliefs and the occasional counterpoint of disagreement.

THREE

"Of God and His Community": Kepler and the Catholic Church

In February 1605 the Italian mathematician Bartolomeo Cristini had good news to report to fellow mathematician and astronomer Giovanni Antonio Magini in Bologna. The papal nuncio to the imperial court in Prague, wrote Cristini, "declares that he hopes to convert Master Kepler to the Catholic religion within the next few days."[1] This imminent conversion would fulfill a dearly held hope of many of Kepler's Catholic friends and colleagues, who had often, subtly or directly, pressed him to consider the religious and professional benefits of becoming a Catholic.

Despite the rabid animosity that seethed between Catholics and Protestants in the years leading up to the Thirty Years' War, there was still a great deal of movement between the two confessions—some of it principled and some of it pragmatic. Kepler's own uncle Sebaldus converted to Catholicism and became a Jesuit in 1576, only to leave the Society of Jesus, take up a life as an itinerant astrologer, and then convert back to Protestantism.[2] Kepler's father, Heinrich, remained a Lutheran but took up arms as a mercenary on the Catholic side of the Dutch Revolt in 1574.[3] Christoph Besold, a fellow Lutheran and close friend of Kepler's—and one of the jurists who helped Kepler defend his mother during her witchcraft trial—shared many of Kepler's own irenic views. Though a subscriber to the Formula of Concord, Besold was sympathetic to many aspects of Catholic doctrine, and in 1630—the year of Kepler's death—he secretly converted to Catholicism, making the conversion public five years later.[4] There was ample precedent even among Kepler's own immediate Lutheran circle of friends and family for switching sides; it might even have made a great deal of sense for Kepler to do so, given that he worked for a Catholic emperor and that his own side ultimately rejected him in 1619.

Cristini's news of Kepler's imminent conversion was, however, wide of the mark; Kepler had no intention of converting to Catholicism (or to the Calvinism of which he was so often accused) and remained committed to his Lutheran faith long after 1605—and, indeed, long after his official excommunication from the Lutheran Church to which he clung so tenaciously. Yet Cristini's report, though false, was just one of many rumors about Kepler's impending conversion that swirled around Kepler throughout his lifetime. These rumors stemmed, in part, from Kepler's close personal and professional relationships with many influential Catholics. These included Bavarian chancellor Herwart von Hohenburg, lifelong friend and patron of Kepler, as well as Jesuit mathematicians such as Paul Guldin, Johannes Decker, and Albert Kurz. These men not only corresponded with Kepler on questions of mathematics, astronomy, astrology, and chronology but also helped Kepler to circumvent some of the troubles that plagued his fellow Lutherans living in Catholic territories in the empire. Herwart, for instance, likely helped Kepler to obtain the exemption that allowed him to remain temporarily in Graz when all other Lutheran teachers and priests were banished by official decree.[5] Guldin helped Kepler with some of his publication difficulties and arranged for a Jesuit telescope maker to provide Kepler with his very own telescope.

The close interest that so many Jesuits took in Kepler, particularly after he rose to the position of imperial mathematician, bolstered the impression that they harbored hopes for the conversion of so well respected a figure. Indeed, in letters to Kepler his Jesuit correspondents emphasized both their respect for his mathematical prowess and their fears for the current state of his soul, fears to which they believed he would be receptive. As Albert Kurz wrote to Kepler, the very fact that Kepler was willing to discuss his theological difficulties implied that he had doubts about his faith, and those doubts would be resolved if he would only trust in the leaders and traditions of the Catholic Church. Kurz wrote: "In other things I will admire you, in other things I will praise you but in that business of your salvation, I ask you to rely on the blood that drips from Jesus Christ rather than on your mental prowess alone, as though you alone could hope to touch on the truth without a guide or companion; rather, you ought to fear going astray amid the light and company of so many great stars."[6] Nearly echoing the Lutheran theologians at Tübingen who warned Kepler about the dangers of using reason to navigate the world of faith, Kurz urged Kepler to rely on the guidance of others skilled in the truths of theology rather than on his own limited understanding of those truths.

The Jesuits believed that they were the ones best suited to offer Kepler the guidance he needed to find his way to the truths of Christ and his church, and they made repeated attempts to do so. Kepler's own willingness to consider Jesuit perspectives on mathematical issues, and even to take the Jesuit side in a chronological dispute against the Protestant one,[7] signaled to the Jesuits that he would be receptive to their message. Still further, at a time when relations between Catholics and Protestants were stridently antagonistic, Kepler's willingness to seriously consider the Catholic position on questions of theology and his repeated declarations that he belonged to the "Catholic Church" made it seem that his eventual move to Catholicism was already well under way.

In this chapter, I consider Kepler's relationship with both specific Catholics and with the larger Catholic Church. I do not offer an exhaustive survey but rather focus on some emblematic episodes that help tell a story about Kepler's approach to religious community and his hopes for the future of the church, understood in very broad terms. In particular, I focus on a theme that Kepler stressed throughout these episodes, and indeed throughout his lifelong attempts to articulate a theory of churchly harmony—the theme of accommodation.

The theory of accommodation has ancient roots in both classical rhetorical theory and the theological discussions of the church fathers. I will consider rhetoric and rhetorical accommodation closely in the next chapter; here, I focus on the theological associations of accommodation and the ways that Kepler invoked them in his encounters with the Catholicism of his day.[8] To understand those invocations, we need to recognize that the church fathers used accommodation in a variety of different aspects, the most common of which was hermeneutical. The phrase with which they asserted this hermeneutical approach, taken originally from Jewish sources, was *scriptura humane loquitur* (scripture speaks in the language of man). The language of scripture, in other words, was accommodated to human understanding and perception.[9] Much of the language of the Bible, in this view, did not reflect the way things really were but rather the ways that humans understood or engaged with them. Anthropomorphic statements about God were just some examples of biblical language commonly understood as accommodated in this way. Church fathers often analogized this kind of biblical linguistic accommodation to the ways in which adults spoke to children in order to be understood by them. Origen, for example, referred to accommodated biblical language as a "children's language" spoken by a schoolmaster.[10] Calvin followed in this tradition of accommodation years later when he spoke of accommodated scriptural language as "lisping" and explained that "God's mind . . . contains nothing corporeal nor any corporeal accident, nor any human emotion. . . . Instead Scripture, by lisping indistinctly with the words of men . . . accommodates itself to men's weakness, like a nurse or a mother using baby language to her child to make herself understood."[11]

A related but broader approach to accommodation was pedagogical. According to this approach, it was not merely specific language in the Bible that was accommodated but God's overall historical interaction with humanity. Since human culture and understanding evolved over time, God accommodated his requests and expectations to the particular stage of his people at any given time. The twelfth-century Jewish scholar Maimonides relied heavily on pedagogical accommodation in his approach to Jewish law, and many of the church fathers—Augustine in particular—saw God's historical dealings with his people through this lens.[12] God had commanded sacrifices early on, according to this perspective, only because he knew that such practices were appropriate for that historical period; as Augustine explained, "the sacrifice which God had commanded was fitting in those early days, but now it is not so. Therefore he prescribed another one, fitting for this age, since he knew much better than man, what is suitably accommodative to each age."[13] This was true, Augustine argued, for God's words and commands to his people throughout the ages; he spoke to them and requested of them things that they were able to understand and appreciate, and these things changed with time, as the people developed. In Augustine's words, "the word of God appears to have dealt with the history, making the capacity of the hearers, and the benefit which they were to receive, the standard of appropriateness of its announcements [regarding him]."[14]

Beyond the hermeneutical and pedagogical approaches to accommodation, there were two additional theological ways in which God accommodated humanity: one due to his transcendence and a related one that was tied to the Incarnation in particular. According to those who emphasized God's transcendence, accommodation went beyond specific scriptural language or even specific pedagogically oriented practices and lay behind the ability of humans to interact with God in any way whatsoever. As the distance between God and man was so immense and—from the human perspective—so utterly unbridgeable, it would be impossible for humans to understand or know God in any way without God's willingness to somehow lower himself, to accommodate his transcendence such that his followers might grasp him, however imperfectly. The Incarnation was often viewed as one of the central aspects of God's accommodation—what greater accommodation could there be to human weakness, after all, than to assume human flesh? Luther referred to both these aspects of God's accommodation when he wrote that "if he should speak to me in his majesty, I would run away—just as the Jews did. However, when he is clothed in the voice of a man and accommodates himself to our capacity to understand, I can approach him."[15]

Still another way that accommodation was often invoked in theological discussions relied on the example of Paul in the Bible. Paul, according to some of the church fathers, took divine accommodation as a personal model, which in turn, he argued, ought to become a model for all Christians in their interactions with each other. As John Chrysostom explained, Paul exhibited contradictory behavior—at times following Jewish law, for example, and at times overturning it—because he hoped to accommodate himself to the needs of his followers and in so doing to help as many of them as possible on the path to salvation. As Paul claimed in 1 Corinthians 9:22, "I have become all things to all people so that by all possible means I might save some." Rather than insisting on the one true way to act, Paul—following the divine model—opted to focus on the specific needs of the various communities he addressed, subordinating the idea of truth to the good of the larger Christian community.[16]

Kepler was clearly familiar with all these forms of theological accommodation, and he relied upon all of them throughout his work. In his interpretation of scripture, Kepler depended heavily on the idea of hermeneutical accommodation. He wrote in the introduction to his *New Astronomy* that "the Holy Scriptures . . . speak with humans in the human manner, in order to be understood by them. . . . No wonder, then, if Scripture also speaks in accordance with human perception when the truth of things is at odds with the senses, whether or not humans are aware of this."[17] In his efforts to argue for the truth of Copernicanism against biblical passages that seemed to suggest a geocentric earth, Kepler asserted that the passages were accommodated; when, for example, the Bible spoke of Joshua stopping the sun, it portrayed the event as men would perceive it, not as it had actually occurred.[18]

Kepler relied, too, on the transcendental and Christological forms of accommodation, though in unique ways. As we saw in chapter 1, Kepler believed that though the gap between God and man was wide, it was not unbridgeable: man could begin to understand God because of the geometric archetypes that linked God, man, and the cosmos together. Tellingly, Kepler at times used the language of accommodation when referring to these archetypical linkages, arguing in the *Mysterium Cosmographicum*, for example, that God had deliberately constructed the universe so that the number, distances, and motions of the planetary bodies were "accommodated" to one another and to man's perceptions.[19] Likewise, in the *Harmony of the World*, Kepler emphasized that God had "accommodated" each planet's speed to the harmonic proportions.[20] Kenneth Howell has emphasized that for Kepler this kind of cosmic accommodation was closely linked to the hermeneutical kind: "in both cases, God has adapted his creative acts (or words) to the capacities of human minds."[21] It may, however, be even more accurate to say that for Kepler, both the cosmos and man were accommodated to the geometric archetypes originating in God himself. Still further, Kepler linked both these kinds of accommodation to the idea of the Incarnation when he claimed that both the geometric sphere and the cosmos represented the Trinity. According to Kepler, God accommodated himself both by taking on flesh in the form of Christ and by representing that incarnation—and the triune nature of divinity—in both the cosmos and the geometric forms that underpinned it.

In this chapter, we will see two more instances where Kepler argues for accommodation, and does so in particularly distinctive ways. Both are linked to the traditional arguments for accommodation, specifically the pedagogical and the communal/Pauline. As we will see, Kepler portrayed astrology itself as a kind of divine accommodation—yet another means by which God might reach out and speak to his people in ways they could understand. In some ways this was akin to hermeneutical accommodation; the heavens were as much God's canvas as the Book of Scripture. Yet Kepler linked it even more to the idea of pedagogical accommodation. Scriptural exegetes had long seen particular aspects of the law—especially the Old Testament law—as accommodations to the historical needs of an evolving people. Kepler replaced the historical element of this tradition with the different theological practices and beliefs of the major confessions. In so doing, he argued that God might use astrological signs to denote the various

Catholic sacraments—even though many of those sacraments might not have any real significance—in order to accommodate his message to the theological ideas embraced by his worshipers. That larger message, and the ways that it might ultimately shepherd his people toward divine truth more generally, was more important than the truth of the terms that it used.

In support of this position, Kepler referenced José de Acosta's theory of priestly accommodation when it came to the idolatrous practices of the New World. Much as priests might rely on the ideas of the native peoples about their own gods in order spread the word of the one God, God himself might rely on Catholic ideas to spread his true message. This linkage—between idolatry and issues of confession—was invoked quite often during the sixteenth and seventeenth centuries, usually to the opposite effect.[22] Idolatry was often invoked as a synonym for false belief, or heresy, and concerns about idolatry in the New World were often linked to the polemics of confessional disputes in Europe itself. All those who believed differently than one's own confession, in this view, were idolaters worthy of punishment. Kepler took the opposite approach and relied upon the idea of idolatry to urge more inclusiveness rather than more intolerance.

Kepler's pedagogical usage of accommodation was thus very much entwined with the Pauline and communal usage, for its purpose was not just to teach but to unite and conciliate. And as we will see in this chapter, Kepler focused directly on the idea of communal accommodation, not just in a general and theoretical sense but in a very personal and specific one. When the Lutherans refused to allow him to receive Communion, Kepler argued that he should be allowed to participate in the Mass in a Catholic Church, with his particular theological objections duly noted. In so doing, he argued—much as he had done when attempting to regain admittance to the Lutheran Church years earlier—that God's accommodation ought to be imitated in very practical ways by those congregations that sought to serve him. This meant reaching out not to the limited audience of those who agreed with one another on every particular but to the broad audience of all those who called themselves Christians. A religious community, Kepler contended, ought to accommodate itself to multiple individual perspectives and beliefs for the good of the whole. Though some positions were truer than others, division would only hurt the communal body, whereas peace would strengthen it. If peace could not be achieved by a unified position on all the doctrinal particulars, then accommodation of difference was the only way to reach any kind of harmony.

Ultimately, when Kepler argued for accommodation, he was really arguing for harmony by another name. The language of harmony had already been associated with the idea of accommodation since at least the time of the church fathers. When Augustine wrote about God's pedagogical accommodation through the sacrifices, he noted that "He, who knows better than man what pertains by accommodation to each period of time, commands, adds, augments, or diminishes institutions . . . until the beauty of the whole history, whose parts these periods are, unfolds like a beautiful melody."[23] History itself was harmonious, according to Augustine, because of the differences that characterized its successive epochs. As Amos Funkenstein has noted, Augustine shows us "that the process of history . . . is as beautiful, if seen as a whole, as the cosmos is, and for the very same reasons."[24] Kepler made a similar argument, though about a time-bound rather than a historical community: accommodation of individual difference lent beauty to the whole. Here, he followed directly in the footsteps of Erasmus, who also linked communal accommodation to harmony, maintaining that "if a moderate accommodation softens the paroxysms of discord, the medicine of the synod will be more effective in producing concord."[25] To be like God was to accommodate, and accommodation meant the active pursuit of harmony on earth, reflecting the harmony that God had embedded in the cosmos.

Kepler, Herwart, and the New Star of 1604

In September 1604, observers across Europe were enthralled by a dazzling new object in the sky, often referred to as a new star. Portentous in its own right, the new star was made doubly significant by the place of its emergence: it appeared close to the conjunction of Mars, Jupiter, and Saturn in the sign of Sagittarius—a conjunction that initiated the Fiery Trigon, a period of great astrological significance. These two momentous events, and their close proximity, resulted in a flood of pamphlets arguing for the new star's earthly significance. Kepler also produced a work that focused on the implications of the new star—*De Stella Nova*—though it was not published until two years later, and it was far more circumspect than many of the texts that preceded it.[26] The majority of the book was devoted to the physical significance of the new star, while only the final chapters addressed what future events it might portend. And though Kepler willingly speculated on some possibilities for its future significance—among them the fall of the Ottoman Empire, the

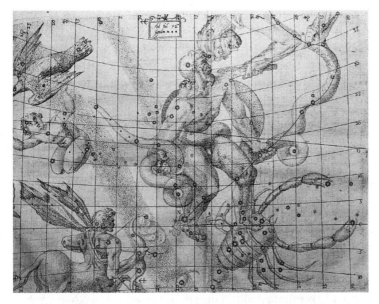

3.1   The location of the 1604 supernova (labeled N) amid the constellations, from Kepler's
De Stella Nova (1606)

Second Coming of Christ, and the conversion of all nonbelievers to Christianity—he did not endorse any option as certain. Reading detailed portents into the heavens was risky business, he argued, because God had provided no formula by which astrological phenomena could be easily interpreted. "If it had pleased God to openly indicate what he wished to men," he wrote, "he would have inscribed it in the heavens with written words; thus, men struggle in vain to conjecture about the divine will."[27] Though he considered it clear that the new star, appearing precisely where and when it did, was a sign of divine providence, Kepler hesitated to proclaim what such providence signified in the realms of communal politics or religion. Instead, he urged his readers to use the new star as an opportunity to examine their own lives, while maintaining a sense of humility about its global significance.[28]

Rather than focus on the arguments in *De Stella Nova*, I want to consider instead an exchange that took place shortly after its publication between Kepler and Hans Georg Herwart von Hohenburg, Bavarian chancellor and friend and patron of Kepler's. Though primarily a statesman, Herwart also had deep scholarly interests in astronomy, chronology, mathematics, and philology, and it was Kepler's detailed answer to a chronological query of Herwart's in 1597 that initiated a relationship that would span fourteen years and more than ninety letters. Over the course of this relationship, Herwart lent Kepler books, supplied him with information about the work of other scholars in the Republic of Letters, and pushed Kepler to refine his own scholarly views, particularly when it came to questions of astrology.[29] Herwart had read Kepler's *De Stella Nova* with some interest. There, Kepler had suggested—though only as possibilities—some religious implications of the new star, yet he had not considered what significance the new star might have when viewed within a specifically confessional context. In March 1607 Kepler received a letter from Herwart asking him to push his ideas further and contemplate the distinctively Catholic significances of the heavenly bodies.

Kepler's discussion of the new star and the birth year of Christ, Herwart wrote, had motivated him to provide Kepler with a brief *judicium* he had written that considered questions of astrology and religion, as he was aware that there were many Catholic theologians who still attributed inclinations or significances to the stars. Herwart maintained a degree of skepticism on the question, noting that he had yet to find a firm foundation on which to base such suppositions; he therefore admitted that "I would well have cause to withhold [my *judicium*], for perhaps it is not worth the effort of writing or of rebuttal." Yet he had read in Kepler's astrological writings a similar uncertainty on issues of astrological inclination or signification, and for that reason he hoped that Kepler would openly communicate his thoughts on the matter. He asked that Kepler keep Herwart's musings secret, since they were, after all, only "bare speculation." Moreover, he emphasized that though his speculations were "drawn out from the tradition of our Catholic Church," with which he knew Kepler was not in full agreement, he hoped that Kepler would take them as nothing less than well intentioned.[30]

Despite the Catholic Church's formal opposition to the practice of judicial astrology (predictions about specific individuals or enterprises), Herwart's contention that many Catholic theologians still accepted its basic tenets was sound. Ugo Baldini describes the paradox of a society in which, up until the seventeenth century, "outright condemnations of judicial astrology coexisted with its widespread and public practice . . . and with its substantial acceptance by social elites, and even by the ecclesiastical hierarchy."[31] Part of this paradox lies in the church's somewhat unsystematic approach to the discipline of astrology, to the distinctions between natural astrology (including large-scale predictions about weather, war, famine, and the like) and judicial astrology, and to the enforcement of its own astrological condemna-tions. Though Pope Sixtus V's bull *Coeli et Terrae Creator Deus* of 1586 formally condemned astrology in fairly restrictive terms, arguing that "God has reserved certain knowledge of future things only unto him-self,"[32] the bull focused specifically on the application of judicial astrol-ogy to the future lives of individuals and did not discuss the implica-tions of natural astrology more broadly. And even after the pope's bull was officially accepted by the Catholic world, judicial astrology still had many supporters; Jesuits still taught private classes in astrology,[33] Catholic courts still had official judicial astrologers,[34] and astrological discourses were still addressed even to cardinals at the papal court.[35]

Though officially on doctrinally shaky ground, then, Herwart's as-trological speculation was in good company, as he indicated.[36] In that *judicium*, Herwart began his speculation by equating the three divisions of the heavens with the three aspects of God. The supreme and invis-ible heaven, according to Herwart, signified the Father; the intermedi-ate heaven, or the fixed stars, signified the Son; and the lower heaven, or the seven planets, signified the Holy Spirit. This kind of association, he would have known from Kepler's letters and books, was very famil-iar to Kepler, who himself had identified the Trinity in the configura-tion of the heavens (though Kepler's own argument for the Trinity was slightly different, as we've seen). Herwart then moved on to the more exclusively Catholic portion of his *judicium*. Since the seven planets, in his view, represented the Holy Spirit, they represented more specifically the gifts of the Holy Spirit, or the seven sacraments. The moon, because of its connection with water, represented baptism. Mercury, wrote a cynical Herwart, represented the sacrament of marriage, as it was "somewhat obscure, wandering, constantly beneath the sun or needing penance." By extension, Venus, "clear, lucid, shining, most beautiful of all," represented the sacrament of Holy Orders. The sun represented penance, as it was the source of all the other planets, which all inclined themselves toward it and depended on its motion, just as all the other sacraments depended on penance. Mars, which recurred every two years, corresponded to childhood and hence to confirmation. Jupiter, recurring every twelve years, corresponded to puberty and hence to the Eucharist. Finally, Saturn's thirty-year cycle corresponded to the sacrament of extreme unction, as Herwart noted that thirty years "con-cludes the age of the perfect man"—likely an allusion to Jesus, who began to preach at the age of thirty, according to Luke.[37]

In his reply to Herwart, Kepler first addressed Herwart's contention that he hadn't clearly articulated his own astrological position, and he took pains to emphasize that he generally considered himself an opponent of the kind of astrology practiced by the majority of its adherents. "What else is the entire little book," he wrote of *De Stella Nova*, "but a near crucifixion of all judicial astrology, with the aspects alone enduring as parts of the natural order?" But he willingly engaged with the linkages that Herwart had posited between the heavens and the seven sacraments and included a *"judicium de speculatione"* of his own at the end of his letter. He first addressed the premise that the heavens represented the church overall and deemed this premise "probable." "Because it is a general belief," he explained, "that God sends us signs with reference to our earthly circumstances and sends these signs from the heavens . . . [and] the most important of our circumstances are those related to the church." Therefore, he concluded, it was sensible to assume that heavenly signs referred to churchly matters.[38]

Kepler then went one step further and conceded it likely that heavenly signs referred to matters specifically relevant to the Catholic Church. His argument for this assumption was based primarily on practicality:

If I concede it plausible that it is not absurd that God speaks with astrologers, who are few in number, and forms his words from the particular principles of astrology, although little certain, it will be much less absurd [to believe] that God says something through celestial signs to those who extend the name of Rome through the whole breadth of the Catholic Church—for they are today the most numerous and most powerful part of the world—and that he speaks to them in their principles and according to their understanding.[39]

God wants to speak to his people, Kepler explained, and to do so he will employ whatever means necessary for them to understand him. Just as Kepler had employed the principle of biblical accommodation in a hermeneutical sense in his interpretation of scripture, Kepler here implied that even if the sacraments of the Catholic Church carried no great weight from a divine perspective, God might employ them as a pedagogical means to transmit his messages, given the significance of the sacraments for so many people on earth.

Kepler had made a similar argument linking astrology to the principle of accommodation in the *De Stella Nova* itself when he discussed the possible effects of the new star. He expressed his hopes that the new star might signal the rise of a leader who would quell some of the warfare plaguing Europe, and he noted that such might have been God's purpose in making the star appear where and when he did. To clarify what he meant, he provided an example from an entirely different realm: José de Acosta's descriptions of his encounters with the native peoples of the New World in his *Natural and Moral History of the Indies*. Kepler focused on Acosta's descriptions of Indian idolatry and the remedies for it, and he quoted directly Acosta's account of the ways in which idolatry was an inseparable part of daily life for the Indians. He highlighted in particular Acosta's suggestion for the extirpation of idolatry and his belief that "salutary rites be introduced in place of noxious ones, and ceremonies be replaced by ceremonies."[40] Rather than trying to remove all shades of idolatry entirely, Acosta had argued, priests should consider the function that idolatrous practices served. Idolatry was essentially about religious ceremony, and hence the best way to remove idolatry would be to substitute religiously appropriate ceremonies for the idolatrous ceremonies to which the Indians were accustomed. Priests might accommodate the Indians' habits, desires, and ways of thinking by channeling them to an appropriate religious end. Still more, the idolatrous practices of the Indians, instead of being a stumbling block, could be used as a means by which to bring them to the church, by showing the Indians the ways that Christian practices could serve similar purposes.

Kepler implied in *De Stella Nova*, through quoting this passage at some length, that God's use of the new star might be conceived of in similar terms to the priestly use of religious ceremony in the New World. Many of the astrological beliefs of the common man might well be baseless superstition, that is, yet that very superstition might serve a purpose. Like the priests who might use the desire for religious ceremony inherent in the practice of idolatry, God might use the astrological beliefs of his people, though faulty, as a means to turn their attention in appropriate directions and effect positive change.

Kepler made this kind of argument once again in his 1609 *Antwort auff Röslini Discurs*, a reply to a work by Helisaeus Röslin in which Röslin supported the entire discipline of astrology against some of Kepler's attempts to limit it in *De Stella Nova*. There, Kepler insisted that his arguments against the natural foundations of certain astrological principles like the zodiacal signs, or even his argument that the nova of 1604 signified nothing "by nature," need not imply that therefore they contained no messages for mankind. God might, he averred, employ mistaken but commonly held astrological beliefs to signify something of future importance. God might, that is, employ the very prejudices of his followers to lead them to truth.[41] Kepler's answer to Herwart's *judicium* was based on a very similar premise. The astrological linkage that Herwart had posited between the sacraments and the planets was warranted, Kepler argued, since it was yet another example of divine accommodation. Since sacramental language was the language of a large majority of his followers—though not, it should be emphasized, of Kepler himself—Kepler believed that it was sensible that God would embrace that language as a means to reach them.

One might suppose that Kepler conceded so much to Herwart because of the value of his patronage, not because of genuine agreement on Kepler's part. Herwart was a courtier with important connections, and Kepler certainly had no wish to alienate him. Yet Kepler, as a Lutheran, clearly and openly disagreed with his Catholic patron on many questions of religion. Here, however, he exhibited no such dissent. "This all seems to me entirely believable," he concluded, so much so that he denied the possibility of producing a better comparison of heavenly bodies with churchly things.[42] Kepler then ventured still further to contemplate the meaning of the new star in the sacramental context Herwart had conveyed. The new star, he speculated, signified a great new bishop, and its coincidence with the Fiery Trigon signified a new doctrine or heresy connected with the sacraments of extreme unction, the Eucharist, or confirmation. He speculated still further that because the new star appeared along the ecliptic, it signified that the new bishop would assume power in the usual way, through apostolic succession. And as the new star was beautiful, the new bishop would entice people with his words but would fall from power quickly, as the new star had disappeared in February or March 1606.

At this point Kepler rather abruptly backed away and ceased elaborating further on the star's significance. Because his speculation had led him to the realm of prophecy, he wrote, he would end his ruminations and be content with this short "prelude," as he called it. He likened this to his attitude in *De Stella Nova*, in which he had briefly speculated on the new star's future significance only at the end of the book, where, as he wrote, "I wanted to add an ending appropriate to the *fabula*." "For truly," he explained, "there is no other method in my predictions than the one that my speculation has used here."[43] His astrological predictions, that is, like the speculation he sent to Herwart, were based not on certain truths but rather on assumptions that needed to be accepted in order for the conclusions to seem valid.

This is not the only place where Kepler highlighted the uncertainty that lay at the heart of astrological speculation. He emphasized precisely this point in *De Stella Nova* when he left the significance of the new star undecided. There and elsewhere, Kepler compared the practice of astrology to the practice of medicine—both were imprecise arts, based on a posteriori observations and experiences, in striking contrast to the certain demonstrability of mathematics. Kepler also emphasized the degree to which the practice of astrology was tied up with cultural assumptions—he included here the signs of the zodiac and their associations with the earthly elements.[44] Should any of those assumptions be false, predictions based on them would also be false. Finally, Kepler emphasized the inscrutability of God's intentions; professing certainty about the future plans of God could only be a sign of hubris.[45]

Kepler's speculative play with Herwart, then, is revelatory on two levels, both of which pertain to his understanding of religious community and confessional identity.[46] First, it illustrates Kepler's very broad conception of God's relationship with his people, be they Lutherans, Calvinists, or Catholics. Kepler himself had a complicated view of his own confessional identity; as noted in the previous chapter, he disagreed with some of the central tenets of Lutheran orthodoxy and aligned himself more closely with the Calvinists when it came to the presence of Christ in the Eucharist. Yet Kepler's consideration of Herwart's Catholic astrology is not indicative, as some have argued, of any close affinities with Catholic sacramental thought.[47] Instead, it reveals Kepler's very practical understanding of God's communications with his people, as well as his expansive sense of who God's people were. Kepler relied on the pedagogical notion of divine accommodation and extended it from the realm of law and worship to the realm of astrology. Just as God could command practices like sacrifice, though they were essentially meaningless, in order to lead men on the ultimate path to salvation, so too could God utilize the bodies of the heavens to refer to faulty theological doctrines of men to achieve the same purpose.

Though Kepler himself had specific opinions about which positions were theologically valid and which were not, this did not mean that God was interested in addressing only the privileged few who had succeeded in divining this religious truth. Unlike many of his time, that is, for whom God's people represented very narrowly their own confessional allies—and for whom members of competing confessions were not merely misguided but were sinful heretics—Kepler understood God's people in a very broad sense. God's people were all Christians, regardless of confession. Though Catholics had certainly erred in adopting specific theological positions, like the belief in the seven sacraments, this did not mean that they had forfeited the right to their place in the larger brotherhood of Christendom. And as God undoubtedly intended to communicate with all his people—among whom

Catholics still encompassed the majority by far—considerations of the manner in which he might do so, and the specific religious language that he might employ, were surely warranted. Indeed, the very fact that Kepler cited Acosta—a Jesuit whom most Protestants would have been loath to embrace—illustrates his own very broad notion of where truth might be found.

The second noteworthy aspect of Kepler's discussion with Herwart lies in the reason he hesitated to speculate too closely on any concrete confessional implications of the planets. In marked contrast to Kepler's earlier insistence (see chapter 1) that the cosmological claims of the *Mysterium Cosmographicum* were demonstrable via reason alone and therefore indisputable[48]—as were any religious conclusions drawn from them—Kepler argued here that his astrological speculations lacked method and certainty, as astrology was a discipline characterized by a posteriori observations and experiences collected over time. It would therefore be impractical, he claimed, to draw any strong religious conclusions based on them. Yet it is likely that Kepler recognized both the impracticality of such speculations and also the danger inherent in them. After all, the problems of confessional divide were tearing Europe apart in direct and deadly ways, ways with which Kepler was already all too familiar. To attempt to speak about the future of an individual was uncertain business, to be sure—but to attempt to do so about religious groups, and in weighted, confessional terms, was a far more dangerous game to play. Kepler was fully aware that his speculations could easily be mobilized by opposing groups to suit their own agendas and could be used to fan the flames of a fire already perilously out of control.[49]

Moreover, much as he knew that astrological predictions were based on cultural assumptions, he knew too that religious conflicts were based on assumptions of their own. Kepler had argued in *De Stella Nova* that David Fabricius, a Lutheran theologian and astronomer who had also observed and interpreted the new star, had allowed his own biases to color his interpretations. Fabricius's predictions, he wrote, were "nothing other than complaints about his neighbors, his opinion concerning the condition of the empire, and his desire for vengeance and improvement."[50] Worried that any predictions he might make would either be interpreted similarly or, still worse, be used to further the grievances of others, Kepler steered clear of making claims with obvious confessional implications when they were based on so uncertain a foundation to begin with. If his work was more likely to aggravate disputes between the confessions rather than settle them, Kepler sought to de-emphasize its confessional relevance. He highlighted what he perceived to be the essential continuity between cosmos and confession only when he felt that such continuity would help to heal the church rather than add to its troubles. And as we saw in chapter 2, he ultimately shied away from emphasizing even those continuities that he felt were clearly warranted by his a priori metaphysical commitments when he felt that he had no audience willing to listen.

Kepler, Guldin, and Religious Community

By 1627 Jesuit mathematician Paul Guldin had already sustained a friendly correspondence with Kepler for many years. The two had discussed innovations in mathematics and the hardships of daily life, and Guldin had even arranged for Nicholas Zucchi, Jesuit mathematician and telescope maker, to gift Kepler with one of his telescopes. In thanks, Kepler sent Guldin copies of his *Harmony of the World* and the second edition of the *Mysterium Cosmographicum*, and he had appended his *Somnium* with a grateful letter to Guldin. There, he asserted that "there is hardly anybody with whom at this time I should prefer to talk about astronomical studies face-to-face than with you"[51] and emphasized that he delighted in both Guldin's friendship and his scholarship.

When in 1627 Guldin finally broached the topic of religion by asking Kepler whether he would consider conversion to the Catholic Church, Kepler's answer was likely not what Guldin expected. "My piety toward God must be considered very poor indeed if I began to be a Catholic only now," Kepler wrote. "[After all,] I was baptized into the Catholic Church by my parents immediately when I first crossed the threshold into life . . . nor did I ever leave the church from that time, nor did I act otherwise than according to the church's particular doctrines."[52] Rather than respond positively or negatively to Guldin's question, that is, Kepler insisted that it was beside the point—why should he convert to Catholicism when he already considered himself a lifelong member of the Catholic Church?

Of course, what Kepler had really done was redefine the question by redefining the meaning of the "Catholic Church." Indeed, he proceeded to explain to Guldin exactly what the concept entailed for him—and what it did not. The Catholic Church could not, he argued, be defined solely in terms of the official hierarchy of that institution as it was constituted at the time. Rather, Kepler went on to clarify— very much in line with the teachings of Luther—that any hierarchy that saw itself as the promulgator of true doctrine was fundamentally unnecessary. The doctrines of the true church, Kepler believed, were available directly to all individuals through the grace of the Holy Spirit, conferred via baptism.

Kepler went on to explain that the arrogance and ostentation that characterized most of the leaders and institutions of the current Catholic Church made it clear that they had strayed from the true teachings of the Holy Spirit. He also claimed that certain churchly practices were clearly in error and pointed to idolatry, such as the worship of the image of the crucifix, of saints, and of Christ in the Eucharist (which Kepler thought characterized the idea of transubstantiation). He argued that the laity should be allowed to take part in the Mass under both kinds, the bread and the wine, and that their exclusion from it was a modern invention. "These are only a few [of the reasons]," he continued, "which inspire me to hold back from fully obeying the church of the Western Metropolitan and Patriarchs: which I reject not out of pride but because I think that Christ, its head, must be obeyed more."[53] Moreover, Kepler insisted that his firm commitment to the true Catholic Church was so strong that "on behalf of the refusal of those things which I do not recognize as apostolic, and thus which are not Catholic, I am prepared not only to abandon the rewards which are now held out to me, and in which his holy majesty [the emperor] most magnificently and liberally shares, but also the Austrian lands, the whole empire, and even something more serious than all those things, astronomy itself."[54] This was no mere boast: Kepler had already suffered much on behalf of his religious principles, and his frequent moves made it increasingly difficult for him to devote himself fully to his astronomical pursuits.

Up to this point, Kepler's answer to Guldin was fairly typical of Protestant attempts to redefine the idea of the church in the wake of the Reformation. From the start of the Reformation, reformers had based their movement on the idea that the Catholic Church was not the "true" church. At some point in the past, they argued, the Catholic Church had strayed from the truth; rather than creating something new, reformers were simply restoring the true doctrines and practices of the ancient apostolic church that had been lost over time. Though Catholics and Protestants alike agreed that some notion of "the church" needed to be continuous over time—after all, Christ had promised that the Holy Spirit would always be with his followers—reformers came to identify the true church not with the institutions of the official church hierarchy or even with the succession of individuals who made it up. Rather, they identified the true church with true doctrine, and with the small, continuous body of faithful people who adhered to it while the majority strayed.[55]

Kepler's answer, at least to this point, adhered fairly closely to this idea. In claiming to be a Catholic, he essentially redefined "Catholic Church" to mean "true church," and the ideas that he associated with it were for the most part Lutheran ones, with some notable exceptions. In beginning to answer Guldin, Kepler essentially responded by saying, "If by Catholic you mean true believer, than I am a Catholic," while implying that the Lutheran Church had established itself as the church that most closely aligned with true belief and practice. Yet in the remainder of his letter, Kepler departed from the typical Lutheran approach, which both asserted itself as the true church and also positioned the Catholic Church as not only false with respect to particular practices and doctrines but false through and through. In this view, the Catholic Church was a community brought to its current state not by weakness or human error but by the workings of Satan and the Antichrist. The true church was distinguished both by its own adherence to truth and by its active resistance and opposition to Rome.

Kepler, by contrast, continued his letter to Guldin by clearly distinguishing between the doctrines and practices of the Catholic Church with which he disagreed and the larger communal structures of the church. Though the Catholic Church clearly suffered from problems within, it was still a community of God, and it ought to be recognized as such even by those who disagreed with it. Still further, he explained that "I remain firmly in the Catholic Church, and I join in the full affection of God and his community, even when it rages and strikes out, as much as any human who suffers weakness. And if indeed I am tolerated, with those few reservations, I am prepared in silence and patience to practice and acquit my art among the common men of the predominant party while abstaining from all insults, sneers, improper interpretations, hyperbolic exaggerations, accusations, and perversions."[56] He argued, in other words, that the doctrinal errors of the Catholic Church would not prevent him from aligning himself to it and participating in its communal activities, with his theological reservations duly noted. Indeed, he was prepared to do just that, and he promised to act respectfully even toward those activities with which he disagreed, as befitted any member of a religious community.

This far, Kepler described a rather passive participation in a Catholic community—he would remain a silent observer, taking what he could and attempting to cause as little offense as possible. Kepler continued, however, to argue for a more active religious participation in Catholic communal activities: "In fact, there is a further condition by which I can both take part in the Mass and join my prayers to the prayers of the available church . . . : if it clearly accepts my protestation and that of all my [family] that we do not agree to those things that we believe to be in error, but only to the general, ultimate, holy and catholic intention of the Mass." If the leaders of his local Catholic community acknowledged his right to believe in his own conception of *communicatio idiomatum* and the presence of Christ in the Eucharist, that is, then there was no reason why Kepler could not participate in the Mass in a Catholic Church. His points of disagreement with Catholic doctrine were merely matters of personal conscience, as he had argued to the Lutherans earlier—matters that should not necessarily rule out participation in the church's communal religious activities. Kepler argued still further that this was a position with which Guldin ought to agree, because it was rooted in the Bible itself. "If you love me, I think that even you ought to support my candor," he insisted, "because you know the rule: he who has doubts is condemned if he eats, because it is not from faith: however, nothing is unclean through itself, unless there is someone who thinks it unclean."[57]

Here, Kepler referred to Paul's discussion in Romans 14 about eating meat deemed "unclean" by the laws in Leviticus. Those who are strong in faith, Paul had argued, would recognize that "nothing is unclean in itself" and therefore that no food could automatically make a person impure; those weaker in faith would continue to avoid eating unclean meat because they believed it would affect their own purity. Yet Paul insisted that two things ought to take precedence for both the weak and the strong in deciding whether or not to eat unclean meat: one's personal convictions along with the larger cohesion of the community. If the weak believed the meat to be unclean, Paul wrote, they should not eat it—they should be ruled by their own personal faith and conscience. Yet the strong likewise should not distress the weak by eating unclean meat in their presence, for, as Paul explained in Romans 14:7, "the kingdom of God is not a matter of eating and drinking but of righteousness, peace, and joy in the Holy Spirit."

With this discussion, Paul emphasized that even when one approach was theologically more justified than another, the harmony and peace of the community were more important than unity in religious practice. By referencing this passage from Romans, Kepler was referencing more broadly the idea of Pauline accommodation; just as Christ accommodated differences among his followers, so too did Paul. And as Paul had urged all those who followed the gospel to be imitators of him, as he was of Christ, other Christians ought to adopt this Pauline accommodation of differences in their dealings with each other. Kepler thus argued that so long as he did not violate his own beliefs, as Paul had held with regard to the unclean meat, there was no problem with him, as a Lutheran, participating in the Catholic Mass. To be sure, this emphasis on personal conscience over doctrinal consensus did not serve him well with his own Lutheran Church, and Kepler likely did not expect Guldin to take his suggestion seriously. Instead, he hoped to give Guldin some insight into his religious sensibilities; Guldin's binary opposition between Catholic and Lutheran was simply not the way Kepler saw the world.

Kepler ended his letter by emphasizing that though he had taken the opportunity to air some of his innermost thoughts, Guldin need not respond; it would be better, in fact, for them to return to the less troubling questions of astronomy. And though Guldin did not respond personally, he—apparently feeling that some reply was necessary— asked a fellow Jesuit to write on his behalf, perhaps hoping that a response from someone without as personal a relationship with Kepler would be more effective. Guldin then sent this reply on to Kepler. This Jesuit, whose name is never mentioned, undertook his task with the assumption that Kepler's letter showed clear signs that he wavered in his Lutheran faith. To refute Kepler's arguments would consequently be at once to prove him wrong on specific points and, at the same time, to potentially win him fully over to the Catholic side. He began his response by casting aside as foolish the claim that Kepler had somehow always been connected to the Catholic Church, "as if there were a necessary connection between baptism in Christ and Catholic opinions."[58] Indeed, he continued, if someone were baptized at infancy but seized on to heretical opinions as he reached the age of maturity, and continued to believe in those errors throughout his life, that person could hardly be called a true Catholic.

He likewise rejected Kepler's claim that the practices of the Catholic Church were riddled with errors and novelties and argued that by contrast, the essential elements of the church had remained constant from its origins until the present day. "Thus to condemn all these things," he wrote, "to reject them as errors and to accuse them of abuses, is to be filled not with the divine spirit but with a malignant and poisonous hatred of the church."[59] He emphasized that Kepler's belief that the Holy Spirit alone would allow him to distinguish between truth and falsehood was belied by the short history of the Protestant Church, a history in which Luther's initial error led to the multiplicity of further errors and sects. As a reliance on personal interpretation clearly led to the arbitrary multiplication of beliefs, a reliance on the long history and traditions of the Catholic Church and on its leaders throughout the ages was to be preferred. He focused on the small number of Protestants compared with Catholics, and on the size and power of the Catholic Church, as well as the might of the Catholic emperor and the Austrian empire. The successes of the Catholic Church proved that God must approve of it, for he would not have allowed it to thrive for so long had it been in error.

As Kepler explained in his letter of reply to Guldin, he was much troubled by this letter of rebuttal, both by its tone and also by the way it appeared to misunderstand his central claims. The writer had done as many "clever men of our time" tended to do, wrote Kepler; to defend himself, he had attacked the words of another and taken them completely out of context. This was true on a general level throughout the letter, according to Kepler, for "while I said what I did for the sake of explaining my innermost spirit, it was attacked as if it were intended as a public offense against the Roman religion." This kind of debate was no true conversation, for if Kepler chose to respond still further, at each step along the way his words would be twisted far beyond their original intent. As an example, Kepler offered the following: Guldin's colleague had argued that the successes of the Catholic Church and emperor proved that they were correct in their faith, for otherwise God would not have permitted them to succeed. Kepler responded: "If I were to insist that the first emperors were clearly infidels and yet they seized the rule of all the lands, the enraged and hostile debater would then interject: so you compare the emperor to an infidel? I would have been ready to reply, but it does not follow: but hatred emanates from closed-up ears." Accordingly, Kepler declared that he could respond to the entirety of the letter by Guldin's colleague simply by asserting that "my refuter has deviated from my viewpoint."[60]

Kepler did briefly enumerate some of the specific instances where he had been misunderstood. More broadly, he emphasized that he had not intended to attack the beliefs of his friend, for he conceived of the church in terms that transcended the time-bound perspectives of both Catholics and Lutherans. "I acknowledge that the church is one and the same at all times," he wrote, "but its members are mortal. The church was [around] before Rome, still more before Wittenberg. Just as quarrels appear among the citizens or factions of a republic, so also among members of the church, separated by time and place, errors appear out of human weakness." He insisted that Guldin and his fellow

Jesuits need not argue with him—rather, he wrote, "dispute with someone who plainly cuts Rome from the church, and who does not know
how to distinguish between the temple of God and the one sitting in it,
and thus concludes that the whole church is the Lutheran Church."[61]

Kepler here clearly distinguished his own views about the definition of the "church" from those of many of his fellow Lutherans.
While central to the Protestant understanding of the church was the
idea that the church had turned down the wrong path at some point
in the past, this was directed very pointedly at Catholics and Catholic
history. The idea that the church was subject to error and decay was
true, many Protestants insisted, not for "our" church but for "their"
church, the Catholic Church—"their" ancestors had erred, not "ours."
Moreover, error was typically attributed not to human weakness or error but rather to the work of the Antichrist—to evil within the part of
the church that had gone wrong. Kepler, by contrast, argued that the
church, like any human institution, was fallible on all fronts, and in all
its variants. Just as in a political community, Kepler reasoned that in a
religious community the passage of time necessarily led to the accumu-
lation of mistakes and wrong turns, ones to which Protestants were as
susceptible as Catholics. Indeed, as we saw earlier, Kepler found fault
with many of the Lutheran theologians of his day and emphasized the
humanity of even those theologians, like Hafenreffer, whom he greatly
admired. And though he continued to identify as a Lutheran, he defined the "true" church not as the Lutheran Church, inscribed backward in time, but rather as something that transcended the particularities of confession. The true church existed before Rome, as it did
before Wittenberg; it was not bounded by the beliefs and structures of
the Lutheran Church, nor did it exclude the Roman Catholic Church,
though it did not correspond exactly to it either.

As a result, Kepler argued that it was necessary "to distinguish
between the temple of God and the one sitting in it"—between the
broader idea of religious community and the particular leaders or doctrines that happened to hold sway in any given time or place. Much as
in his earlier discussion with Herwart, Kepler emphasized that while
particular confessional communities were limited, God's community
included all of Christendom, Catholics and Protestants alike. And he
implied that as no one religious community could have a monopoly on
the truth, confessional disputes and partisan exclusivity were typically
only destructive forces. Where no one principle could be agreed upon
by all the confessions, personal conscience ought to decide the particularities of faith, while the general outlines could be shared by those

who might continue to differ on the particulars. Religious community more generally, Kepler insisted, ought to be characterized by harmony and freedom rather than unanimity of practice and belief.

Kepler's position here—his willingness to align himself with the Catholic Church and its religious practices—is noteworthy when compared with his approach in a letter to Johannes Pistorius, father confessor and adviser to Rudolf at Prague. Pistorius had been raised as a Lutheran but had later converted to Calvinism and eventually to Catholicism, in which he studied for the priesthood and rose through the ranks to become imperial confessor. Kepler and Pistorius generally respected each other's learning; as imperial mathematician, Kepler was asked to report his findings to Pistorius periodically, and the two carried on a lively and wide-ranging conversation during their years together in Prague. In 1610, two years after Pistorius's death, Kepler referred back to those conversations, hailing Pistorius as "that polyhistor of all the sciences."[62] Yet in 1607 Kepler had harsher words for his colleague. Pistorius, suffering from serious illness, had written to Kepler that he was certain the end was near. He added that this knowledge did not trouble him, for he rejoiced in the possibility of going to heaven and leaving behind the "inanities" of the world.

In response, Kepler expressed his strong sympathies for Pistorius and noted that though the thought of Pistorius's death filled him with sadness, he was "comforted by your upright courage and your unwavering desire for the eternal life of which your letter bears witness." But he also took the opportunity—in particular, Pistorius's use of the word "inanities"—to reflect on what he considered to be one of the primary causes of the world's inanities: the practices of the Catholic Church, with which Pistorius was affiliated. He wrote: "Of the inanities of the world . . . I think a great part stems from the fiery zeal of the factions, who believe that they have a special right to blessedness, decide that the only gates to heaven are at Rome, and [are filled] with complete contempt for those who embraced their liberty." The Catholic Church had completely skewed the teachings of the apostles, leading to "a Roman monarchy or church tyranny." In the face of this tyranny, Kepler insisted that he remained "in that freedom in which I was born with God's approval, and which insists that I not voluntarily bend myself to the Roman yoke of those who not only force Christians into following indifferent ceremonies [i.e., adiaphora] . . . but also interpret the words and commands of Christ and the apostles in a most dangerous way and claim the right of interpretation only for themselves and put human understanding in chains, so that one cannot judge otherwise but the interpretation here." Kepler claimed that "if one admits to this strict interpretation, then in the end the Antichrist will lack nothing . . . in order to erect his empire in the church and overthrow the dominion of Christ."[63]

This letter is nearly always cited by historians who have considered Kepler's approach to the Catholic Church. Some have viewed it as paradigmatic; for this reason, M. W. Burke-Gaffney labels Kepler generally as "anti-Catholic" and notes that his "anti-Catholic views were exposed plain and unvarnished" in his letter to Pistorius.[64] Others have simply found the letter puzzling. Max Caspar writes that "Kepler, otherwise so peace-loving in religious matters," in his letter to Pistorius leveled "such a sharp attack on the Catholic Church, as scarcely seems possible from him and of a kind which he makes nowhere else."[65] Caspar speculates that "memories of specific earlier conflicts" played a part in the harshness of Kepler's response—perhaps his expulsion from Graz. He also emphasizes that the harsh attitude in Kepler's letter to Pistorius was likely tempered over time by Kepler's experiences with the Lutheran Church, experiences which made him realize that "the Württemberg consistory was more popish than the Pope, and did not possess that tolerance with which Kepler was treated by the Catholics in Prague and also later in the Hapsburg lands."[66]

It is certainly true that Kepler's approach in the letter to Guldin was motivated in part by the personal circumstances in which he found himself in 1628. When Kepler replied to Guldin, it had already been nine years since he had received the final word about his excommunication from the Lutheran consistory in Stuttgart. Adrift, no longer fully part of any religious community, and his faith in the principles of his own church shaken, Kepler was highly motivated to find a way to participate in religious communal life in any way he could, while still preserving his own particular religious beliefs. His views on the Catholic Church likely did become somewhat milder over time, as his personal motivation for tolerance increased.

Yet it is a mistake to view the letter to Pistorius as indicative of his larger attitude to the Catholic Church in 1607. Instead, the letter to Guldin, personal motivation notwithstanding, should be taken as more paradigmatic of Kepler's attitude throughout his life. After all, his letter to Herwart, also written in 1607, evidenced a very broad conception of religious community, one that included Catholics and took seriously their theological perspectives. To be sure, twenty years separated the letters to Herwart and to Guldin, and in the former Kepler's openness to the Catholic faith was largely theoretical; he did not evidence a willingness to personally align himself with a Catholic community but rather expressed a belief that God included all confessions in his larger communal embrace. As Kepler was forced from his own Lutheran community, his openness to Catholicism moved from the realm of the theoretical to the realm of the personal and practical. Yet throughout that time, he sustained friendly conversations with Catholics other than Herwart, and with Jesuits other than Guldin, some of which also openly touched on questions of religion.

Hence, instead of arguing for the letter to Pistorius as representative of an earlier, "anti-Catholic" Kepler, and the letter to Guldin as representative of a later, more tolerant Kepler, the letter to Pistorius should be viewed as a very specific letter directed toward a very specific person in very specific circumstances. Pistorius was, after all, a master polemicist, and his polemical writings were notably strident and abrasive. His famous *Anatomia Lutheri*, for instance, was a personal attack on the reformer and claimed that Luther was filled with the seven evil spirits and was an abomination. Moreover, Kepler knew well that Pistorius had moved through the different confessions and that he had argued strongly, at different times of his life, for all of them. Kepler clearly respected Pistorius, yet he debated with him as someone who knew the details of Lutheranism intimately and had chosen not only to convert to Catholicism but to become one of its most prominent and belligerent spokesmen. It is likely that their conversations on topics of religion, despite their friendly personal relationship, were always somewhat aggressive; indeed, Kepler ended his letter to Pistorius by adding that "I hope you receive these thoughts from my writing as you customarily do when we speak and debate about the same subject."[67] For his part, Pistorius responded dismissively in his next letter with only one sentence devoted to religion: "I wish you would avoid theology; clearly, you understand nothing about it."[68] The tenor of their debates likely lent itself to the tone of Kepler's letter because of who Pistorius was, not because of who Kepler was.

Kepler, the Jesuits, and Jesuit Science

Protestant attitudes toward Catholics in the late sixteenth and early seventeenth centuries, harsh as they typically were, paled in comparison to the vitriol leveled at the Jesuit Order in particular. Martin Chemnitz, the famed Lutheran theologian and follower of Melanchthon, was not more strident than many of his fellow Lutherans when he wrote in his 1572 pamphlet *Concerning the New Order of the Jesuits* that the Jesuits were "no other than perjured, treacherous, oath-breaking, dishonorable, desperate villains, against whom the German land ought to be well on its guard."[69] The widespread consensus among Protestants was that the Jesuits were by nature deceitful and hypocritical; according to one common refrain, they were wolves dressed in sheep's clothing.[70] In this view, their missionary work and their focus on education were merely ways for them to insinuate themselves in the minds and hearts of unwary Protestants and lure them toward the Antichrist in Rome, or toward whichever Catholic ruler they decided to support; as everything they did was guided by this motivation, their public words and image could never be trusted. As Antoine Arnauld, lawyer and councilor to Henry IV, wrote of the Jesuits in 1594, "all their thoughts, all their designs, all their actions, all their sermons, all their confessions have no other aim than to subjugate the whole of Europe to the domination of Spain."[71] The English diplomat Henry Wotton—a correspondent and acquaintance of Kepler's—likewise criticized the Jesuits and their ambitions: "Never," he wrote, "were men more gripping after lands and possessions, more imperious over consciences and families."[72]

This kind of invective was by and large an extension of the general confessional dispute between Catholics and Protestants; Jesuits were, after all, among the most visible and successful arms of the Catholic Church and hence an easy target for Protestant attacks. Yet it quickly spilled over from the realms of theology and politics to the realm of scholarship. If the Jesuits simply could not be trusted, many believed, their scholarship too must be suspect. Joseph Justus Scaliger and Isaac Casaubon generously shared their scholarly vituperation with many, to be sure, but they leveled their highest scorn at Jesuit scholars in particular. The Jesuit mathematician Cristoph Clavius was "smug and shameless" and even "rustic and illiterate," wrote Scaliger; his books were "foul and pedantic" or "mangy and feverish."[73] Jesuit Martin Del Rio was the "*stercus diaboli*" (filth of the Devil).[74] Casaubon wrote that Cardinal Bellarmine and his Jesuit colleagues were "ignorant," "untrained and laughable" and that Jesuit Andreas Eudaemon Johannes was one of "my enemies, wretched heretics, teachers of assassination, Cretan Greeklings, and fuzzy thinkers," "most assuredly a servant of Satan."[75]

Most Protestant members of the Republic of Letters never descended to this level of vituperation. The Republic of Letters was, after all, a community in which religious beliefs were supposed to be sidelined in favor of the interests of scholarship, and where religious affiliation was supposedly irrelevant.[76] Yet despite the supposed religious neutrality of the Republic, most Protestant scholars in the late sixteenth and seventeenth centuries continued to view the scientific activities of the Jesuit Order with suspicion, and even disdain. This was due in part to the perceived backwardness of certain aspects of the Jesuit approach: their commitment to Aristotelian natural philosophy and their condemnation of Copernican theory and subsequent treatment of Galileo. Yet it stemmed even more from the belief that Jesuit science was itself polluted by its linkages to the religious goals of the society. While affirming their own ability to separate their religious and scholarly interests, many Protestants insisted that the Jesuits were uninterested in, and indeed incapable of, a similar separation; these Protestants were filled, writes Ann Goldgar, with "the prejudice . . . that the Jesuits were unwilling to abandon their own prejudices."[77] Jean Cornand de la Crose merely summed up a belief that had been popular for many years when he wrote, in 1691, that "the Character of a Jesuit . . . doth not well agree with that of a sincere Writer."[78]

The perceived theological motivation behind the Jesuit pursuit of science—and of scholarship more generally—indicated to many that the results of Jesuit scholarship and science were not to be trusted. John Beale emphasized this in a letter to Henry Oldenburg, where he noted that the Jesuits "are to be suspected in point of candor and severe truth."[79] Oldenburg agreed, adding that though the global Jesuit network made the Jesuits seemingly ideal candidates for scientific correspondence and specimen collection, their reports could not be trusted, "considering the principal end of such men's voyages, which is, to propagate their faith, and to greaten and enrich themselves by their craft."[80]

The new forms of scientific association that arose over the course of the seventeenth century accentuated this suspicion of the close linkages between Jesuit scientific and religious pursuits. Like the supposedly nonsectarian Republic of Letters before them, many of the new academies and societies of the seventeenth century were established in order to deliberately segregate the realm of natural philosophy from the realms of politics and religion—at least in theory. These spaces would provide, it was hoped, an impartial, apolitical forum where religious disputes could be sidelined and in which "reasonable men from a wide range of ideological positions . . . could collaborate in gathering information which they hoped that all would be able to accept as undeniably true."[81] Rhetoric and reality often parted ways, of course, and much like the Republic of Letters, the academies and societies of the seventeenth century were never truly free of political or religious overtones. This same disjunction between official rhetoric and reality characterized the Jesuit pursuit of science as well. Though Ignatius of Loyola had maintained that the official aim of all the activities of the Society of Jesus was "to aid our fellowmen to the knowledge and love of God and to the salvation of their souls,"[82] many Jesuits were drawn to the society specifically because it allowed them to pursue their own natural philosophical and mathematical interests, without thought to any religious applications of such pursuits. Moreover, Mordechai Feingold has emphasized in his study of Jesuit science that "while scholarship often served partisan goals in the charged religious atmosphere of the early modern period, Jesuit scientific practitioners as a group seem to have resisted the temptation to yoke science to other ends as well as did practitioners of any other religious denomination."[83] In practice, that is, the members of seventeenth-century scientific academies and societies were often as likely to invoke political or religious concerns as the Jesuits were to avoid them. Yet the official positions of these different groups highlight why many members of both the Republic of Letters and the new scientific societies, so invested in the formal separation of scholarship from partisan interests, would view the Jesuit linkage of science and religion with suspicion and concern.

Given the common Protestant perspective on Jesuit science, Kepler's own views make him something of an outlier. As noted earlier, he had some close relationships with his Jesuit correspondents, whose scientific prowess he clearly respected. This, of course, was also true of other members of the Republic of Letters; despite the reservations of many Protestants about Jesuit "prejudices," many Jesuits attained respected places in that community of scholars. Yet in addition to his personal relationships with specific Jesuits, Kepler differed from many of his fellow Protestants in his vocal opposition to all those who spoke dismissively of Jesuit science and who rejected any particular scientific endeavor simply because it originated with the Jesuits. He supported the chronological work of some of his Jesuit correspondents against the work of Scaliger, a fellow Protestant, and bemoaned the fact that both sides—and he excoriated the Protestants, in particular—allowed their religious affiliation to color their evaluation of each other's work. "Everyone is preoccupied with prejudices and hatred," he wrote. "If a Jesuit writes, it is considered as though not written by those among whom [Scaliger] reigns."[84]

Why was Kepler so willing to openly and favorably consider the work of the Jesuits, at times even supporting them against his fellow Protestants? The answer relates in part to Kepler's irenical stance more generally: his refusal to limit the larger community of Christians to merely one confession and his lifelong pursuit of churchly unity and harmony. Yet in this case, something still more underpinned Kepler's attitude. Kepler's fellow Protestants were troubled both by the religious beliefs and goals of the Jesuits and, more broadly, by the close alliance between their scientific and religious pursuits. Though surely influenced by religious agendas of their own, many Protestant natural philosophers of the early modern period increasingly tended to argue that natural philosophy or experimental science ought to occupy a realm separate from all partisan interests—and here they included both politics and confessional theology. This was particularly so as religious tensions escalated and confessional disagreements grew more heated; scholars invested in the ideals of the Republic of Letters increasingly argued that any linkage between natural philosophy and religion stood in tension with the need for civility and community that the new science demanded. Jesuits, who upheld the old linkage, were therefore viewed with an increasing degree of distrust.

Yet as many began to assert that the new science ought to be theologically neutral, Kepler continued to openly proclaim, in both his letters and his published books, that he viewed his astronomy as a means to bring people closer to God. Like his medieval predecessors, he understood this, at the very start of his career, to apply not simply to the broader outlines of a religious worldview that any Christian would necessarily accept but more specifically to the linkages between his cosmological theories and the very religious issues that were being hotly debated by the competing confessions. He stressed that his astronomy served a theological purpose and that he saw himself as an interpreter of the messages that God had built into his creation. "I wanted to be a theologian," he asserted, and "for a long time I was distressed: behold, God is now celebrated too in my astronomical work."[85] He continued to emphasize throughout his career that he saw himself as a "priest of God with respect to the Book of Nature"[86] and that he linked his astronomical and religious pursuits in direct and explicit ways.

In many ways, that is, Kepler conceived of his own scholarly pursuits along similar lines to those of his Jesuit counterparts, though the details of their theological stances certainly differed. Indeed, he explicitly connected his own attitude with that of the Jesuits in an analogy that at first seems merely a rhetorical flourish but betrays something deeper. In a letter to Maestlin in which he discussed his goal of reforming astrology, Kepler argued: "Do I not act rightly if I dedicate this work to persuading learned men and philosophers about the distinct operation of the heavens? And thus I act as the Jesuits do: who improve many things in order to make men Catholics. Or on the contrary, I do not act [as the Jesuits do]: for those who defend all the nonsense are like the Jesuits. I am a Lutheran astrologer, who rejects the chaff and retains the kernel."[87] On a simple level, Kepler invoked the Jesuits because of their skill at persuasion; much as they used their learning to persuade men to become Catholics, Kepler argued, he used his learning to persuade men about the true workings of the heavenly bodies. Still further, he invoked the Jesuits as a trope to represent the old, traditional approach, while his description of himself as a "Lutheran" astrologer alluded to his new, reformed approach to the discipline of astrology. This accounts for Kepler's association of Jesuit learning and "nonsense," a linkage that he surely did not believe when applied beyond this analogy. Yet the choice of this analogy—and Kepler's claim that "I act as the Jesuits do"—was not simply happenstance. Kepler implied, through this comparison, that his work was in many ways as closely tied to a religious agenda as that of the Jesuits. He was a Lutheran astrologer, both in the sense that he hoped to reform astrology and in the ways that his astrology, and his cosmological work more broadly, served a religious and—at least at the start—a specifically confessional purpose. Through his work, he elucidated the patterns of creation, brought people closer to God, and pointed the way to a reunified church.

Though at first it may seem paradoxical, Kepler's willingness to openly link his own religious and astronomical pursuits made him far more inclined to consider the religious and scientific claims of his Jesuit colleagues apart from one another. When men like Oldenburg asserted that natural philosophy or experimental science occupied a realm separate from religion or politics, and consequently condemned any explicit linkages between them, they automatically found any science that was tinged with religious overtones to be suspect, especially if the religious source it came from was suspect. In the instance of Jesuit science, they furthered the very linkage they sought to avoid by condemning the science on religious grounds. By contrast, because Kepler argued for no such separation, and believed that religious motivation did not automatically invalidate scientific pursuits, he was willing to consider the two—the specifics of the religious motivation and the specifics of the science in question—apart from one another. He knew that his own religious beliefs should not invalidate his astronomical claims; on the contrary, he tried to be more rigorous in his demands for accuracy and precision than many of his contemporaries.

Kepler's own religiously inflected science, in other words, seems to have resulted in his greater ability to separately evaluate the religious and scientific claims of others than those who openly argued for the necessity of such a separation. He insisted that calling something "Jesuit science" should not necessarily invalidate the science in question. Similarly, he counseled one of his Jesuit correspondents not to dismiss the work of Scaliger simply because Scaliger was a Protestant. "He may be a heretic, an innovator, a critic of the Holy Fathers; he may even have lied about his lineage. What, pray tell, do all these things have to do with the present question? Can it not be discussed without disparaging him?"[88] In questions of scholarship, as in questions of theology and religious community, Kepler struggled against the divisiveness he saw around him and defended an approach characterized instead by inclusiveness and harmony.

Conclusion: Adiaphora and Accommodation

Kepler's attitude toward Catholics, I argued at the start of this chapter, can be understood through the larger lens of accommodation. Kepler believed that God accommodated his message to human language and perception and that he did so both in the words of scripture and in other realms, like astrology, in order to lead humanity toward some greater truth. Much like Paul in the Gospels, he argued as well that just as God had accommodated himself to humanity, so must Christians accommodate one another, allowing for diverse practices and beliefs within the larger, united religious community. Yet what precisely should unite the community, and on what differences was it permissible to accommodate and compromise? Most Christians, even during the divisive confessional era, would surely have been able to identify *some* issues on which it was permissible to disagree and still remain acceptably orthodox. The question was where to draw the line, which means that discussions of accommodation were often very much linked to debates about adiaphora—matters not considered essential to faith.

Originally rooted in Stoic philosophy as well as in the statements of Paul in Corinthians and Romans, discussions of adiaphora became particularly relevant in the political environment in the Holy Roman Empire after the Reformation.[89] After Charles V defeated the Schmalkaldic League, he issued the Augsburg Interim of 1548 as a way to manage theological differences between Catholics and Lutherans in the empire. The Interim, while binding on all Protestants, was supposed to represent a kind of compromise with them; its authors had endeavored to keep it aligned as closely as possible with the basics of Luther's teachings, so that it would be theologically acceptable to Protestants. Yet it also required Protestants to return to the old, Catholic forms of religious practice and worship in many respects.

In response to Lutheran opposition, some Lutheran theologians, Melanchthon among them, crafted a Leipzig Interim in 1549, modifying the Augsburg Interim. There, they attempted to clarify the essentials of Lutheran doctrine while adopting some of the requirements of worship from the Augsburg Interim, among them episcopal rule, confirmation, and extreme unction. They justified these compromises by saying that the elements of worship required were minor and not essential to faith—they were adiaphora. Because they were inessential, they were negotiable and could be used as the basis for compromise. And compromise was essential, these theologians argued, not merely because it was important to obey the political authorities but also because it led to peace: "our concern and our intention are always directed not toward causing schism and complications, but rather toward peace and unity."[90]

Many Lutherans harshly condemned this kind of compromise, among them Melanchthon's former pupil Matthias Flacius, and argued that any form of worship that was practiced under compulsion was no longer a matter of adiaphora and must be resisted. Melanchthon, by contrast, insisted that even if particular forms of worship were ordered by the political authorities, it was foolish to forbid Lutherans from practices that were inessential and would not harm their faith, if such practices would lead to peace. Against the Lutherans who refused to compromise, he claimed that "the violent persons who want to compel everybody to hold the opinions they hold are instituting a new kind of popery."[91]

The debate about adiaphora centered, in part, on the question of just how similar to each other the varying confessions actually were. How much that divided the confessions was minor and inessential to faith and how much was fundamental? Those who argued for compromise on issues of adiaphora tended to argue that Protestants and Catholics had more that united them than that divided them. And it was on this basis that irenicists and conciliators often sought to reunify the church: by identifying a fundamental core of shared beliefs that proved that the divisions separating the confessions were not insurmountable. In fact, the attempt to identify commonalities among religious groups was not unique to the post-Reformation era. In his *De Pace Fidei* of 1453, Nicholas of Cusa had ambitiously applied this idea to religions beyond Christianity, suggesting that all religions shared a common core and that tolerance and harmony might be achieved on that basis. What God really wanted, he wrote, was that "henceforth all the diverse religions be harmoniously reduced, by the common consent of men, unto one inviolable [religion]."[92] In the post-Reformation era, Marcus Antonius De Dominis, the theologian whom Kepler heralded as a prophet, had likewise used the idea of adiaphora—in his words, "unity in necessary things; liberty in doubtful things; charity in all things"—as a fundamental part of his plan to reunite the churches in his *De Republica Ecclesiastica* of 1617.[93]

Like De Dominis and others, Kepler insisted that the warring confessions were far more similar than their bitter polemics suggested. "I acknowledge that the church is one and the same at all times," he wrote.[94] He too claimed that compromise and tolerance were important and rested his arguments on the very passages in Paul—the eating of unclean meat—that often formed the basis for discussions of adiaphora. Yet Kepler made these arguments not about those aspects of worship that might easily be deemed inessential but rather about the central and highly contentious issue of the Mass itself. And he did so not by claiming that the proper approach to the Eucharist was inessential to faith but by claiming that the question of faith could be accommodated by a personal objection. He maintained, that is, that the question of communal participation should be considered separately from questions of faith. The act of taking the Mass might accommodate multiple interpretations—so long as a person held true to the interpretation he deemed correct, what matter where he participated in the sacrament and what those around him believed to be true?

In fact, Kepler asserted neither that all the confessions essentially agreed on the question of the Mass nor that they all essentially differed; rather, he claimed that it was the Catholics and Calvinists who shared an essential core of beliefs, and the Lutherans who had departed from this essential—and correct—core.[95] Kepler did not argue that doctrine was unimportant or that the proper approach to the Eucharist was inessential to faith. He did not assert that all the confessional positions were equally valid when it came to the presence of Christ. Rather, he argued that even when it came to this contentious question, there could be a path to peace and unity; so long as practice and belief were decoupled to some degree, those with different beliefs might still practice together. He would still, after all, have participated in Communion with his fellow Lutherans—indeed, he tried very hard to regain the ability to do so after his excommunication—even though this was the one confession that, he believed, had clearly departed from the true interpretation he associated with antiquity.

In many ways, Kepler's attitude aligns most closely with Paul's stance in Corinthians. While in Romans Paul argued for compromise with regard to a very specific practice—the eating of unclean meat—in Corinthians his description of accommodation was far more sweeping. When Paul claimed, in 1 Corinthians 9:20–22, to "become all things to all people," he meant this quite literally: "To the Jews I became like a Jew, to win the Jews. To those under the law I became like one under the law (though I myself am not under the law), so as to win those under the law. To those not having the law I became like one not having the law (though I am not free from God's law but am under Christ's law), so as to win those not having the law. To the weak I became weak, to win the weak." In making this kind of sweeping argument for accommodation, Paul argued that as an apostle he had the right to live freely under the gospel and to insist that his way was the right way. Yet he willingly surrendered that individual right, he claimed, for the greater good of the community. In this same section of Corinthians, he redefined freedom as slavery to everyone, since the greater good of the Christian communal body came first, and it necessitated compromise and accommodation.[96]

The kind of sweeping compromise and changeability epitomized by Paul here could easily seem weak, insincere, and self-serving. Plutarch had described such a man, "so indefatigable, so changeable, so universally adaptable, that he can assimilate and accommodate himself to many persons, . . . reading books with the scholarly, rolling in the dust with wrestlers, following the hunt with sportsmen, getting drunk with topers, and taking part in the canvass of politicians, possessing no firmly founded character of his own."[97] Paul insisted that his own accommodation stemmed not from self-interest but from concern for the salvation of others. Kepler, too, lamented that his urge to compromise led others to view him negatively. "People think that I am someone who agrees with all the parties, not out of a conscientious heart, but rather to garner favors from everybody. . . . They think that I am a godless scorner of God's word and of the Holy Communion. . . . They consider me a doubter . . . and also unstable, siding first with this side, and then with another."[98] Yet he, too, asserted that his tendency to accommodate came not from a desire to help himself but from a desire to help others. "I do not like to be looked upon as a man apart," he continued. "It hurts me in my heart that the three great factions have among them torn the truth so badly that I must gather it piecemeal where I can find it."[99]

Kepler's arguments for accommodation, first to the Lutherans and then to the Catholics, all rested on his desire for harmony and for a unity that he believed could be restored, with effort, to the greater church. He contrasted his own attitude with those who insisted on seeing the church through the narrow lens of one confessional perspective. "I am pleased either by all three parties, or at least by two of them against the third, in the hopes of agreement. But my opponents are pleased by only one party, imagining eternal irreconcilable division and quarrel. My hope, so help me God, is a Christian one; theirs, I do not know what."[100] Kepler, unlike so many of his contemporaries, hoped for agreement when possible and for unity above all. But this was a unity that allowed for difference, and even for error; the church as it existed on earth, after all, was only a human construct, in Kepler's view. The best its members could do was model themselves, as closely as possible, on the real truth of the heavenly church and the cosmic harmonies that celebrated it.

FOUR

"An Ally in the Search for Truth": Kepler and Galileo

Kepler did not have many fond memories of his mother, whom he described as "small, thin, dark, sarcastic, and obstinate, with a negative character."[1] Yet one memory in particular stood out for him in later years. One night in 1577, when Kepler was turning six, his mother shook him awake, took his hand, and led him up a small hillside in Leonberg. There, they gazed together at the sky, to watch a comet so bright that it was visible throughout Europe and beyond. When Kepler later recalled this moment fondly, he was likely thinking primarily of his mother's presence at his side, for he probably could not have seen the comet itself as anything more than a blur of light. At the age of three he had suffered a particularly bad bout of smallpox, which irreparably damaged his eyesight, leaving him nearsighted in both eyes and with double vision in one. Kepler would lament this lack of clear sight throughout his life, especially given that his life's work focused on the movements of the distant stars. Kepler's poor vision is no doubt one of the reasons he was so overjoyed to learn of the invention of the telescope, a miraculous object that could make comets and other celestial objects visible even to him. "O telescope, much-knowing and more precious than any scepter!" wrote Kepler in his *Dioptrics* of 1611. "Is not he who holds you in his right hand made king and lord of the works of God?"[2]

Kepler rejoiced in the telescope of Galileo not only because it revealed the wondrous secrets of the heavens for all to see but also because it further supported the Copernican cause to which Kepler was so devoted. Even before Galileo's remarkable telescopic feats, Kepler had learned of Galileo's Copernicanism and had eagerly welcomed Galileo into the Copernican community that he struggled to create. In this chapter, I focus on Kepler's conception and construction of that community, through a focus on the relationship between Kepler and Galileo from their initial correspondence in 1597 to the aftermath of Galileo's telescopic discoveries in 1610. In so doing, I focus in particular on the ways that Kepler understood and marshaled a number of rhetorical techniques that he and his fellow late Renaissance scholars had inherited from antiquity. Rhetoric, I argue, was a particularly important tool in Kepler's arsenal as he struggled to spread the Copernican worldview, and the rhetorical techniques on which Kepler drew varied greatly depending on his audience and purpose.

Those rhetorical techniques can be traced back to particular representatives of the ancient Greco-Roman rhetorical tradition, whose work had been passed down through the generations; Aristotle, Cicero, and Quintilian were the most visible of these. These ancient rhetoricians had recognized that reason without rhetoric was powerless to effect real change. The vast majority of people, according to these rhetoricians, were simply not able or willing to follow the logical conclusions of their reason if those conclusions implied any real change from the status quo. They needed to be *persuaded* to change, and this was the job of rhetoric.[3] "Wisdom in itself is silent and powerless to speak," argued Cicero; therefore, "wisdom without eloquence cannot do the least good for cities."[4] Language was an active tool, according to the rhetoricians, and not just a passive medium for conveying ideas; it could be used "to prove, to please, and to sway or persuade."[5] It could impress, deceive, transform, and subvert, and—if the rhetorician were skilled—it could do all these things without being obvious, making it seem like reason alone was at play.

To be effective, however, rhetoric needed to be adapted to the particular needs and dispositions of the audience toward whom it was directed. And here is where the tradition of accommodation, which I discussed in the previous chapter through the lens of theology, intersects with discussions of rhetoric. Just as the church fathers had argued that God's word was accommodated to the needs of his human audience, so too did rhetoricians argue that the skilled speaker needed to accommodate his own words to the specific audiences that he addressed. In

Cicero's words, orators "who desire to win approval have regard to the goodwill of their auditors, and fashion and accommodate themselves completely according to this and to their opinion and approval."[6] And as in the case of the Bible, where this meant a kind of "baby talk," Cicero emphasized that in the vast majority of cases rhetoricians, too, needed to accommodate their speech to "popular intelligence," focusing less on precision than on probability and persuasion.[7] In fact, the obvious connections between the rhetorical emphasis on accommodation to one's audience and the theological emphasis on divine accommodation in scripture led some, like Melanchthon, to directly link the two. In his reform of the German universities, Melanchthon had highlighted the discipline of rhetoric and emphasized its importance in scriptural exegesis. Since God had spoken "through the Word," he claimed, "the nature of speech must be known." What else had Paul meant when he insisted on the importance of the gift of tongues in Corinthians, Melanchthon asked, if not rhetorical eloquence?[8] God, according to Melanchthon, was the perfect rhetorician, and it was incumbent on his audience to fully immerse themselves in the techniques of rhetoric in order to understand the true import of his prose.

What, then, were the most important rhetorical techniques? Traditionally, argumentative speech was divided into five categories, also a legacy of the ancients: *inventio*, which meant determining the argument itself; *dispositio*, ordering the argument appropriately; *elocutio*, the style of the argument; *memoria*, the mnemonic architecture; and *actio/pronuntiatio*, the argument's delivery.[9] In the Renaissance, scholars immersed in the rhetorical tradition stressed in particular those aspects of speech that were conducive to persuasion, typically the *inventio* itself and the subsequent delivery. Both required a focus on the ways that the content and meaning of an argument related to its audience and considered oratory not in terms of its abstract technical merit but in terms of its specific social effects. In terms of the *inventio*—the crafting of the basic argument—the speaker interested in effective persuasion needed to focus on gaining the audience's confidence and trust. It was not enough to merely convey the bare facts of the matter; the audience needed to believe so much in the words of the speaker that they might agree to positions that they would otherwise scorn.

There were multiple ways to accomplish this; one was a focus on *ethos*, on the person conveying the message (whose authority and character were emphasized), rather than on the message itself. The advocate needed to convince the audience that he himself, or his client, was trustworthy and possessed a character that couldn't be doubted. And it was less important to theorists like Aristotle and his successors that he actually *possess* these characteristics than that he *project* them.[10] *Ethos* was not necessarily an inherent characteristic but an aura acquired deliberately through skilled manipulation. It could be achieved by emphasizing that one's argument agreed with accepted wisdom; to this end, rhetoricians often emphasized *auctoritas*, the citing of authoritative exempla. It could also be achieved by stressing the speaker's, or his client's, lack of bias and general desire for the public good. A related rhetorical device depended on *fides*, alternatively translated as "credit," "faith," or "belief"; this was the method by which ancient rhetoricians evaluated the credibility of witnesses and their testimony.[11] It was also a standard applied in a number of early modern disciplines, among them the *artes historicae*,[12] and also in the discipline of law. It, too, depended on the persona of the speaker or client, and orators often sought to develop an aura of respectability and a reputation of trustworthiness in other matters, which the audience might then transfer over to the matter at hand.

Since a primary goal of rhetoric was to persuade an audience to believe, rhetoricians justified certain deceptions, so long as they helped convince the audience of the larger argument. Persuasion, manipulation, and "moving" an audience in ways beyond reason all suggested, after all, that in accommodating himself to the desires and prejudices of his audience the orator might need to maneuver a bit with respect to the details of the truth itself. The same was true, of course, of divine accommodation in scripture; theologians had long recognized that any theory of accommodation would imply a kind of deception on the part of God. In a minimal sense, this was simply a linguistic limitation—there was no language that would encompass the divine perspective, since language itself was a human tool. If one took this accommodated language at face value, one would be deceived. Yet some theologians went far beyond this in arguing for God's deception. Origen, for example, declared that God would sometimes deliberately mislead people for their ultimate good; as an example, he cited God's prophecy of Nineveh's destruction, a prophecy that was never intended to be true.[13] God, according to Origen, was like a doctor who deceived his patient if the truth might harm the patient's ultimate recovery. Deception itself, that is, was a pedagogical tool employed by God to lead his people to salvation; it was just another form of accommodation.

The idea of accommodation as deception thus linked both rhetoric and exegesis and was adopted, in the sixteenth and seventeenth centuries, by one group in particular—the Jesuits. Jesuits hoped to win their audiences over to the truths of the Catholic faith, and as their audiences ranged far and wide across the known world, they developed a theory of cultural accommodation that might help them do so. According to this theory, Jesuits need not condemn but might instead accommodate false beliefs if such accommodation enabled them to win new souls for the true church.[14] These false beliefs ranged from the caste system in India to the cult of ancestors in China; all were accommodated in order to lead their adherents to the ultimate Christian truth.[15] If cultural differences were a matter of truth and falsehood—as nearly all Jesuits believed—then this kind of cultural accommodation was a deliberate acceptance of untruth when it led to a greater good. Jesuits relied on the very same passage in Corinthians that underpinned the Pauline idea of communal accommodation, where Paul had argued that he "became all things to all men." Christians who modeled themselves after Paul had typically understood this to refer to accommodation of fellow Christians who might differ from them in some respects, but Jesuits extended this far beyond the Christian community. As Jesuit Louis Le Comte wrote, echoing the words of Paul, "one must be barbarous with the barbarians and polished with civilized people; one must lead an ordinary life in Europe and a deeply austere life among the penitents of India; one must be well dressed in China and half naked in the forests of Madurai; in this way the uniform and unchanging gospel will be more easily insinuated into the spirits."[16] At the same time that they emphasized this kind of cultural accommodative deception, Jesuits developed an elaborate theory of casuistry, based on earlier discussions of the subject by the Dominican theologian Domingo de Soto and the canonist Martín de Azpilcueta (better known as Doctor Navarrus). This theory allowed for deception in the form of mental reservation or equivocation when that deception would be helpful in morally complex or dangerous situations, particularly ones that might threaten the life of oneself or one's allies.[17] The proper rhetorical training, moreover, could make all the difference in navigating the tricky boundaries between justified equivocation and outright imposture.

Rhetoric, accommodation, and deception were all very much part of the cultural legacy that Kepler inherited in sixteenth-century Europe. Kepler himself was well attuned to rhetorical traditions and techniques, having studied rhetoric at Tübingen with Martin Crusius and then having taught rhetoric himself in Graz. He knew well that rhetoric, a central feature of the classical curriculum, was at its core a tool that focused on the art of persuasion. And he did not hesitate to employ his rhetorical skills in the service of his campaign to construct a community of scholars who would accept the new Copernican cosmological system. In particular, as we will see, Kepler marshaled the ideas of *ethos* and *auctoritas* to persuade all those who hesitated to adopt Copernican theory that their hesitations were misguided. When his initial *ethos*-based campaign failed in rather dramatic fashion, Kepler turned instead to a rhetorical focus on the idea of *fides*, or the power of legal testimony, particularly in advocating for the truth of Galileo's recent telescopic discoveries.

In all these cases, Kepler took the techniques of rhetoric and deployed them as means by which disputes over the new science might be adjudicated. He was not alone in this disciplinary move; a number of recent scholars have highlighted the process by which rhetoric was increasingly emphasized in the sixteenth and seventeenth centuries not only in the human and moral disciplines but also in the natural sciences, where certain proof came to be seen as an elusive possibility and credible testimony a more accessible tool for the discovery of natural truths.[18] Other scholars have likewise highlighted a similar move in which legal techniques and ideas, particularly credibility, evidence, and testimony, were deployed in the evaluation of natural philosophical matters of fact, with men like Robert Boyle emphasizing moral certainty as the only standard truly attainable when it came to the physical world.[19] Mario Biagioli has focused similarly on Galileo's emphasis on the ideas of credit and trust, as well as his rhetorical construction of his own authority; Galileo's personal authority, in this view, was a rhetorical resource for the production and dissemination of his work rather than an a posteriori result of that work.[20]

Following this emphasis on the early modern melding of rhetoric and natural philosophy, I argue not only that rhetoric was an important tool in Kepler's arsenal but also that the rhetorical techniques on which Kepler drew varied greatly depending on his audience and purpose.[21] I do not claim that Kepler negated the possibility of certain knowledge or that he emphasized credibility and moral certainty at the expense of logic and demonstration. Rather, I underscore the fact that the early modern scholar had a broad range of techniques of rhetorical persuasion at his disposal; in reading Kepler's correspondence with Galileo, I argue, we see Kepler suggest and then deploy these techniques, with the goal of supporting and promulgating Copernican theory.

One of those rhetorical devices, rooted in the idea of accommodation, was the use of deception as a form of persuasion. Kepler attempted, as we will see, to base the early stages of his Copernican campaign on claims that were not themselves true but that might convince his audience to adopt the ultimate truths of Copernicanism. As God had done in scripture, as Paul had done for the Corinthians, Kepler would do in his Copernican campaign: he would lie in order to save. In this, Kepler's approach to Copernicanism was remarkably similar to the approach of the Jesuits to their missionary activity. As I noted in the previous chapter, Kepler was well acquainted with Jesuit strategies for propagating the Catholic faith and at times deliberately likened his own Copernican campaign to their religious campaigns. In his attitude toward dissimulation, we see Kepler once again mirroring the Jesuits; by accommodative dissimulation, he hoped to win his audience over to the truth of Copernicanism and create his own kind of community of believers. Kepler, it is clear, thought of Copernicanism not merely as an astronomical truth but as an all-encompassing one to which he needed to devote the full strength of his persuasive energies. "Because I am completely persuaded by the opinion of Copernicus," he wrote, "I am religiously prevented from proposing anything different. . . . This glory suffices, that with my discovery I can guard the gate of the temple in which Copernicus celebrates at the high altar."[22] The words Kepler used to describe Copernican theory were rife with theological overtones. And when it came to truths that were this fundamental, dissimulation was warranted. Kepler hoped that he and Galileo could be at once allies, working to establish the truth of Copernicanism, and missionaries, striving to bring that truth to as broad an audience as possible.

Kepler and Galileo: First Contacts

After completing his first book, the *Mysterium Cosmographicum* of 1596, Kepler eagerly worked to publish it as quickly as possible.[23] He sought an audience for the text among both his own friends and patrons and the broader community of scholars throughout Europe with an interest in astronomy—the "mathematical Republic of Letters,"[24] as his professor Michael Maestlin described it. He strongly believed that the book offered a more powerful defense of the Copernican system than any that had been offered previously—more powerful, indeed, than Copernicus's own, as Kepler had constructed an a priori argument from geometry rather than an a posteriori argument from the observations alone.[25] By constructing an archetypal explanation of the structure of the heavens, one that was both mathematical and aesthetic and that had resonances in both astrology and music, Kepler thought that he might truly persuade those who hesitated to abandon the idea of a geocentric universe. Indeed, he felt his ideas to be so obviously true that the effort he had expended in working them out might be easily overlooked; he likened his discovery in the *Mysterium Cosmographicum* to the "egg of Christopher Columbus,"[26] a story that proved that something that seemed so obvious in retrospect might seem impossible at the start. It had taken "the aid of divine inspiration"[27] for Kepler to arrive at the ideas of his *Mysterium Cosmographicum*, and he was eager to swiftly share those ideas with others who might benefit from them.

Kepler arranged with his printer to send copies of his book to the Frankfurt Book Fair, while he himself distributed copies to friends and acquaintances. In particular, he sent two copies of the book to Italy with friends traveling in that direction. One of these copies came into the hands of Galileo in Padua, though Kepler had not sent it to him specifically; indeed, it is likely that Kepler had not heard of Galileo until this point.[28] Upon receiving the text, Galileo immediately wrote a quick letter to Kepler, praising him for his achievement. "I received your book, most learned man, sent to me by Paul Homberger not days but mere hours ago," he wrote. Since Homberger was set to return to Germany immediately, Galileo had decided to write to Kepler on the spot, having read only the preface of the book. From the preface alone, he wrote, he could discern the general argument of the book, and he thanked Kepler for that argument: "I am extremely grateful," he wrote, "that I have an ally in the search for truth." He greatly lamented that "those who are devoted to truth are rare and that those who are not pursue a perverse method of philosophizing."[29]

Galileo went on to clarify that the specific truth to which he referred was the truth of Copernicanism, a truth that many of his generation were still unwilling to accept. He himself, he explained, "came to the Copernican belief many years ago, and on the basis of this set of propositions I discovered the causes of many natural effects, which certainly could not be explained through the common hypotheses."[30] Galileo added that though he had assembled many proofs for Copernicanism and refutations of its detractors, "thus far I have not dared to bring them to light, having been frightened because of the fate of Copernicus, our teacher, who was given immortal fame among some but was nevertheless ridiculed and rejected by an infinite many others (for such is the number of fools)." If he truly believed that there were others of the same opinion as he and Kepler, he continued, he might disclose his true beliefs, but "since there are not, I will refrain from business of this kind."[31]

Kepler replied at once; he was thrilled to receive Galileo's letter, he wrote, both "on account of the friendship begun with you, an Italian, [and also] on account of our consensus in the Copernican cosmography." He hoped that Galileo might have since had the time to finish reading his book, for he eagerly awaited Galileo's opinion. "I prefer the censure of one judicious man, however harsh," he wrote, "to the thoughtless applause of all the masses." Still further, in response to Galileo's more subdued initial letter of support, Kepler argued impassionedly for the vocal defense of Copernicanism by its adherents, claiming that only if all Copernicans banded together could the rest of the world be swayed. He acknowledged that Galileo had the noble precedents of Plato and Pythagoras behind his belief that "we must yield to universal ignorance, nor must we rashly throw ourselves before or oppose the rages of the learned masses." Yet Kepler insisted that the opposition that Galileo feared was no longer universal and that theirs would not be the first voices to argue for a new way of understanding the cosmos. With Copernicus and several others having already paved the way, wrote Kepler, "moving the Earth should not be a new thing anymore." Kepler argued still further that since the ideas of Copernicus were not well understood, learned Copernicans like Galileo and Kepler needed to guide the world to the correct views cautiously and deliberately rather than letting the opponents of Copernicanism gain the upper hand and watching from the sidelines as things spiraled out of control.[32]

Kepler then outlined an ambitious plan with which he and Galileo, and with them other Copernicans, could set about actively changing the minds of the public, bringing the world slowly around to the Copernican cause. His picture of the gradual dissemination of Copernicanism involved a strong emphasis on authority and the use of small deceptions, justifiable because they led the public to the ultimate larger truth of Copernicanism. The *vulgus*—the ignorant public—reasoned poorly and needed guidance, Kepler declared, and it was incapable of judging arguments on their merits. The only way for Kepler and Galileo to convince the public of the truth of Copernicanism, then, was to rely on the idea of authority, overpowering the public with the supposed number of experts who came down on the side of the Copernican system. "Let us begin to overwhelm the public more and more with authority," he wrote, "if, perhaps, we can guide it to the same knowledge of truth through deception." This would serve two purposes. First, those who already believed in the ideas of Copernicus but, like Galileo, were frightened to publicly state their opinions would receive comfort in hearing of others who shared their views. Coming forward,

Kepler urged, "would help all allies laboring against unfair judgments, long enough for them to either take solace from your agreement or protection from your authority." Indeed, though he still insisted that the Copernican question had already been broached by others before, he acknowledged that "it is not only you Italians who are not able to believe that they move unless they feel it but also we here in Germany do not find ourselves in the best favor with our enemies because of this dogma."[33]

Second, while the vocal support of committed Copernicans would help all those who already believed in the ideas of heliocentrism, that support was also important, Kepler argued, to counter the cries of the ignorant masses who opposed it. Those who disagreed with the ideas of Copernicus would be far less likely to voice that opposition when overwhelmed by the presence of experts who supported those ideas with the voice of authority. The *vulgus* would be forced to concede to the authoritative voices of Kepler, Galileo, and all those who stood with them, or they would be too overwhelmed to say anything to the contrary. Yet Kepler did not just hope to convince an unskilled and ignorant *vulgus*; in fact, on many occasions he seemed ready to discount the *vulgus* entirely. Rather, he hoped to focus most of his energies on the *docti mediocriter*, those moderately learned men who dabbled in the scholarly debates of many fields—erudite patrons, perhaps, or scholars not well acquainted with the rudiments of mathematical astronomy. To convince such men of the truth of Copernicanism, Kepler advocated a rhetorical campaign of persuasion, anchored in the idea of scholarly expertise and dependent upon the structures of the Republic of Letters itself.

Kepler drew, in particular, on the rhetorical techniques that he had studied with Martin Crusius at the University of Tübingen when he outlined to Galileo the method by which they might convince their audience of moderately learned men to judge their case favorably. In particular, he emphasized the idea of *ethos*: he and Galileo should highlight their own reliability rather than the details of the system that they propounded. And they should do so, Kepler argued, through the use of *auctoritas*; they needed to provide examples of reliable and authoritative sources on whom their readers could rely. In so doing, he and Galileo would be providing them with persuasive grounds for assent. Yet there remained a problem: there were, in fact, few Copernicans whom Kepler and Galileo could legitimately hold up as authoritative witnesses to support their cause. This was where the "deception" to which Kepler referred came in, when it came to convincing both the *vulgus* and the *docti mediocriter*. Kepler and Galileo, and other Coperni-cans along with them, needed to marshal their own authority in the absence of external support and to project an image of such unques-tionable expertise as to make it *appear* that the weight of *all* authority was on their side. The unlearned *vulgus* would simply follow those they believed to be in the know. For the moderately learned, Kepler and Ga-lileo would need to provide something a bit more elaborate; then they would be "bewitched by the authority—I speak from experience—of those with mathematical expertise."[34]

How would Kepler and Galileo target the moderately learned, in particular? First, Kepler emphasized the production of astronomical tables by Copernicans. When people, he wrote, "hear what ephemeri-des we already have, constructed from the hypotheses of Copernicus, [they will hear that] all those who write ephemerides today follow Copernicus." When, moreover, they are provided with mathematical demonstrations of the phenomena from a Copernican perspective, he said, it will seem to them that "the phenomena cannot be estab-lished without the motion of the Earth." Neither conclusion, of course, was justified—ephemerides were often constructed on the basis of the older, Ptolemaic system, and mathematical demonstrations could bol-ster Ptolemaic, as well as Copernican, theory. Yet Kepler argued that this deceptive strategy was justifiable because of the ultimate truth of Copernicanism: "since [the claims of Copernican theory] are true," he wrote, "why should they not be forced upon [others] as irrefutable?"[35]

Second, Kepler emphasized that Copernican mathematicians like himself and Galileo should write frequently and supportively to one another. Most university mathematicians, he noted dismissively, were lost causes because they demanded certain proofs and were hostile to the Copernican cause. If Copernicans isolated themselves from other mathematicians and then wrote supportively to one another, they could show these letters to the public, a practice that would make people think that Copernicanism was the accepted norm in scholarly, mathematical circles—that, as Kepler wrote, "all professors of math-ematics everywhere agree."[36] The sharing of letters, according to this strategy, would create not only a bond between like-minded individu-als but also the illusion of universal assent. And this illusion would make the claims of individual Copernicans much more readily ac-cepted by an audience easily swayed by the specter of authority.[37]

After outlining this ambitious plan, Kepler ended his letter by backtracking. "Is there truly need of this deception?" he asked. "Be confident, Galileo, and go forth. If I conjecture well, few of the distinguished mathematicians of Europe will want to separate from us: such is the force of truth."[38] Having only just heard from Galileo, one of the few fellow Copernicans with whom Kepler was familiar, Kepler could not have intended this statement as anything more than a rhetorical flourish. Or perhaps that was precisely the point—in ending with a forceful statement about the power of truth to unite all learned experts, Kepler was demonstrating the way he hoped to marshal rhetoric to achieve, with effort and skill, the very goal that he presented as artless and inevitable.

Kepler's Strategy Enacted with Duke Frederick of Württemberg

The strategy that Kepler articulated to Galileo was not simply a hypothetical suggestion. Even before Kepler's *Mysterium Cosmographicum* ever made its way to Galileo, Kepler had already enacted some of his own suggestions in his attempt to solicit princely patronage for the book, in the person of Frederick, the Duke of Württemberg. Kepler recognized that Frederick would have little reason to support an unknown mathematician who advanced hypotheses that seemed to contradict accepted scholarly wisdom. He pursued a two-pronged strategy in his attempts to convince the duke that his discovery was worth supporting. On the one hand, he argued in a letter to the duke that his discovery was novel and unprecedented, sure to set its readers alight with wonder; on the other, he asserted that Copernican theory, which formed the foundation of his text, was accepted by all those with real expertise and hence was neither dangerous nor controversial. Moreover, not only was Copernicanism accepted by modern astronomers, but it had an ancient pedigree, which hearkened back to Plato and Pythagoras. Kepler provided a context out of which his work emerged, while highlighting the ways that it advanced beyond that context and provided a valuable contribution.

Positing Copernicus as the theorist "whom all the famous astronomers of our time follow, instead of Ptolemy and Alfonso,"[39] Kepler briefly summarized Copernican theory and noted that no one had yet ascertained why things were as Copernicus had described them. He then described his own theory, in which he had proven geometrically that the universe was structured around the five Platonic solids. Kepler was right to stress the novelty of his own contribution, yet his contention that Copernicanism was accepted by all experts—indeed, by "all the famous astronomers of our time"—was dubious at best. Kepler had not yet received Galileo's letter, and aside from himself and Michael Maestlin, his teacher of mathematics and astronomy, there were few astronomers whom he could legitimately claim for the Copernican cause.

Kepler also proposed a unique and tangible materialization of the ideas of the *Mysterium Cosmographicum* for the duke's *Kunstkammer*, his cabinet of curiosities. He would fashion "a credenza-goblet . . . with a real and actual image of the world and a model of the creation, up so far as human reason can reach, and the likes of which has previously neither been seen nor heard by anyone."[40] Here, he appealed to Frederick's respect for the authority of expert astronomers as well as to traditions of patronage and courtly culture, which placed great value on novelty and collectability. Kepler sought to entice the duke to support both his own career and the ideas of Copernicus by embodying his discoveries in the credenza-goblet, something concrete and collectible.

At Kepler's request, Maestlin also wrote to Frederick, adding his own authoritative voice to Kepler's. Like Kepler, Maestlin argued that Copernican theory had an ancient pedigree and was the standard from which all expert astronomers operated, although he was a bit more circumspect in his language. "Kepler follows the new or newly revived hypotheses of Copernicus," wrote Maestlin, "which some time ago Aristarchus and other very wise philosophers taught: that the Sun sits at the center of the world, immobile, but the Earth moves, etc." Maestlin noted that Copernican theory had not yet been perfectly formulated and was more complicated than Ptolemaic theory; for this reason "the common and ancient hypotheses [here, Ptolemaic theory] are kept for the youth, and since they are easier to understand (as is only fair), they are taught: but all practitioners remain convinced by the demonstrations of Copernicus."[41] Though it is possible that Maestlin referred to the technical use of Copernican theory for the production of tables (such as the *Prutenic Tables*), his language was ambiguous and most likely conveyed to the duke the impression that Copernicanism in its entirety was adopted by all experts.

Duke Frederick, reassured by Maestlin's letter and intrigued by the idea of the goblet in the form of the cosmos, agreed to accept the goblet and patronize Kepler's book. In addition to setting to work on the construction of the goblet, Kepler acknowledged Frederick's support by dedicating the opening diagram of the book—a depiction of the goblet—to him. Unfortunately, the credenza-goblet was never finished; the project was a difficult one, and Kepler ran into trouble dealing with the artisan responsible for its construction, particularly from a distance. Likewise, Kepler did not immediately receive any monetary gifts for his work—he had to wait until several years later for his 250-gulden reward from the estates of Styria. Still, Kepler did benefit from the dedication to the duke, and in ways he had not fully anticipated. As the Tübingen theologians began to express their reservations about Kepler's book, they were restrained by the power of the important patron Kepler had secured for himself. Indeed, when Maestlin wrote to Kepler describing the complaints of the theologians, he noted that "the book somewhat offends our theologians, though moved by the authority of our prince (to whom the principal diagram is dedicated), they leave it undecided."[42] Likewise, Kepler himself acknowledged that his book likely could not have been printed in the first place had he not secured Duke Frederick's support. He was pleased, he wrote to Maestlin, that "with the authority of the prince having been indirectly won over, the book was able to be printed at Tübingen, which certainly could not have been done if the court had not fortified me against the hidden thoughts of many people. They perhaps thought they would have gotten in the way of the prince's pleasures if they had impeded my little book. At any rate, I truly did not consider this when I first approached the prince."[43] Though he may not have turned the duke into a Copernican, Kepler realized that the support of the duke had been crucial to the spread of his Copernican message. By impressing on the duke that his theories were both novel and authoritative, and then by offering to make those theories the concrete property of the duke in the form of the credenza-goblet, Kepler had secured himself against some of the "rages of the . . . masses" that Galileo had so worried about.

In their dealings with Duke Frederick, Kepler and Maestlin thus performed the very rhetorical maneuvers that Kepler outlined to Galileo shortly thereafter. Kepler contended that Copernicanism was a theory to which all experts gave their ready consent, and he provided the *Mysterium Cosmographicum* as an example of a great achievement made possible by the ideas of Copernicus. He then had Maestlin send a supporting letter, emphasizing precisely these points and reinforcing Kepler's contention that the weight of authority was on his side. And he imagined an object that the duke could hold up as a material embodiment of that authority. As Kepler would later emphasize to Galileo, the slight deceptions involved in this campaign were justifiable, for Frederick's support would do wonders for the spread of Kepler's book and hence the Copernican cause more broadly.[44]

Kepler's Strategy Enacted: The *Dissertatio cum Nuncio Sidereo*

Despite Kepler's best wishes, Galileo never replied to Kepler's letter of encouragement. But thirteen years later, Galileo published his famous *Starry Messenger,* which contained reports of his telescopic observations of the moon and stars and of his discovery of the moons of Jupiter, which he called the "Medicean stars" in honor of his Medici patrons. After publishing the book, Galileo renewed his contacts with Kepler, writing to the Tuscan ambassador in Prague with a request for Kepler's opinion of the work. Though Kepler had no telescope with which to

IOANNIS KEPLERI
Mathematici Cæfarei
DISSERTATIO
Cum
NVNCIO SIDEREO
nuper ad mortales miffo
à
GALILÆO GALILÆO
Mathematico Patavino.

Alcinous.
Δεῖ δ' ἐλευθέριον εἶναι Τῇ γνώμῃ τ μέλλονα Φιλοσοφεῖν.

Cum Privilegio Imperatorio.

PRAGÆ,
TYPIS DANIELIS SEDESANI.
AnnoDomini, M. DC. X.

4.1 Title page of Kepler's *Dissertatio cum Nuncio Sidereo* (1610)

verify Galileo's claims, he responded with a long letter of support, published with few changes under the title *Dissertatio cum Nuncio Sidereo* (*Conversation with the Starry Messenger*).[45]

When we consider Kepler's *Dissertatio* in light of his 1597 letter to Galileo, his quick show of support makes perfect sense.[46] Not only was Kepler excited about the telescope with which Galileo made his observations in the *Starry Messenger*, but he also could not have failed to note that in the *Starry Messenger* Galileo had, for the first time, publicly supported the Copernican system when he remarked, almost as an aside, that the four Medicean stars "execut[e] with one harmonious accord mighty revolutions every dozen years about the centre of the universe; that is, the Sun."[47] When Kepler received the request to evaluate Galileo's telescopic work, he saw it as a prime opportunity to support a fellow Copernican. Indeed, when asked for his opinion of Galileo by Giovanni Magini, he responded at the outset that "we are both Copernicans" and noted, echoing Erasmus, that "like rejoices with like."[48]

Moreover, the strategy that Kepler marshaled within his *Dissertatio* has much in common with the outline for the dissemination of Copernicanism that he had first laid out in the letter of 1597—both in the emphasis on expertise and authority as grounds for assent and in the idea of communication as a means to convey the impression of a community of believers. First, Kepler wrote of his support in a quick letter to Galileo, which he published thereafter as the text of the *Dissertatio*, with very few changes. He emphasized at the opening of the text that the *Dissertatio* represented a private conversation between two like-minded experts, one that he was sharing only to make clear to everyone his agreement with Galileo. In his dedication to Giuliano de Medici, he described the book as a text that had been "private and particular to Galileo" and was only now being made public in printed form,[49] and he remarked in his notice to the reader that in creating the *Dissertatio* he had simply adopted "this shortcut: I have had printed the letter that I sent to Galileo (which was composed with great speed, within the prescribed time, among other urgent occupations)."[50] The publication of the *Dissertatio*, literally "conversation," was, for Kepler, equivalent to the supportive sharing of letters that he had outlined so many years earlier.

Second, within the *Dissertatio*, Kepler emphasized Galileo's expertise and the expertise of those who transmitted his message as reasons why his claims must be accepted, also a theme of the 1597 letter. Kepler noted that Johannes Matthäus Wacker von Wackenfels, who had first brought him the news of Galileo's discovery, had told him that "very illustrious men, exalted by their education, dignity, and constancy far above the foolishness of common people, report these things about Galileo."[51] The authority of these reports influenced Kepler, he wrote, as did "the authority of Galileo . . . acquired by the rectitude of his judgment and by his skilled nature."[52] Kepler defended his immediate concurrence with Galileo's writings: "perhaps I shall seem reckless, since I trust your assertions so willingly, bolstered by no personal experience. But why should I not trust a most learned mathematician . . . ?"[53] He contrasted Galileo's expertise as a user of the telescope with his own inexperience: "Should I, dim-sighted, disparage someone with keen sight? Or someone equipped with optical instruments, while I myself, bare, lack this equipment?"[54] Thus, Kepler presented Galileo's expertise as incontrovertible and implied that his claims were similarly unassailable.

Kepler also noted external factors that bolstered the credibility of Galileo's testimony. Galileo had publicly made his views available, so that they could be verified by others. He was, moreover, a "Florentine noble,"[55] one who was backed by one of the most respected families in Tuscany, the Medici. "Would it have been a small thing," wrote Kepler, "for him to trifle with the family of the Grand Dukes of Tuscany and to affix the Medicean name to his own inventions while on the other hand promising planets?"[56] Kepler pointed out that if Galileo had wanted to perpetrate a fraud, he would have done so in a way that was more logical and intuitive, one that would more easily gain the assent of his victims. Why, asked Kepler, would Galileo have invented planets only around Jupiter and not around any of the other planets? And why would he have chosen four planets specifically, a number not paralleled elsewhere in nature? With this line of reasoning, Kepler argued that Galileo's claims were credible precisely because they were unusual, and even illogical, of the sort that no one would deliberately fabricate.

Finally, a key element of Kepler's strategy of persuasion, as outlined in his 1597 letter, involved his conveying the illusion of wide-ranging assent. To this end, Kepler devoted much of the *Dissertatio* to an illustration of the many people who anticipated Galileo or whose ideas and discoveries coincided with those of Galileo.[57] Kepler first noted that although the idea of so powerful a telescope may have seemed unbelievable, "it is by no means impossible or new."[58] In fact, not only had it been produced recently by the Dutch, but it had been discussed even earlier, when "already many years ago it was revealed by Io. Baptista Porta in his book of natural magic."[59] Likewise, Kepler himself had anticipated the telescope in his *Optics*.[60]

Kepler also cited authoritative predecessors who agreed with Galileo's claims about the nature of the Moon's surface, the possibility of additional planets, and the mass of stars in the Milky Way—predecessors who included Pythagoras, Plutarch, Bruno, Maestlin, and, of course, Kepler himself. "You confirm those who previously asserted the very same thing as you," Kepler wrote.[61] Galileo, in Kepler's telling, followed in the tradition of many innovators who progressed by walking in the footsteps of illustrious predecessors: "that which you say you recently discovered with your own eyes, they had predicted long before you, as is required."[62] In citing the members of a preexisting community of believers, Kepler hoped to bolster Galileo's claims: if not to present them as foregone conclusions, then at the very least depicting them as commonly held beliefs rather than novelties.[63] The weight of authority, he hoped to show, was on Galileo's side.

Kepler's Strategy Backfires: The Dangers of an *Ethos*-Based Campaign

Before Kepler published his *Dissertatio*, a number of interested friends and fellow scholars wrote to him requesting his opinion of Galileo's new discoveries. Martin Horky, a young Bohemian living in Bologna, was one such friend. "It is a wondrous thing, and an astounding thing," wrote Horky to Kepler of Galileo's story of the moons of Jupiter in late March 1610. "Whether it is true or false I do not know."[64] He wrote two more letters in the next two weeks again asking for Kepler's opinion of the Galilean discoveries, though by the last he was starting to doubt Galileo and wondered whether the moons might be simply a *fabula*, a tall tale.[65] Finally, by April 27, he had decided that Galileo was simply not to be trusted. Indeed, he wrote to Kepler, Galileo had visited him and his mentor, the mathematician and astronomer Giovanni Antonio Magini, in Bologna only recently and had brought his telescope with him to show them "those made-up planets." At night, they had tested the telescope out, with very disappointing results. "On earth it works miracles," he wrote, but "in the heavens it deceives." In particular, he noted that while he and all those with him were able to see four very small stars around Jupiter, as Galileo had claimed, "all acknowledged that the instrument deceived." According to Horky, "Galileo became silent, and on the twenty-sixth, a Monday, he sorrowfully departed from our most illustrious master Magini very early in the morning. And he gave no thanks for infinite kindnesses and reflections [he had received], because he sold a tall tale." Horky added, in a German post-script at the very bottom of the otherwise Latin letter, that he had sur-reptitiously "made an impression of the spyglass in wax," so that he might construct one better than Galileo's upon his return home.[66]

Rather than immediately replying to Horky, Kepler wrote his *Disser-tatio* to Galileo as an open letter for all to see. He hoped that Horky and others might read his letter and see in it his clear support of Galileo and alter their own views accordingly. In June 1610, after publishing the *Dissertatio*, he finally wrote back to Horky and held up the *Disserta-tio* as both his response to Horky and his true opinion of Galileo. "You are still caught up in your original doubts about the stars of Galileo," he wrote to Horky. "I do not marvel, nor do I blame you: it is proper for the opinions of philosophers to be free." Yet Kepler hoped that if Horky truly respected his opinion, as he had claimed, he might read Kep-ler's *Dissertatio* with an open mind, and his doubts might be resolved. Horky had mentioned to Kepler that he hoped to write a book attack-ing Galileo, and Kepler hoped that he had not yet done so; if he had, Kepler asked, "please, to gratify me, who loves both you and truth: free me from concern, and write to Galileo that, having read my *Dissertatio*, you began to believe that which previously seemed untrue to you."[67]

Horky had managed to read Kepler's *Dissertatio*, but unfortunately he did not receive Kepler's June letter before going ahead and publish-ing his own book, the *Brevissima Peregrinatio contra Nuncium Sidereum* (*Brief foray against the Starry Messenger*). And while Kepler had hoped that Horky might read his *Dissertatio* and decide to judge Galileo more favorably, Horky learned exactly the opposite lesson from Kepler's pub-lished letter. He believed that Kepler had intended his *Dissertatio* as a critique—not an endorsement—of Galileo's text. Kepler's *Dissertatio*, to Horky's mind, had only bolstered his initial impressions of Galileo's deceit. Indeed, Horky specifically cited the *Dissertatio* of Kepler in his *Peregrinatio* as supporting evidence for his claims of Galileo's trick-ery. So convinced was Horky that he had read Kepler correctly that in May 1610 he sent Kepler a copy of his own book in which he wrote that he owed Kepler a great debt, for "I know where the deception comes from, as you most learnedly discovered in your *Dissertatio*."[68] (This letter evidently took some time to reach Kepler, for Kepler's June let-ter to Horky, asking him to read the *Dissertatio* and write positively to Galileo, betrays no knowledge of Horky's use of the *Dissertatio* in his *Peregrinatio*.)

While Horky may have simply been a poor reader, he was not the only one to understand Kepler's *Dissertatio* as a critique of Galileo rather than an endorsement. Michael Maestlin—himself already a Copernican, and not someone who would rashly and unthinkingly oppose Galileo's new discoveries—also praised Kepler for taking Galileo down a notch: "You have deplumed Galileo,"[69] he wrote, alluding to Aesop's story of the jackdaw that dressed itself in the borrowed feathers of other birds. Georg Fugger, the imperial ambassador in Venice, used this reference yet again. He praised Kepler for his trenchant critique of Galileo and wrote that Galileo was "accustomed to collecting . . . the feathers of others, in order to decorate himself as Aesop's crow."[70]

Why was Kepler's *Dissertatio* so drastically misunderstood?[71] It seems that while Kepler had noted all those who had anticipated Galileo's claims in order to highlight them as supporting authorities and to create the impression of a community of experts all in agreement, others interpreted Kepler's work as a negative statement about Galileo's character. Kepler had intended to convince people to believe him and his fellow Copernicans through a rhetorical campaign based on the idea of *ethos*, focusing on their perceived characters alone; and in a way he had not foreseen, he did just that. In highlighting the many thinkers who had anticipated Galileo, Kepler unwittingly created the impression that Galileo had claimed credit for the discoveries of others—hence, the two references to Galileo as Aesop's crow. And if Galileo could not be trusted to name those individuals whose ideas had led to his own, perhaps his new claims should be distrusted as well.

Maestlin clearly articulated this sentiment. "You have deplumed Galileo," he wrote, "by showing that he was not the first author of this new telescope nor the first to notice that the Moon has a rough surface nor the first in the world to show more stars in the heavens than we have up to now found listed in the writings of the ancients." And on this basis, Maestlin argued, should not Kepler be pleased that Martin Horky had taken the final, logical step, showing that not only was Galileo not credible as a person but also that his claims themselves—specifically, the satellites of Jupiter—were false? Horky, according to Maestlin, had completed Kepler's work when he had "noticed the deception . . . in the telescope of Galileo himself and therefore slain [Galileo] with his own sword."[72]

Kepler was far from pleased—he was furious with Horky's public misuse of his text. On the same day, he quickly wrote to both Horky and Galileo, struggling to undo the damage. In his letter to Galileo, Kepler tried both to distance himself from Horky and to show that unlike both himself and Galileo, Horky had incorrectly assumed the mantle of expert. He had only recently received Horky's book, and he marveled "at the temerity of this youth—with the native scholars only murmuring that he, foreign and ignorant, [alone] contradicts, with the matter not yet proven." Perhaps, Kepler guessed, it was his very youth, foreignness, and inexperience that made him speak so boldly—they served, Kepler wrote, "like the mask of an actor," allowing him to avoid true responsibility for his words. Though Kepler believed that Horky's words were so worthless as to be a waste of Galileo's worry or time, he had decided to write to Galileo immediately to explain the specific ways that Horky had "abused" his letter.[73]

As soon as he had learned, Kepler explained, that Horky was one of Galileo's critics, he decided to take action immediately—not merely because Kepler himself was a supporter of Galileo personally but because "I knew how [these sorts of people] oppose all new discoveries [and hoped] to forestall it on this occasion." In other words, Kepler had interpreted Horky's dismissal of Galileo's claims as a dismissal of novelty in astronomy more generally—and of Copernicanism in particular. He had hastened to convince Horky to change his mind, in keeping with his plans from years before to remain a vocal and committed supporter of Copernicus and his adherents. For that reason, he had sent Horky a copy of his own *Dissertatio* so that Horky might learn from it either to think more wisely or at the very least to restrain his overly hasty attack. Much to Kepler's chagrin, Horky had learned neither lesson— instead, he had "strangled my friendship most slanderously."[74]

Though Kepler had claimed earlier that expert arguments would easily sway an inexperienced public, he conceded now that the idea of expertise was itself problematic, as the public was incapable of judging accurately who was an expert and who was not, so long as one invoked the general idioms of science. Horky had taken advantage of the ignorance of the *vulgus*, which, "unskilled in optical methods, listens freely to critics speaking of optical things. It cannot distinguish between a blind man and a seeing one and rejoices in any spokesperson of its ignorance whatsoever." To remedy this, Kepler suggested that actual experts in optics address the questions concerning the telescope, so that Horky could be revealed as a fraud. Of course, one could suggest to the *vulgus* that they study optical writings themselves, in order to rebut Horky's "thoroughly stupid little book" on their own, yet this was not their way. "They prefer to follow this author when he says that a curved line is straight, in order to be able to run riot against philosophy, rather than to take up the work themselves." He hoped that they would be swayed from their ignorance if "a scholar knowledgeable in this science would put pen to paper to contradict such nonsense."[75]

Indeed, when neither Galileo nor any other such scholar was forth-coming, Kepler did just this himself in 1611: his *Dioptrics* addressed the optics behind the telescope specifically to explain and bolster Galileo's claims. Simple statements of support and general allusions to expertise would not be enough, it seemed, to draw an impressionable public to the side of truth.

Kepler then paused to ponder why it was that so many people de-nied the actual objects that the telescope revealed, even those, like Horky, who had actually had the chance to use a telescope. He at-tributed this to the fact that vision in general could be deceptive and could vary from person to person; "I see it is not impossible," he wrote, "that one person sees what thousands of others do not." Kepler him-self had found that sometimes the telescopes he used were problematic for others, while the ones praised by others were blurry for him. Yet certainly, he insisted, there were others who could bolster Galileo's ac-count of what he had seen. Kepler's attempt to cite as witnesses those who had anticipated Galileo had backfired in dramatic fashion, so he asked instead for witnesses who could verify Galileo's claims using the telescope itself. "Though I no longer have any doubts," wrote Kepler, "nevertheless, it has pained me that all this time I have lacked other witnesses who would confirm the belief of others." Since Galileo had written that he had witnesses who could testify on his behalf, Kepler asked him to name them; at present, wrote Kepler, "I can produce no one other than you . . . by which I may defend the reputation of my letter: in you alone rests all the authority of the observation."[76] His own reputation was now entangled with the reputation of Galileo, and it would take credible witnesses beyond the two of them to repair the misreading of their texts and convince their readers to judge them—and, more importantly, their claims—favorably.

Kepler's request for additional witnesses, and his post-*Dissertatio* let-ter to Galileo more generally, are evidence of a shift in the focus of Kepler's rhetorical techniques—a shift that can be implicitly detected, at times, in the *Dissertatio* itself and that Kepler made explicit only af-terward, in light of the *Dissertatio*'s misreading. Kepler had focused his *Dissertatio*—much as he had planned in the 1597 letter to Galileo—on the ideas of authority, expertise, and a scholarly community all in agreement. He had cited all those who had anticipated Galileo in or-der to overwhelm his readers with the weight of authority, blinding them with the image of a group of mathematical experts who all held to the same irrefutable views. At the same time, however, he empha-sized that his audience should accept Galileo's discoveries on the basis of the credibility of Galileo as a person rather than simply the credibility of his claims. To this end, he highlighted Galileo's status as a gentleman with princely patrons. This latter, more legal standard, which focused on credible testimony and trustworthiness, was one that Kepler adopted more unambiguously when his initial *Dissertatio* failed to achieve its purpose.

In explaining the ways in which his *Dissertatio* had been misunderstood, Kepler emphasized in his later letter to Galileo that his goal in citing the work of others had been not to criticize Galileo but the opposite—to support him. "I did not enumerate that earlier, similar things had been observed in order that he might disparage you but rather so that others might believe the testimony of many." The mere fact that others had said things similar to Galileo in the past should in no way imply that Galileo had copied from them, as Horky and others had suggested; indeed, Kepler insisted, "often people come together to the same target in different ways." Moreover, if he had wanted to make that kind of accusation, he would have said so clearly and directly rather than insult a fellow scholar in an underhanded way. Horky should not "think that I am so sleepy that I do not know how to speak openly. He should let me [speak for] myself."[77]

In bringing the testimony of additional witnesses to support the claims of Galileo, then, Kepler had never intended to accuse Galileo of deception or of borrowing the ideas of others; instead, he had hoped to "add to the *fides* of the *Nuncius*."[78] Ancient rhetoricians evaluated the credibility of witnesses and their testimony according to *fides* ("credit," "faith," or "belief"),[79] and this method was applied in a number of early modern disciplines, including the discipline of law. Kepler relied on this linkage, and in the face of opposition to Galileo's discoveries and of the misreading of his own text, he explicitly reframed the entire debate as a legal one. "Truly," he wrote, "this is not a philosophical problem but a juridical question of fact: whether Galileo deliberately deceived the world."[80] He emphasized here that despite their philosophical implications, Galileo's claims were not *themselves* philosophical and could not be refuted using philosophical reasoning. Rather, they had the status of legal questions of fact—a phrase that anticipates the "matter of fact" that underpinned the program of the Royal Society years later.

Kepler named himself as witness, testifying to the character and credibility of Galileo as author. "This is a contest between virtue and vice," he wrote. "I, as an honest person considering Galileo's claims, judge that such evil does not lie within him." At the same time, he sought to discredit Horky by referring to his general dishonesty: Horky was someone "who has no sense of honesty, and since he therefore does not deem it important, he probably judged others according to his own character." Kepler acknowledged the possibility that he was wrong in his positive assessment of Galileo. In this case, Kepler would be pitiable but still honest, while Horky, who would have been right only coincidentally, would still be a scoundrel who had twisted the facts of the matter. Moreover, Kepler insisted that there was no way to avoid the possibility that his own support of Galileo was misguided, "for this is the way of the law, that anyone is presumed to be good until the contrary is proven—how much more so if the circumstances have created *fides*?"[81]

Kepler concluded by noting that he hoped that this letter, unlike his earlier *Dissertatio*, might speak clearly for itself and plainly convey his opinions about the status of Galileo's claims. He was pained, he wrote, only by the way his own words had been twisted, not by the general scorn of the *vulgus*: "I am not so stupid that I am swayed by the negative attitude of the crowd or by their listlessness and inability to argue against the experience and skill of an astronomer." Though he had earlier urged Galileo to set aside the Pythagorean model of scholarly silence, he now acknowledged that such an approach clearly had its merits. Yet he was still very pleased that Galileo had seen fit to "cast the die . . . [and] open the sanctuary of heaven." All those who refused to look at the wonders Galileo had revealed could only be pitied; the *vulgus* "is punished for its contempt for philosophy with everlasting ignorance." Kepler added finally that this letter, like his *Dissertatio* earlier, was written directly to Galileo but intended for public consumption. Galileo should feel free to share it widely—not, as he had indicated earlier, to convey the sense of a shared community of believers but rather to establish it as an *epistola publici juris*.[82] Although this phrase could simply imply publication, Kepler likely intended it to bear some legal overtones as well, following the focus on law throughout his letter.

Kepler and Horky Revisited

On the same day that Kepler sent his letter off to Galileo, Kepler drafted a second letter and sent it to Horky. Here, Kepler made clear that he was greatly displeased with Horky's misuse of his *Dissertatio*. While he knew, he wrote, that his harsh words might well cause Horky to end their friendship, he hoped at the very least not to make an enemy of

Horky, largely out of respect for Horky's father, a diplomat at the court in Prague and a friend of Kepler's. He warned Horky that he had written a letter to Galileo ("a letter of the sort you can imagine," he added) and that he had granted Galileo the authority to print it at his own discretion. He noted that Horky's father worried about him and would be still more worried "if he knew about your *Peregrinatio* and my foray against you."[83] Kepler advised Horky to avoid the public stage and leave Italy soon, if possible.

As it happened, Horky did leave Italy—but he did so before Kepler's letters to him arrived. Consequently, he left with no idea that Kepler opposed his criticism of Galileo and his reliance on the *Dissertatio*. In fact, on his journey home he specifically stopped at Kepler's home to further discuss their supposed mutual disapproval of Galileo's "deceptions." Kepler, at the same time, had no idea that Horky had not received his letter and assumed that Horky knew of his disapproval. As Kepler related to Galileo afterward: "It was a strange and remarkable meeting, when he with an exultant expression and as if triumphant over Galileo spoke to me as though I agreed, while I responded as I had done in my letter, in which I had renounced my friendship with him. It threw each of us into a great confusion, because he did not know about my renunciation (since my letter was delivered after his departure from Bologna), nor did I think otherwise than that he had read the letter in which I had explained myself."[84] After a great deal of awkwardness and confusion, the two finally realized their mistake, and Horky immediately endeavored to regain the favor of Kepler, whom he greatly respected. He clarified to Kepler the reasons for his suspicion of Galileo, while Kepler relayed to him his own observations, ones that refuted Horky's claims. "It hurt him badly," Kepler wrote to Galileo, "when I recounted to him what I had written to you."[85] Horky was particularly distraught that Kepler had accused him of treachery; even if he had misrepresented Kepler's views, he insisted, he had done so with only the best intentions, fully convinced that he had accurately represented Kepler's own point of view.

Horky tried not only to prove his good intentions to Kepler but to make clear to Kepler why he still opposed Galileo's discovery of the moons of Jupiter, even as Kepler held up his own observations, as reported in his *Narratio de Jovis Satellitibus* (1611), as proof. To bolster his own position, Horky argued that "he had followed the public opinion of many of the most learned professors at the University of Bologna and others throughout Italy."[86] He said that the absence of any rebuttal by Galileo to the claims of these university professors proved that his conjectures were unsound, for if he had real proof of his claims, would he not, as a man of honor, have immediately taken it upon himself to supply it? Indeed, Horky added, many scholars in Italy had wanted to publicly debate the matter with Galileo, but Galileo had deliberately avoided them and had even changed his residence because he feared the confrontation, allowing the authority of his prince to protect him from even the deserving queries of fellow scholars.

Horky argued that it had therefore been only a matter of time before someone had come forth with a book like his—he had simply happened to be the first. In the end, Kepler confessed to Galileo, Horky had convinced him—not that Galileo's claims were flawed or problematic but rather that Horky's own behavior had not been entirely unjustified. Horky had "persuaded me," he wrote, "and I forgave him; we became friends once again." They ended their meeting cordially, but Kepler promised that "as soon as he, with me demonstrating, has seen and recognized the satellites of Jupiter, he will change his mind."[87] As Horky needed to leave to travel to his father, this joint telescopic adventure would take place the next time that Horky came to Prague.

Because of Kepler's newly reached understanding with Horky, Kepler also asked Galileo to refrain from publishing the letter denouncing Horky that he had sent to Galileo earlier and that he had given Galileo permission to broadcast widely. Even if Galileo felt that he could improve his reputation and counter the attacks against him by publishing Kepler's letter, Kepler assured him that "the glory of your triumph will be greater if, as I hope, I send to you a voluntary confession of your enemy." Moreover, since sending Galileo that letter, Kepler had also published his *Narratio de Jovis Satellitibus*, recounting his own personal experiences with the telescope that he had procured in the interim, and he noted that this might be the best way for Galileo to refute those who denied his discoveries. Regardless, Kepler urged Galileo to have some compassion for poor Horky and to avoid addressing Horky's work directly. "Consider his youth," he wrote. "Nothing is more common at this age than to agree passionately with one's teacher's ideas, and from them to dash ahead in a fit of rash courage as if from some rampart, in order to engage in battle with an enemy." With this in mind, Kepler argued, Horky's antics should be dismissed as the excesses of youth, nothing more—and certainly nothing worth openly combating. Not only would it be beneath Galileo's dignity to address the intemperate claims of someone so far beneath him both intellectually and professionally, but it would only make others give more credence to those claims. Moreover, it would open the floodgates for other similarly meaningless attacks, by making it clear that Galileo deemed them worthy of reply. While Galileo ought to support those truly interested in understanding and verifying his discoveries, Kepler argued, baseless condemnations were best ignored. Kepler noted that having left the university, Galileo had thankfully left behind the world of the disputation and need reply only to the questions that had merit, not to every meritless critique. "Leave their manners to the schools that you have left," he wrote.[88]

If Galileo did decide that it was necessary to respond personally to Horky and to invoke Kepler's own letter against Horky, despite Kepler's advice and requests to the contrary, Kepler begged that he refer only generally to the theme of that letter, while leaving out the particulars. In particular, Galileo should "remove the personal names and insulting words," such as "when I attributed petulance to him, called him a sputum of a man, accused him of destruction, called him a sycophant and a buffoon . . . [or spoke of] ignorance, temerity, stupidity, and other most unfortunate turns of phrase."[89] Kepler noted that the other terms he had used were more acceptable; these extreme words had been appropriate only given Horky's seeming exploitation of Kepler's own words. Yet once Kepler realized that Horky had meant no harm and had thought himself true to Kepler's intentions, Kepler felt that his own insults were no longer appropriate. The Republic of Letters ought to be characterized by civility, after all. If Horky had truly twisted Kepler's words on the public stage, he would have placed himself outside the boundaries of the Republic of Letters, and Kepler would have been justified in responding in kind.[90] But Kepler had come to realize that Horky's words stemmed from, not "vice or the disgraces of life, but rather the faults of character or age."[91]

Kepler's change in attitude toward Horky stemmed in large part, it is clear, from his realization that Horky was simply a well-intentioned and injudicious youth and not someone who had deliberately set out to wrong Kepler or Galileo. At the same time, it stemmed from his growing frustration with Galileo himself. After the initial letter that Kepler had received from Galileo in 1597, in which he learned of Galileo's Copernicanism, Kepler had rejoiced to meet a fellow scholar with whom he might honestly share his ideas and opinions. Yet while Galileo had ultimately not disappointed him when it came to support for the Copernican enterprise, he had disappointed him as a fellow member of the scholarly community. After the first letter, there had been no word from Galileo at all for thirteen years. When Galileo finally reached out to Kepler again, it was with thrilling news—a new instrument that revealed new bodies in the heavens, bodies that themselves seemed to support the theories of Copernicus. Kepler, once again, wrote to Galileo in full support, immediately publishing his thoughts for all to see. Yet Galileo continued to be less than forthcoming. Kepler repeatedly begged Galileo for a copy of his telescope. "You have inflamed me with a great desire to see your instrument," he wrote in one letter to Galileo.[92] The telescopes available in Prague were far inferior to Galileo's and made it impossible to verify Galileo's discoveries. The Tuscan ambassador to Prague himself asked Galileo to make a telescope available to Kepler, but Galileo claimed that none was to be had. He wrote to Kepler that he had already given some to his patrons and that the construction of others was "very laborious"[93] and simply not possible at the present time. In a postscript to his next letter to Galileo, Kepler asked him to "free us at the first possible moment from the desire for your new discovery."[94] Sensing that Galileo hesitated to reveal his secrets to others—indeed, aware that this was the very reason that so many had challenged Galileo's results to begin with—Kepler insisted that "there is no one whom you should fear as a rival."[95] Even so, Kepler's telescope was to come not from Galileo but from the elector of Cologne; he was one of the noble patrons to whom Galileo *had* given a telescope, and he temporarily loaned it to the court in Prague during a visit there.

Similarly, even when Galileo made new discoveries with his telescope, he did not share them immediately. Twice he sent to Kepler anagrams that contained hidden within them the nature of his discoveries—one about the triple form of Saturn and one about the phases of Venus—which Kepler tried desperately to solve. To the second anagram, Kepler sent Galileo back a full eight guesses, none of which hit the mark. "I implore you," he wrote in his letter, "do not keep the matter hidden from us any longer. See that you are dealing with real Germans. I come away impatient from your various literary secrets. Do you see the misery in which you cast me with your silence?"[96] Kepler continued to speak highly of Galileo in all his writings—particularly in the *Dioptrics*, which he published the same year and which contained the solutions to the anagrams that Galileo had finally revealed to him. Yet he never felt comfortable with Galileo's secretiveness. He surely recognized that Galileo had crafted his work "to maximize the credit [he] could expect from readers while minimizing the information given out to potential competitors"[97] and knew that Galileo put all his fellow scholars into this class.

Kepler, by contrast, viewed his own ideas and innovations not as personal products to be guarded secretively but rather as contributions to the general store of public knowledge. If Galileo saw himself writing not to any coherent community but rather to a diverse and disconnected field,[98] Kepler's goal was the deliberate construction of just such a community. This perhaps explains Kepler's insistence in 1597 that Galileo share his thoughts with a much broader audience, along with Galileo's own silence after that lone initial letter. Indeed, until the publication of the *Starry Messenger* in 1610, "Galileo seemed content to limit his audience to small groups of Paduan academics, Venetian patricians, and Florentine courtiers with whom he discussed philosophy, music, mathematics, and literature."[99] And when he did reach out to a broader audience, it was to secure his status as an inventor, not because he had agreed to join the community that Kepler struggled to create. Part of the reason that Kepler's *Dissertatio* was seen as an attack on Galileo lay precisely in the fact that Kepler strove to draw attention to that broader community and to highlight the larger efforts amid which Galileo had made his own discoveries.

Kepler believed that the open exchange of information and discoveries was the only way that a community of scholars might succeed; if he shared freely, he hoped others would share freely with him. As he once wrote to Magini (later to be Horky's host in Italy, as it happened), "I am so passionate for the astronomical arts that I cannot refrain from communicating my ideas to expert practitioners, so that through their advice I may accomplish subsequent achievements in this divine art."[100] The question of priority and personal credit had already been raised for Kepler years before with regard to Galileo himself. When, in 1603, Kepler's friend Edmund Bruce wrote to him that Galileo was supposedly claiming Kepler's ideas as his own,[101] Kepler urged Bruce to let it go. "I am not worried that Galileo claims my ideas," wrote Kepler. "There are those who emphasize truth and the glory of God the creator rather than their own reputations. Let the Garamantes and Indians hear these and other mysteries of God, let my enemies make them known, let my name also perish in the meantime, so long as the name of God, the Father of minds, is thus promoted."[102] Kepler knew how to act the courtier, to be sure: he fashioned his ideas into products tailored to his patrons, as we saw earlier with Duke Frederick, as surely as Galileo fashioned the moons of Jupiter into emblems of the power of the Medici.[103] Yet as with his deceptions of the *vulgus* and the *docti mediocriter*, he reserved this behavior for those *outside* the Republic of Letters. Fellow scholars, he felt, should be treated with openness and candor, not subterfuge or cunning. Likewise, the kind of insults he had leveled at Horky were appropriate only for those who had deliberately placed themselves outside the Republic and its guiding precepts; for those who adhered to the norms of scholarly decorum and civility, only civility could be returned in kind. Both Horky's earnest attempt to justify his actions and Galileo's continued evasiveness moved Kepler to reevaluate his condemnation of Horky's behavior toward Galileo. Though he continued to approve of Galileo the Copernican, his evaluation of Galileo the courtier was less positive.

Ultimately, Kepler's ideal vision for the astronomical community that he hoped to create was modeled on the larger Republic of Letters. Kepler reached out to Galileo as a fellow scholar, one who he hoped would welcome his ideas, criticize them honestly, offer his own ideas in turn, and contribute to his attempts to strengthen their communal enterprise. Though he was thrilled to have Galileo join the ranks of Copernicans and delighted that Galileo's telescopic discoveries further supported the truth of heliocentrism, he found himself troubled by Galileo's secretive behavior. For Kepler, who labeled his very first book with the word "mystery," or "secret," nature's secrets meant its inner workings. These were to be plumbed and then shared, not kept hidden; nature's surface secrecy did not correspond with any kind of secrecy or concealment on the part of its investigators. Kepler, that is, firmly believed in the ideal of public knowledge that was to characterize many early modern scholars of his time, in contrast with the hermetic exclusivity of many of his predecessors.[104]

While Galileo, too, had no interest in limiting his discoveries to a select few initiates, he emphasized secrecy in the sense of privacy and priority. Always concerned that others might take advantage of him and claim credit for his discoveries, Galileo's refusal to share his telescopes or his plans for their construction greatly frustrated Kepler, as did his sending of ciphers rather than clear and immediate news of his discoveries. In the end Galileo's secretiveness was among the reasons Kepler withdrew his condemnation of Horky. Though Kepler continued to support Galileo and to argue on his behalf, he also believed that since Galileo had not been truly open about spreading or defending his ideas, he could not expect his work to go unchallenged. Likewise, Kepler tried to model his behavior toward Horky in accordance with the guidelines of civil behavior that characterized both the larger Republic of Letters and the mathematical Republic of Letters within it. When Horky's own behavior placed himself outside those guidelines, Kepler's response was harsh and his break with Horky decisive. Yet when Horky convinced Kepler that his actions had been justified, Kepler immediately retracted his negative words about Horky and sought to ensure that they would not be publicized, despite his continued disagreement with Horky's stance. While the ideas of Copernicanism that undergirded Kepler's conception of scientific community were very important to him, the ideals of civility and shared knowledge were critical as well.

The interactions between Kepler and Horky also reveal that those ideals could be difficult to uphold in the face of the practical constraints of an imagined Republic spread across geographic boundaries. The communications that enabled such a union of scholars might be slow in coming and easily misunderstood. The situation in which Kepler and Horky found themselves was due as much to the fact that their letters kept crossing paths as it was to their intellectual disagreement. Horky's negative letters to Kepler about Galileo were not answered quickly enough to prevent his misconstruing Kepler's own feelings on the subject. And Kepler's angry condemnation of Horky's work completely missed Horky, who arrived at Kepler's home eager to revel in their mutual disapproval of Galileo. Likewise, Kepler's own efforts to support Galileo through his *Dissertatio* were complicated by the fact that readers often completely misunderstood his tone. The problems of communication at a distance—delay, crossed paths, and misconstrued tone—were as much a challenge to Kepler's attempts at establishing community as Galileo's emphasis on secrecy and priority.

Conclusion: Astronomical Community versus Religious Community

Kepler's vision of a community of science rested on an ideal of public knowledge. Yet the public to which he referred in this case was clearly a scholarly one. When it came to the unlearned or moderately learned public, Kepler was willing to be less than open. This larger public, Kepler felt, would not be swayed by the open sharing of knowledge; other methods would have to suffice, methods that might be secretive or even deceptive. And here, Kepler, as a trained humanist, had a large rhetorical arsenal with which to build his persuasive campaign. Kepler, like Galileo, could be the consummate courtier when he so chose. The rhetorical techniques on which he drew could be used alongside traditional logical or mathematical arguments to further bolster one's case, and they could be targeted at scholars and nonscholars alike. Even the letters of the theoretically open and transparent Republic of Letters were themselves crafted with great care and attention to rhetoric; letters were, after all, as much self-conscious attempts at the fashioning of one's persona as they were methods of communication.[105] Kepler's *New Astronomy* was targeted specifically toward the scholarly reader and was itself a masterpiece of rhetorical persuasion.[106] Yet the specific strategies that Kepler labeled as deliberately deceptive were aimed at a nonscholarly audience in particular and relied heavily on the manipulation of opinion through an emphasis on authority. These deceptions were appropriate, in Kepler's view, only because they were addressed to outsiders; he faulted Galileo for playing the courtier even to his intellectual colleagues.

Interestingly, when it came to the *vulgus* and the *docti mediocriter*— the unlearned and moderately learned with respect to natural philosophy and mathematics—Kepler's attitude clearly differed from his attitude toward the theological layman, the *laicus* or *idiota*. As we saw in chapter 2, Kepler insisted, like Nicholas of Cusa before him, that the layman often had *better* access to truth than professional theologians and could interpret doctrine on the basis of his own knowledge of scripture and tradition without blindly following the interpretations of the professionals. When it came to laymen in the astronomical realm, however, Kepler was dismissive of their reasoning abilities and insisted that they needed guidance and persuasion, even if that persuasion was rooted in well-intentioned deceptions. Here, we see a remarkable inversion of the traditional Augustinian paradigm for God's two books. Augustine had argued that it was the Book of Scripture that required knowledge and training, and the Book of Nature that was open to all. "The page of the divine scripture is open for you to read, and the wide world is open for you to see," he wrote. "Only the literate can read the books, but even the illiterate can read the book of the world."[107] Kepler, too, saw the universe as a book written by God, but he did not have any confidence in the abilities of the general public to read it correctly. Though they disagreed about much, he concurred here with Galileo's argument that the Book of Nature "cannot be understood unless one first learns to comprehend the language and read the letters in which it is composed. It is written in the language of mathematics, and its characters are triangles, circles, and other geometric figures, without which it is humanly impossible to understand a single word of it."[108]

Kepler thus inverted the older understanding of which of God's two books was easier to access. At the same time, he co-opted the pedagogical metaphor describing God's attempts to aid his people in understanding the Book of Scripture and applied it directly to the Book of Nature. The metaphor itself—accommodation described as a parent helping a child progress—has a parallel in ancient rhetoric, with Cicero's claim that the student of rhetoric needed to be taught by someone who would "put into his mouth none but the most delicate morsels—everything chewed exceedingly small—in the manner of wet nurses feeding baby boys."[109] In applying this rhetorical idea to God's words in scripture, Calvin described God's kindness in "chew[ing] out every word and syllable" for us, so that we might read and understand the difficult messages of scripture.[110] Kepler certainly saw the Bible as accommodated in language, but he used the metaphor of a parent carefully helping a child when describing not knowledge of scripture but rather knowledge of the Book of Nature. "Does God the creator," he wondered in the very same book in which he tried to support Galileo's telescopic discoveries, "lead mankind, like some growing youngster gradually approaching maturity, step by step from one stage of knowledge to another? . . . How far has the knowledge of nature progressed, how much is left, and what may the men of the future expect?"[111] Kepler believed that only the mutual efforts of a community of Copernicans might bring the world to truly understand the Book of Nature that God had written for them.

One reason that rhetoric was essential for community building, then, was that it reached out to those who would not—or could not—respond to reason alone and brought them into the fold. At the same time, Kepler recognized—as had many before him—that rhetoric had another important social value. It functioned as a kind of civilizing force in itself, by making men more than mere beasts and persuading them to care for one another. Cicero, for example, linked it to virtue and security and argued that the orator, who "excels men in that ability by which men excel beasts," performed an especially valuable social function.[112] Petrarch, too, felt that the social body couldn't function properly without proper speech; "disordered speech," he claimed, damaged "the right proportion and coherence of things."[113]

Though Kepler did not make the connection between his rhetorical pursuits and the larger social order quite this explicit, he did clearly link civility and social order to a discipline remarkably analogous to rhetoric: music. In fact, rhetoric and music had long been linked, and the analogies between the two realms were particularly ubiquitous in the seventeenth century.[114] In his *Harmonie Universelle*, for example, Mersenne claimed that musicians were orators;[115] Kircher likewise wrote about "*musurgia rhetorica*" in his *Musurgia Universalis* and emphasized the many analogies between rhetoric and music.[116] To Kepler, this linkage would have been obvious: rhetoric could move the passions and so too could music. In his *Harmony of the World* he quoted Proclus to prove that mathematical and musical harmonies were essential to virtuous living;[117] he likewise argued there that music, like rhetoric, could transform the behavior of an audience: "everything is lively while the harmonies persist," he wrote, "and drowsy when they are disrupted."[118] It is in this sense that he follows in the Ciceronian/ Petrarchan tradition linking rhetoric to the social order; music was simply the form of rhetoric with which he made this argument most clearly. Ultimately, then, with his emphasis on rhetoric as an essential tool for crafting a philosophical community in his discussions with Galileo, Kepler circled back to the theme that dominated his approach to all forms of community: the theme of harmony.

For Kepler, the harmonic model represented unified diversity; as we've seen, it pointed to a kind of social cohesion that embraced difference and disagreement within it. In this, Kepler's conception of musical harmony had much in common with Erasmus's conception of rhetoric itself. Erasmus saw in rhetoric both a general civilizing power and, more specifically, a means for avoiding war and achieving the conciliation that he deemed so necessary in sixteenth-century Europe. With the proper rhetoric, the varying sides in the confessional disputes might come to an agreement, Erasmus believed. Still further, he saw in rhetoric a model for communal negotiation that held tolerance as a primary value; the orator, after all, tried his hardest to persuade others to join his cause, but he never imposed his will on them.[119] It was ultimately the choice of the audience to accept or reject his claims. As Gary Remer notes, the rhetorical tradition may have been an important resource for all those humanists who argued for toleration, for it helps to explain their "preference for persuasion over force; their skepticism and toleration in nonessentials; and their emphasis on ethical living over dogma."[120]

This linkage between rhetoric and tolerance brings us back to the start of our discussion in this chapter by way of a question. Tolerance, of course, was what Kepler advocated when it came to questions of theology. Yet it was very clearly *not* what he advocated for questions of cosmology. When it came to the conflict between the Ptolemaic and Copernican worldviews, Kepler insisted (in the face of strong opposition from most of the astronomers of his day) that only the Copernican approach represented the cosmological truth. Indeed, he insisted that the astronomical counterpart to his religious irenicism—the idea that the Ptolemaic and Copernican theories were equally valid tools for considerations of planetary positions and motions—was unsatisfactory. Astronomical theories ought to represent the physical truth of the universe, Kepler argued, and there was only one such truth—Copernicanism. He devoted much of his life's work to proving the truth and superiority of Copernicanism and to converting as many as possible to his way of thinking. Why was Kepler so insistent on establishing an astronomical community that adopted one worldview, yet so willing to embrace a religious community characterized by multiple perspectives and practices?

A related question circles back to the issue of rhetorical dissimulation. If dissimulation was warranted in Kepler's view when it came to Copernicanism—if, that is, Kepler was willing to lie about Copernicanism—then why did he refuse to dissimulate when it came to questions of theology? Kepler was insistently forthright about his religious beliefs, even when it brought him grief; he emphasized that "I have not learned to dissimulate: I deal seriously with religious matters."[121] He never lied in order to convince others to accept his theological positions, nor did he lie to continue participating in the Lutheran Church—even a lie of omission, which is essentially what his signing of the Formula of Concord would have been. The accommodation he called for in the religious community was one in which people were open about their religious objections, not one in which they hid them. But particularly given that forthrightness meant excommunication for Kepler, why did Kepler reject the possibility of theological dissimulation if he accepted dissimulation with respect to the promulgation of Copernicanism?

The answer to both these questions—the question of dissimulation and the question of communal diversity—rests in large part on the relative degrees of certainty with which Kepler believed questions of astronomy and questions of theology could be settled. As we saw in chapter 1, Kepler believed that certain kinds of mathematical and cosmological claims could be demonstrated a priori, via reason alone, in ways that were certain and brooked no dissent. In such instances, Kepler felt comfortable emphasizing the certain truth of a particular cosmological theory, like Copernicanism. Yet if mathematics and cosmology were realms where certainty (and hence unanimity) were theoretically possible, Kepler believed that theology was a realm where certainty could not be achieved on a large communal basis. Even when it came to specific issues, like the Eucharist, that might be analyzed with a greater degree of certainty, Kepler came increasingly to believe (as we saw in chapter 2) that the theologians around him were unable to take those lessons to heart.

This conclusion was one Kepler drew both from his personal investigations and experiences and from the very different trajectories of the history of astronomy and the history of the church. As Nicholas Jardine has noted, one of the supports for the astronomical skepticism common to the sixteenth and seventeenth centuries was historical; men like Ursus, for instance, argued that the abundance and absurdity of competing past astronomical hypotheses proved that astronomers were unlikely to ever arrive at one true hypothesis.[122] Kepler, by contrast, relied on the history of astronomy to argue exactly the opposite. Most famously in his *Apologia for Tycho against Ursus*, Kepler recounted a history of astronomy that proved that though there were still unanswered questions in the field, many of the old astronomical questions were now settled beyond any doubt. He believed that clear astronomical progress had been achieved over time, via improved theories, improved technical procedures, and improved instruments. And he used this history of progress as one of the bases for his realist approach to astronomy—his belief that astronomy could and would discern the truth of the cosmos. Though astronomy was beset by conflict and though it was a human and fallible enterprise, Kepler argued that the clear history of astronomical progress over time indicated that the discipline moved slowly toward a true and complete picture of the world.[123]

The history of the church, by contrast, revealed something else entirely. As confessional disputes grew only more violent and entrenched, history itself became a weapon in the confessional battle, with historical arguments used on all sides to show that one confession or the other had unique access to the truth. Yet even as confessional propagandists used history to further their own orthodox dogmas (by projecting their own views back onto their vision of the original, apostolic church), that same history, when viewed in its totality, made it increasingly difficult to argue for the possibility of a world united by one religious truth. Doctrinal disagreements between the three main confessions, as Wilhelm Schmidt-Biggemann has noted, resulted over time in the "paralysis of theology's claim to truth. . . . The polemical debate which dragged on for generations . . . discredited not only the respective opponents but also the entire area of confessional theology."[124] As Kepler saw it, the history of the church revealed it to be not a timeless and unchanging institution but rather a human one, susceptible to error and decline. He sympathized with its travails, he wrote, "as much as any human who suffers weakness";[125] he likened confessional disputes to "quarrels [that] appear among the citizens or factions of a republic" and linked them to the "errors [that] appear out of human weakness" over the passage of time.[126] And unlike in the case of astronomy, he was highly skeptical that those errors could ever be universally resolved in favor of one agreed-upon religious truth. In arguing for churchly unity, Kepler emphasized peace and harmony rather than complete agreement and outlined ways in which disagreements could be accommodated in a united religious community. Kepler did not privilege cosmological truth over religious truth; rather, he saw the former as universally attainable on a communal level and believed that the latter was in many cases attainable only at the level of the individual. In portraying himself as an astronomer-priest—a priest of God with respect to the Book of Nature—he aligned himself with a theology independent of the confessions, a way to reveal God's hand in the world that had some hope of offering universal truth claims to which anyone could assent.[127]

The same reasoning lay behind Kepler's refusal to dissimulate theologically despite his willingness to dissimulate when it came to Copernican theory. If the only way to achieve churchly harmony was through negotiation and compromise, dissimulation was counterproductive—it would only make things more difficult. By contrast, as we've seen, Kepler felt that the Copernican question *could* be conclusively settled, in ways that were inarguably evident. Indeed, many of his books were attempts to do just this. The evidence provided in the books themselves, he hoped, would convince all scholars capable of truly reasoned thought. For the rest, if dissimulation was required to bring them around, then it was surely warranted in his view. If God's church could not be aligned with any one confession, so be it—after all, as Kepler had written, "Christ the Lord who spoke this word . . . neither was nor is Lutheran, nor Calvinist, nor Papist."[128] Yet God's church might still be a church of Copernicans, and Kepler did his best to see that it would be. For God had imprinted the world with his harmonic blueprint, clear for all to see—and God was a Copernican.[129]

FIVE

"Political Digression(s)": Kepler and the Harmony of the State

In 1623 Kepler contemplated writing about politics by way of fiction. He was deeply troubled by the brutal disputes ravaging the continent, yet he worried that expressing his concerns too strongly might not sit well with those who wielded political power. How, then, to portray just how terribly things in Europe had gone awry, without drawing too much princely ire? Kepler had already drafted a short fictional piece, titled *Somnium* (*Dream*). It was the story of a young man's journey to the moon, and Kepler had not yet published it. It might be interesting, he mused, to use that story to reflect on the pressing political problems of his day. Perhaps the moon could be portrayed as a utopia, a place whose political perfection threw the troubles of Europe into stark relief? As he wrote to his friend Matthias Bernegger, "Campanella wrote a *City of the Sun*. What if I wrote one of the moon? Would it not be a wonderful deed to paint the barbarous habits of this time in vivid colors, but for the sake of caution to leave the earth with such a work and to withdraw to the moon?" Yet Kepler ultimately decided against such an approach. Politics was contentious enough, he believed, that taking even the circuitous route of fiction could be a dangerous undertaking. "How would it help to hold back in this way?" he continued. "Neither More in his *Utopia* nor Erasmus in his *Praise of Folly* was safe, but each needed to defend himself. Therefore, let us thoroughly leave this pitch-black political terrain and remain in the pleasant greenery of philosophy."[1]

While Kepler may have preferred to eschew political discussions, he could not avoid the political sphere entirely. He knew well that his post as imperial mathematician and adviser to the Holy Roman Emperor was a highly political one, and he referred to himself alternately as both a *mathematicus* and a *politicus*.[2] He discussed political questions in his published astrological works, in letters to friends and patrons, in nativities that he was commissioned to produce, in reports that the emperor requested of him, in the prefaces and dedications to many of his astronomical or cosmological books, and even in a "political digression" to his *Harmony of the World*. In the previous chapters, I focused on Kepler's conception of himself as a Lutheran, a mathematician, and an astronomer. In this chapter, I turn to Kepler's conception of politics, of himself as politician, and of the relationship between his political thought and his (natural) philosophical beliefs more broadly. In particular, taking a cue from Kepler's brief consideration of the *Somnium* as political fiction, I consider the ways that Kepler applied the idea of *fictio* to politics, ways that went far beyond a single whimsical text.

The idea of political fiction has, of course, an ancient pedigree, one that influenced the ways that Kepler perceived and invoked it. In looking backward, we might turn first to Plato's *Republic*. Plato's Socrates had little use for poetic mythmaking and believed that Homer's legendary tales had largely corrosive effects. Yet he made an exception for political uses of myth, arguing that "if it is appropriate for anyone to use falsehoods for the good of the city, because of the actions of either enemies or citizens, it is the rulers."[3] To lie to a ruler, for Plato, was the worst kind of crime; yet a ruler was allowed, and at times encouraged, to lie to his citizens, if such lies furthered the public good. Those lies, for Plato, were poetic, mythic ones; the myth of the metals, which encourages all citizens to accept their place in the social and political hierarchy, is one prominent example.[4]

The mere labeling of something as fiction, poetry, or myth need not imply that it had no truth-value. Deception and truth often went hand in hand. Even theology itself could deceive, as we saw earlier when we explored the idea of accommodation. Augustine noted, along these lines, that it was important to distinguish between false *fictio* and *fictio* that was "truth under another aspect."[5] Theology and poetry, according to Augustine, could be both true and deceptive at the same time. For some followers of the Platonic tradition, this was because truth was something reserved for the wise; many believed that God had disguised the truth in densely complex texts in order to protect it from the foolish ignorant, and that the pagan ancients had hidden the truth under layers of myth such that only the truly wise could decipher it.[6] For others, this was because the ultimate, essential truth that underpinned both the scriptures and the pagan myths was more important than the contours of the route that one took to get to it. Duns Scotus, Petrarch, and Boccaccio all referred to theology as "the poetry of God" for this reason.[7]

Fiction was often far more useful than simple truth, particularly if one's purpose was not merely to state but to guide, shape, or transform. *Fictio*, after all, comes from the Latin *fingere*, meaning both "to feign" and also "to fashion." *Fictio* thus encompasses the idea of "forging" or "fabricating" in either the deceptive or the constructive senses of the terms.[8] And for political theorists in early modernity, interested in devising ways to successfully navigate and govern a world that didn't conform to simple rational rules or conventional moral guidelines, *fictio* became essential. As Victoria Kahn notes, Machiavelli's approach to politics embraced both senses of *fictio*; at times Machiavelli endorsed strategically useful deceptions, and at times he stressed the political importance of imposing order, constructing "an artifact understood as a new state of affairs." In both cases, "Machiavelli's idea of making emphasizes the ability to respond to contingency, individually or jointly, with an ever-flexible art of invention."[9]

The Machiavellian approach to statecraft came, by the 1570s and 1580s, to symbolize for many an immoral kind of politics, one unmoored from all religious and ethical standards.[10] Yet Machiavelli's influence was pervasive, and the civil wars, bloodshed, and intrigues that seemed ubiquitous in the sixteenth century made it difficult for theorists to simply return to what now seemed the trite moral maxims of old. It was in this context that Tacitus rose to preeminence as an ancient who had unique insight into the world of politics. The world that Tacitus described—a sprawling imperial one—seemed closer to the lived reality of the late sixteenth century than did Machiavelli's world of "small principalities, independent republics and politically active citizenries."[11] But like the smaller world of Machiavelli, this sprawling imperial one was filled with ruthless and treacherous political maneuverings, chaos, and strife. Consequently, Tacitus's words reverberated with remarkable resonance to those who read him in the sixteenth and seventeenth centuries—so much so that commentaries on his *Histories* and *Annals* came to form "the common currency of political discourse" at the time.[12] The sheer number of books published on Tacitus in the sixteenth and seventeenth centuries bears witness to his popularity: more than one hundred commentaries—mostly political—appeared between 1580 and 1700, along with sixty-seven editions of the *Histories* between 1600 and 1649.[13]

For many theorists, Tacitus became "a surrogate for Machiavelli: a more respectable authority since, unlike the infamous Florentine, he did not advocate the amoral behavior he described."[14] Yet this was not the only way to read Tacitus; because he merely related, rather than advocated, his writings were interpreted in widely disparate ways.[15] The predominant way to read Tacitus was to place him squarely within the newly flourishing reason-of-state tradition. Initiated by Giovanni Botero (who referred to Tacitus forty-four times in his treatise),[16] this tradition posited that the ruler should be above the law, in certain instances, and able to override it. There were typically deemed "good" and "bad" reasons of state: good reasons of state allowed the ruler to serve the public good, while bad reasons of state served self-interest alone. Alongside this focus on the power of the ruler to serve the common good, theorists also emphasized the idea of political prudence, which applied to both rulers and their courtiers and demanded a degree of (warranted) dissimulation and state secrets (*arcana imperii*). Additionally, adherents of the reason-of-state tradition argued that politics was a learned skill, or art, which could be acquired via experience, whether gleaned personally or through a careful study of history. Tacitus was usually read within this political tradition; this was the case for Justus Lipsius, the Flemish scholar most responsible for spreading the words and ideas of Tacitus through his definitive edition of Tacitus's works and his own *Politica* of 1589.[17] Lipsius lived by the words of Tacitus, quite literally: as Anthony Grafton has noted, he offered "to recite all of Tacitus while his listener held a dagger to his belly—and to allow the listener to plunge the dagger in if Lipsius stumbled on a word."[18]

Tacitus, like Machiavelli, was a particularly good source for considerations of political fiction in its various guises. Tacitus had used the phrase "*fingere et credere*" (fabricating and believing) to speak of the inescapable effect that human imagination—and the deceptions it engendered—had on political life. Humans were both devious and credulous, prone both to deceiving others and to being deceived in turn. Tacitus's histories were, in many ways, "a model of merciless introspection, capable of alerting the reader to the insidious role played by language, writing, and the shifting representations of mankind's inner worlds."[19] Given this state of affairs, Tacitus suggested—and his early modern devotees insisted—that it was incumbent on successful politicians both to avoid the fictions of others and to marshal deception as a weapon to suit their own political purposes. Lipsius, for example, devoted much of his *Politica* to an account of various kinds of deceptions and their political uses. He maintained that deception was a central component of political prudence; indeed, it was just "mixed prudence" by another name.[20] Distrust and dissimulation, according to Lipsius, were also essential for the effective ruler, who needed to (deceptively) encourage his people to believe in him while refusing to trust any of them. Francis Bacon, an avid reader of Tacitus who found him to be a better philosopher than either Plato or Aristotle,[21] similarly argued that "in all wise human government, they that sit at the helm do more happily bring their purposes about, and insinuate more easily into the minds of the people, by pretexts and oblique courses, than by direct methods; so that all sceptres and masses of authority ought in very deed to be crooked in the upper end."[22]

If the Tacitean model of politics focused heavily on *fictio* in the form of the deceptions so central to political life, it focused on *fictio* in its other meaning as well—as something fashioned or made. Tacitus, and the reason-of-state tradition more generally, emphasized the importance of fashioning political reality rather than passively responding to it; politics, in this view, was a form of art. As Vera Keller has pointed out, Giovanni Botero portrayed princes and politicians as artisans, who might use their cunning and prudence to perfect the political world. Art, or *fictio*, was not "only a tool for politics; it also presented a model for politics."[23] In part this was because Botero believed that politicians, like artisans, needed to be empiricists who based their ideas and opinions on information gleaned either from the ancient histories or from personal experience. Botero used the metaphor of a mine to argue that cunning politicians should mine their Tacitus—along with other ancient sources—to uncover the *arcana imperii*, the secrets of empire, necessary to successful governance. Yet the mining metaphor was not simply about the search for information itself but also referred to the ways that the state—conceived as an artifact—might be remade on the basis of the mined information. Botero and the Tacitists of his time believed that politicians could use the secrets they uncovered to fashion the world anew.

Several of Kepler's frequent correspondents—among them Christoph Besold and Matthias Bernegger—were noted Tacitists of their time. Like them, Kepler read Tacitus politically. While serving in Prague as imperial mathematician, he even produced his own German translation of the first book of Tacitus's *Histories*. He knew his Tacitus well, and he used stories from Tacitus to bolster some of his own political views and decisions, as we will see in the course of this chapter. More broadly, Kepler infused the entire enterprise of his political astrology with Tacitean overtones. First of all, his political astrology was heavily empiricist; Kepler argued that while astrologers often offered political predictions, these had a hope of accuracy only if they were grounded in political experience and personal discretion. He emphasized that his own astrological reports often drew on his political awareness as much as his knowledge of the motions of the heavens. Second, he highlighted the importance of deception to the political astrologer, who needed to read every situation carefully and determine whether a true astrological report would be helpful or harmful. As an important player on the imperial stage, Kepler claimed that at times it was more important to hide or alter the message of the stars, if that message seemed politically dangerous, than to relay it truthfully.

Beyond the idea of astrology as a political art requiring deception, Kepler focused on the importance of fashioning politics after an appropriate model. Much as Botero portrayed politics as an art akin to mining, Kepler, too, believed that one needed to search deeply in order to reveal the *arcana* necessary for the successful fashioning of the state. Yet for Kepler, unlike for Botero, Bacon, or others, these *arcana* were not just *arcana imperii* of the sort to be found in Tacitus. They were also *arcana mundi*, secrets of the cosmos, like the cosmological secret that he spoke of in the title to his very first book. In the dedication to his *Harmony of the World*, Kepler linked *arcana imperii* and *arcana mundi* and argued that the secrets of nature themselves revealed the fundamental political secret, the model by which his patrons might truly refashion a world worth living in. The mathematical harmony of nature, he insisted, could provide a blueprint for political harmony, particularly in a world as characterized by discord as his own. He saw in King James, in particular—the patron to whom he dedicated the *Harmony of the World*—someone with the desire and the power to embrace and implement such a model.

In this way of conceiving of political fiction, Kepler hearkens back to an older model of human creativity. Machiavelli and the Tacitists, much like Hobbes after them, conceived of political fashioning as an art detached from divinity and, consequently, from nature—an art that was, in the words of Leo Strauss, "a model-less, sovereign invention."[24] For Kepler, however, political fashioning, like all human creativity, rightly took the divine creativity as its model. As Aristotle had argued that all art was based on the imitation of nature, and as Plato

187

had portrayed divine creation as the paradigm for human making,[25] so too did Kepler seek to anchor his conception of political making in the divinely created world. He would have agreed with the canon in *Don Quixote*, who asserted that "the more it resembles the truth, the better the fiction" and who insisted that good fiction ought not to depart "from verisimilitude or from the imitation of nature in which lies the perfection of all that is written."[26] God had created humans in his image, after all, and this meant that in their own artistic endeavors they ultimately followed him.[27] This kind of *fictio*, or fashioning, was far from the *fingere et credere* of Tacitus; not imagined or falsified, it was a fashioning of the political world after the ultimate truth, the archetype whose harmonies would allow even humans bent on self-deception to see things clearly.

Yet while Kepler believed that the mathematics of harmony might lead to a politics of harmony, he did not provide a specific formula for political organization and avoided the model of Jean Bodin, the French political theorist who tried to bolster absolutist, monarchical government with the claim of mathematical certainty. Much as in his discussions of astrology, in his engagement with the political writing of Bodin Kepler emphasized the subjective nature of politics and of political harmony. Though there were certain broad principles upon which he insisted—in particular, the preeminence of the public good and the welfare of the state—Kepler implied that multiple kinds of political configurations were equally valid, so long as they agreed on these principles. While the harmony of nature could point the way to political harmony on earth, it needed to do so indirectly, for politics was too complex and subjective an art to be grounded on the certain truths of geometry.

The Astrologer as Politician, and Kepler as Tacitist

Kepler's most immediate connection to the political sphere was his role as imperial mathematician, which required him to act as astrological adviser to the Holy Roman Emperor. The idea of astrology as a deeply political field was far from new in Kepler's time; on the contrary, astrology had long been "profoundly involved with action and power in the world."[28] The emperors in ancient Rome had relied on the advice of astrologers, and French and English rulers of the late Middle Ages had paid astrologers to construct their natal charts.[29] Yet the value that rulers assigned to astrological advice and the potential for advancement

that astrological expertise afforded its practitioners were both particularly great in sixteenth- and seventeenth-century Germany.[30] Kepler's
own employer, Rudolf II, placed a high premium on astrology and astrological advice because of the tense political climate that characterized his reign. The empire was divided, both by faith and by increasing political disputes; Rudolf felt increasingly threatened by his brother
Matthias, who gained the crowns of Austria, Hungary, and Moravia in
1608—and who would force Rudolf's abdication in 1611. The papacy
had begun an attempt to regain some of its old positions in the empire,
while Spain started to reassert its own imperial interests, particularly
in Bohemia. In 1591 the Ottoman Empire had also begun a prolonged
assault on the Hungarian front, which lasted well into Rudolf's reign.
Faced with both the internal political rivalries of the empire and the
outside threat from European and Turkish forces alike, Rudolf increasingly relied on advisers who were independent, without ties to any rival power.[31] Astrology, in particular, seemed to offer an objective and
more certain source of advice and guidance, and Rudolf, following in
the footsteps of imperial rulers before him, turned to astrology with
high hopes and often relied on astrologers to assist him in the difficult
political decisions he faced.

While for Rudolf, as for his imperial predecessors, astrology had an
important place at court, the astrologer could occupy a range of positions. Some astrologers, like Kepler, held permanent official positions at
court, while others with distinguished reputations, like Cardano, were
consulted as outside experts or given temporary appointments at court.
These men provided more than abstract theoretical explanations of the
heavenly bodies that might be used by the rulers to infer their own
political conclusions. Astrologers typically offered detailed and specific
advice, and often even propaganda, to support an individual ruler's
claim to the throne, to ascertain the result of a forthcoming battle, or
to advance a particular political agenda. Astrology thus had "an immediate and almost palpable role in politics."[32]

When Rudolf II designated Kepler his imperial mathematician, he
expected him to provide astrological advice that was specifically political in nature. He expected this advice to take full account not only
of the positions of the heavenly bodies but also of the earthly situation both in the empire and abroad and to include tangible conclusions about the best course of action. Kepler therefore wrote reports for
Rudolf on subjects ranging from the dispute between Venice and Rome
in 1605, the political situation in Hungary, and the pressing Turkish
threat. He provided horoscopes and astrological advice to Rudolf as

well as to other powerful figures who sought his expertise, including Rudolf's rival and eventual successor, Archduke Matthias; Count Wallenstein, both before and after he became a general; and Chancellor Herwart von Hohenburg of Bavaria.[33] Likewise, even before his promotion to imperial mathematician, Kepler composed annual predictions as part of his calendars, which often included political components.[34]

Kepler knew well, in other words, that the *astrologus* was in many ways also a *politicus*. In fact, in his treatise "On the More Certain Foundations of Astrology" of 1601, he argued explicitly that "in matters of politics and war, the astrologer clearly has an opinion to express."[35] Yet he quickly limited this statement, and explained that

> it is really the height of folly to look for predictions about specific matters. . . . Nothing can be looked for from Astrology except the prediction of some excess in the inclination of souls. What this inclination will lead to in future realities is determined by man's choices in political matters—for man is the image of God, not merely the offspring of Nature—as well as by other causes. Thus whether there will be peace or war in some particular region is a matter for the judgment of those who are experienced in politics, for their power of prediction is no less than that of the Astrologer.[36]

Kepler himself had to offer the very predictions that at first he dismissed as "the height of folly," and hence he argued not that such predictions should be avoided but instead that they should be based as much as possible on political knowledge and experience. Though the astrologer did have some special insight into political affairs, Kepler maintained, that insight was limited—as was astrology itself. The advice of a true *politicus* ought to be based on his understanding of war, peace, and political affairs rather than on the study of the heavens alone.

Kepler tried to take his own advice, and his astrological reports to the emperor were often based on his personal political assessments of current events rather than on the heavenly positions and trajectories alone. When Rudolf requested that his imperial mathematician supply him with a report on the situation in Hungary, Kepler emphasized in the official report that "since I was asked about the particularities of Hungary, I cannot do otherwise but focus on earthly causes, with as much knowledge as I have of them. For I have often made clear to the emperor my opinion that it is a bad and groundless thing to judge by the heavenly signs alone what will happen to the condition of his lands—rather, heavenly signs should be subservient to common perception."[37] Though Kepler's own position hinged on both his astrological expertise and his ability to offer sound political counsel, Kepler sought to emphasize his political skills and give them precedence over his astrological expertise. Astrology was often unsuited to interpreting the specifics of a messy political situation, particularly one with so many unknowns. Along similar lines, Kepler stressed in the same report that as he lacked the nativity of the Turkish emperor, he would have to rely on political causes in his analysis of the Turkish threat as well. A good *astrologus*, Kepler made clear in this report, did not automatically make a good *politicus*. If an astrologer wanted to provide good counsel for his sovereign, he would have to broaden his perspective and take account of the local political forces in any given situation.

In a later "Discourse on the Great Conjunction of 1623," Kepler again maintained that *"ein gueter Politicus"* should acquire a detailed knowledge of earthly circumstances, and that such knowledge provided far more accurate predictions about *"particularfragen"* than the motions of the heavens did. He further specified that "the more expert a *politicus* is, the more he will keep secret his suspicions—not only because even he cannot offer a determinate truth about future contingencies but also because he knows that it does no service to the butcher for someone to tell the ox when it will be slaughtered."[38] With this, Kepler went beyond his earlier assertion that astrologers should base their claims on political knowledge. To be a good *politicus*, Kepler argued here, an astrologer who hoped to offer helpful political counsel might need to *conceal* his astrological knowledge when that knowledge was potentially damaging to his sovereign. Astrology, in this view, was not simply an uncertain art when it came to the political sphere but also a potentially dangerous one. Since it purported to offer independent guidance that stemmed directly from the heavens, it had the power to do more serious damage than political advice from more mundane sources.

This claim echoes one that Kepler made at far greater length in a private letter earlier in his career. In 1611 Kepler wrote worriedly to an anonymous nobleman (whom Barbara Bauer has tentatively identified as Johann Anton Barwitz, secretary and close confidant of the emperor)[39] about the dangers that astrology posed in the fraught political times leading up to Rudolf's abdication. Kepler began the letter by mentioning his belief that, like him, the recipient was a man loyal to the German empire and the emperor rather than to the Austrian or Bohemian factions who opposed him. He then boldly stated the main purpose of his letter: to convince its recipient that "astrology could do great harm to a monarch if some clever astrologer wanted to dupe him because of his gullibility." Kepler hoped to join forces with the recipient of his letter so that "this does not happen to our emperor, [since] . . . the emperor is gullible."[40]

Kepler explained that he referred not simply to popular astrology, of the sort that he believed had little basis in nature and that could easily be twisted to please either party in a dispute. Rather, he felt that even the astrology that he had placed on more certain natural foundations in his own work could cause political damage. People placed too much stock in the word of astrologers, Kepler believed. While malicious astrologers might twist an astrological report to suit their own purposes, even the report of an honest astrologer could lead to undesirable political outcomes. Kepler clarified what he meant by referring to his own recent experience. He had been asked by both the supporters of Matthias and the supporters of Rudolf for an astrological analysis of the political situation and their respective fortunes. Since Kepler took the supporters of Matthias to be enemies of the emperor, he deliberately altered his astrological interpretation, as he worried that the real astrological forecast would only damage the emperor's cause by inspiring those who opposed him. Instead, he had given them a report that supported the emperor's position and undermined their own, in the hope that it might forestall an imminent escalation of the political conflict:

> When asked about the decrees of the stars by those whom I know to be enemies of the emperor, I responded, not with that which I assessed to be of some importance, but rather with that which would demoralize the credulous: I said that the emperor would live a long time, and that there were no bad directions. . . . On the other hand, for Matthias I predicted imminent unrest. . . . I said this to the enemies of the emperor, because even if it does not instill fear in them, certainly it will not inspire confidence either.

Likewise, he explained that he preferred to give the emperor no report at all, since the actual astrological indications were "not of such importance that one ought to place faith in them," and they might cause the emperor to ignore the ordinary means by which he might improve his situation. "Astrology," he wrote, "would put him into a much worse situation than he is in now."[41]

Kepler summarized the actual astrological prognosis for his correspondent, so he would have all the information at hand. He emphasized that though he felt it was insignificant, most astrologers would interpret it by placing their confidence in Matthias and deeming the position of the emperor to be very alarming. Accordingly, he insisted, "I believe that astrology ought to be banished entirely, not only from the Senate, but also from the minds of those who want to daily advise Caesar of what is best for him: indeed, it must be fully hidden from the emperor's sight."[42] Even those who have the emperor's best interests at heart might be negatively influenced by astrological conjectures, Kepler believed. The only way to safeguard the emperor and ensure his best interests would be to eliminate astrology from the political sphere entirely.

In his emphasis on the benefits of political dissimulation, of course, Kepler was embracing his role as *politicus* rather than discounting it. The astrologer who lacked political acumen would offer his interpretation of the heavens without worry of—or perhaps without knowledge of—its potential consequences. That astrologer, however, would be doing a political disservice to his rulers. Likewise, the astrologer who based his predictions on the heavens alone would miss much of the political texture of the times and offer predictions that were incomplete or simply wrong. Kepler did not deny that astrology could offer some political guidance, though he did insist that that guidance was general at best, on the level of "inclinations" rather than necessities. Moreover, he asserted that the guidance afforded by astrology needed to be weighed heavily against the potential damage it could cause. In all this, Kepler stressed that the successful court astrologer needed to be more of a politician rather than less of one. And in his conception of the scope and possibilities of politics, Kepler drew on a political tradition that had become particularly popular in the decades before he assumed his post as imperial mathematician: the tradition of Tacitism.

Kepler highly valued the work of Tacitus, whom he described as the "most venerable" of the pagans for his political acumen.[43] He even went so far as to share the work of Tacitus with his children. While serving in Prague as imperial mathematician, he produced his own German translation of the first book of Tacitus's *Histories*. Like most of his work, Kepler probably intended this for eventual publication, but he first used it as a tool in the Latin instruction of his son, Ludwig. Every week for three years, beginning when he was six years old, Ludwig read through sections of Kepler's German text and translated them back into Latin. He and Kepler then compared his Latin translation with Tacitus's original.[44] At the age of eighteen, Ludwig lent a copy of Kepler's translation to Duchess Herberstorff, who—evidently finding much that was useful in it—asked him for a copy for her son, Gottfried Heinrich von Pappenheim, who later became an officer in Wallenstein's army.

Instead of simply lending her a copy, Ludwig decided—with Kepler's agreement—to publish the translation in 1625, with a dedication to the duchess. He described it much as Kepler seemed to understand it—as a book "full of admirable discussions of regiments and wars, which are no less useful for the present time to read as the general comparison of the old word to the new."[45]

In a letter to an anonymous woman, Kepler described himself as a *"mathematicus, philosophicus,* and *historicus." Historia,* he explained there, "concerns itself with old histories and church histories and also with providing instruction to people of all times"—instruction that, he continued, "today's preachers take to heart less than is fitting, else there would be fewer conflicts."[46] Kepler clearly felt that history could provide lessons for his own time—lessons that would demonstrate ways to eliminate conflict and were consequently particularly useful in the realms of politics and religion. He intended his Tacitus translation to be read at court, by those with an interest in modern-day governance. Though Kepler approached Tacitus with the sophisticated and contextualizing eye of a humanist scholar, and though he "warned the reader not to look for simple, schoolboy lessons about political prudence,"[47] his highly political focus on Tacitus—much like his astrological reports for the emperor—marked him as a *politicus,* one interested in and conversant with the political trends of his day.

Barbara Bauer has astutely noted, however, that Kepler focused on Tacitus both in his translation of book 1 of the *Histories* and in his *Tertius Interveniens*—and that the latter discussion may reveal even more about Kepler's conception of himself as *politicus.*[48] In *Tertius Interveniens,* Kepler responded to a wholesale repudiation of astrology by Philip Feselius, physician to Margrave Georg Friedrich von Baden. Kepler, who had previously argued at length against Helisaeus Röslin's wholesale *adoption* of astrology, here positioned himself as a mediating figure who rejected both the complete embrace of astrology and its complete rejection. In the book, he claimed that while much of astrology was foolish, much of it still had value: "no one should consider unbelievable that out of astrological foolishness and godlessness . . . can be found also useful wit and holiness."[49]

Toward the end of *Tertius Interveniens,* Kepler reflected on the prohibition in Leviticus against consulting *magi* and *arioli,* or seers and astrologers, and asked what kind of astrology this biblical prohibition encompassed. In answer, he referred broadly to the central role of the astrologer in the Roman Empire—for instance, the fact that astrology guided every step of the army. This, Kepler argued, was the sort of thing that the Bible forbade. He then referred specifically to the story of Otho and Galba, as relayed by Tacitus in his *Histories*. According to Tacitus, astrologers had inspired Otho to murder Galba and seek the throne of the empire himself by telling him that his future foretold glory in the coming year. Tacitus had therefore labeled astrologers, whom he called "*mathematici*," "a group of men disloyal to those in power and treacherous to those who aspire to it"[50]—a sentiment on which, Kepler noted, Feselius had based his own anti-astrology tract.

Kepler, however, wanted his book to distinguish between good and bad uses of astrology, and therefore he paused to consider what broader implications this Tacitean story truly had. He concluded that if someone were to approach him with specific yes or no questions about the coming year, such as whether a friend would recover or die from an illness, and if he were to respond by using a natal chart to offer specific answers to those questions, "then I would be an *ariolus*, and a violator of God's law." According to Kepler, the entire process here—the decision of specific future actions on the basis of a horoscope—was groundless and unnatural. If, however, argued Kepler, taking up Tacitus's story, Otho had approached him and simply asked how his horoscope currently looked—without posing any specific questions and without revealing anything of his future plans—then Kepler, or any astrologer, could have told him of the "good revolution" (or good year) and lucky tidings without compunction. In this instance, Kepler wrote, "I would not be guilty in [his plans], because I had told him only that which was natural, with as good a conscience as if he had brought me the urine of Galba"[51] and asked for a medical prognosis on its basis.

Kepler's reflections here are significant for his differentiation between the kind of astrology that was "natural" and the kind that was groundless. He highlighted the fact that astrologers who tried to predict specific events in the future were not practicing an art based in nature but were instead creating foolish fantasies—or, worse, violating the word of God. By contrast, he insisted that the positions of the heavenly bodies, and natal charts in particular, did have more limited uses and could reveal general patterns for the coming year. More importantly, however, Kepler's reflections reveal a great deal about his conception of the role of the astrologer as politician. As Bauer notes, Kepler invoked Tacitus's irresponsible "*ariolus*" as a means for him to relay his own sense of professional ethics. It was clear to Kepler that astrologers had the power to cause great political upheaval, as in the case of Otho and Galba. The ethical astrologer who wanted to avoid Tacitus's charges of disloyalty and treachery had two choices. He could choose the path that Kepler outlined in the *Tertius Interveniens*: he could, essentially, opt out of politics entirely. As Kepler argued, if the astrologer were to base his words solely on that which was naturally warranted, and if he were fully unaware of their potential political ramifications, he could not be held accountable for whatever followed. He could insist not only on his political neutrality but also on his political naïveté and in this way absolve himself of all political responsibility.

Yet this choice was clearly unsatisfactory to Kepler. He wanted not just to avoid the charge of irresponsibility but to effect positive change, not merely to watch from the sidelines as the political situation deteriorated on its own but also to try to help improve it. He wanted to be a *better* politician, not to deny his political role. In this case, the ethical astrologer, fully aware of the political situation and the potential ramifications of his words, needed to think carefully about what he said and to whom he said it. If telling Otho of his good revolution might lead him to murder Galba, perhaps it would be better to tell him something different. Kepler took this approach in his 1611 letter to the anonymous adviser to Rudolf. There, aware of the tense situation between Matthias and Rudolf, he knew that a good astrological report to Matthias and a bad report to Rudolf could have disastrous results. To avoid exacerbating the conflict, Kepler felt that it was most appropriate to conceal his astrological knowledge and dissimulate to Matthias, while trying to convince Rudolf to avoid astrology entirely. Moreover, to make the right decision about when to dissimulate and when to speak the truth, the responsible astrologer needed not to lessen his political involvement but to broaden it—to learn as much as possible about the political situation and to gather experience, whether personal or historical, that would enable him to act wisely and appropriately, for the good of both the ruler and the state.

Kepler's Tacitism thus manifested itself on two levels. On the one hand, like many others of his time, Kepler found in Tacitus a font of wisdom relevant to modern times, particularly to the troubled empire of the early seventeenth century. Like other Tacitists and adherents of the reason-of-state philosophy, Kepler conceived of politics as an art grounded in a rich context of personal and historical experience. Generals tended to make better politicians than astrologers, Kepler believed, because their experiences on the ground made them experts; absent this kind of experience, Kepler relied on his interactions at court and on the historical insights and experiences of Tacitus. On the other hand, like many other Tacitists, Kepler believed that politics had an ethics that underpinned it. That ethics privileged the public good over personal interest and held that the public good could not be pinned down by prewritten laws but needed to be judged on a case-by-case basis. Moreover, it admitted—even required—deception, if such deception furthered the public good.

Kepler's Patrons and the Usefulness of the Harmonic Model

Kepler's role as a politician, as I've described it so far, was restricted primarily to his astrological activities. Yet what of his astronomy, especially his cosmological work? Astronomy itself had clear political uses; astronomical tables could be used to improve the accuracy of navigation, of central importance to rulers whose power and funding relied increasingly on voyages of discovery and conquest.[52] Such tables were also essential for the construction of accurate calendars, and the debates over the calendar were still far from settled, as we will see in the next chapter. These were some of the reasons why Rudolf II agreed to sponsor the new set of planetary tables—begun by Tycho Brahe and finished by Kepler—that would ultimately bear his name. And these reasons could be applied too to cosmological works, if their descriptions of the foundations of planetary motion would enable better calculation of the planetary positions. This is, in fact, precisely what Michael Maestlin argued to the University of Tübingen Senate when he urged them to accept Kepler's first manuscript, as we saw in chapter 1. "There is no doubt," he wrote in 1596, "that those who collect observations are going to find these foundations, given a priori, of the greatest help in the reformation of the motion of the heavenly bodies."[53] The Senate accepted Kepler's book on precisely this ground, as Matthias Hafenreffer, representing the Senate, later wrote to Kepler. "The Senate," he noted, "finds this discovery of yours to be as admirable as it is useful to all readers."[54]

Yet despite all this, Kepler himself declared in the book's dedication to the nobles of Styria that his theories were neither practical nor politically useful.[55] In fact, this was precisely why Kepler argued that the book was worthy of elite patronage. "Why is it necessary to value the use of divine things in cash, as though they were food?" he wrote in the 1596 dedication. "For how, I ask, is knowledge of natural things useful to a hungry stomach, and how is the whole rest of astronomy? Nevertheless, wise men do not listen to that barbarity which shouts that these studies must therefore be abandoned. There are painters who delight our eyes, and orchestras our ears, although they bring no benefit to our affairs. And the pleasure received from either of these works is judged not only civilized but also distinguished."[56] Common people focused on the useful and the commonplace, Kepler insisted, while nobles were interested in the rare, the beautiful, and the pleasurable. "The worth of this material will be all the greater, the fewer people are found who praise it, if they are intelligent," he wrote. "The same things are not appropriate to the public and the princes; nor are these heavenly things fodder for everyone indiscriminately, but rather for a noble mind. . . . Therefore, let these meditations and others like them be scorned by anyone at all, as much as they want, and let them search everywhere for their advantage, wealth, and treasures. For astronomers, this glory will be sufficient: that they write for philosophers, not for the rabble, for kings, not shepherds."[57]

This position was a common one at the end of the sixteenth century, and in it Kepler drew on two separate traditions. The first was the Aristotelian tradition that sharply distinguished between natural philosophy and the mechanical arts, or between nature and art more broadly. The first was contemplative and aimed at knowledge; the second was practical and aimed at utility and control. The first, moreover, was the realm of scholars and the intellectual elite, while the second was the realm of artisans, whose products were neither intellectual nor authoritative. Over the course of the early modern period, this dichotomy was challenged in multiple ways. Men like Francis Bacon highlighted the importance of utility and claimed that knowing meant making, and artists and artisans, interested in raising the status of their disciplines, argued for a new epistemology centered on practical, embodied knowledge.[58] Yet Kepler here relied upon the earlier dichotomy of the contemplative versus the practical arts and argued for lack of utility as a sign of his work's greater value and authority.

Kepler also drew on the traditions of patronage and courtly culture, which emphasized the collection of novel, unique artifacts. According to these traditions, the prestige of a patron was demonstrated both by political conquests and by the men who populated his court, and by the value and uniqueness of the objects he assembled to fill his cabinet of curiosities. By acquiring objects that embodied learning and beauty, patrons confirmed themselves as erudite and cultured.[59] Drawing on this idea, as we saw in chapter 4, Galileo named the moons of Jupiter, seen for the first time with his telescope, the Medicean stars, fashioning his discovery into an object that could be owned and collected by the Medici court.[60] In a similar fashion, Kepler sought to embody his own discovery in a credenza-goblet that could be presented at court—

though not, in keeping with the philosopher/artisan distinction, to make it himself.[61] This concrete embodiment as an object of beauty, and the fact that it had no practical applications, meant that all those who appreciated Kepler's discovery must be great men—philosophers or kings.

Kepler likewise highlighted, as the ideal patron of his work, the philosopher-king. In his depiction, this meant, not a monarch engaged with the real world but informed by his philosophy, but rather a man whose concerns had moved *from* the social and political to the solitary and contemplative. As he wrote in the dedication: "Some Charles will appear again, who, . . . tired of empire, will be captivated by the narrow cell of the monastery of Yuste; and who, among so many spectacles, titles, triumphs, so many riches, cities, kingdoms, will be so pleased by a . . . Copernico-Pythagorean planetarium alone that he will exchange the round world for it and prefer to rule circles with his finger rather than peoples with his empire."[62] As Charles V had retired to a monastery and given up on the tangled political struggles that had dominated his days and nights until then, so Kepler suggested that the truly noble ruler would value his book over and above any political contributions. Esoteric, speculative knowledge—the secret of the universe that he had revealed in this book—was truly the kind to be savored.

By the time Kepler rose to the position of imperial mathematician, he had modified his views on the practical applications of his work. This change may, in part, have been forced early on by the ultimate of practical concerns: continued employment and patronage. As religious tensions slowly escalated in Graz—Kepler's first place of employment—and as most Lutheran priests and teachers (with the exception of Kepler) were expelled from the city,[63] Kepler began to hear rumblings of discontent from some of his patrons. Kepler noted in a 1599 letter to Maestlin, back in Tübingen, that some of his patrons continued to be impressed by his mathematical work and "want me to use this philosophical leisure of mine toward illuminating the mathematical disciplines: and if it remains in their hands, my salary will not be abolished."[64] Yet increasingly there were other voices—those who looked down on the mathematical arts because they were not useful or relevant. Those men, wrote Kepler, "shout against me today (or certainly mutter) that now is a time, not for learning, but for waging war."[65] The timing, Kepler emphasized, was the crucial factor in his patrons' change of heart—while mathematical erudition and collectible novelties were valued in times of peace, surely, his patrons wondered, there were other uses to which learning could be put in times of war.

In particular, they suggested that Kepler study medicine—a noble and learned profession but one that had clear and immediate practical applications. "They ordered," Kepler wrote to Herwart von Hohenburg, "followed up with threats of dismissal, that I put aside astronomy for the present and embrace medicine and that I transfer my attention from speculations of beautiful magic to the care of public utility, which is beneficial in this difficult time."[66] The problem was compounded by Kepler's recent trip to Prague to pursue further patronage opportunities with Tycho Brahe; Tycho had offered Kepler a post with him in Prague, provided his patrons in Graz continued his stipend. This trip made an impression on Kepler's patrons, but not the one he had hoped. They focused on the fact that Kepler did not mind journeying from home—why go to Prague and focus on astronomy when he could instead go to Italy and study medicine? Kepler was reluctant to give up his work in astronomy, yet he acknowledged that he needed to earn a living, and if study medicine he must, so he would. "The sum of my deliberation is this," he wrote. "I want to spend a few years still in astronomy, on account of rich Tycho, the possessor of these things [observations], whether that happens in Bohemia or whether here in Graz, which I would prefer. Meanwhile, I shall advance slowly in medicine. Toward this end, I need a continuation of my salary, which I have from the orders of this province, and I need security in the business of religion, which the prince can supply, of the sort that is enjoyed by Tycho beneath the emperor."[67] At this stage, Kepler still indicated that he wanted to pursue his astronomical studies *despite* their lack of use.

Over time, however, Kepler came to articulate a very different vision of his work, one that emphasized its close relation to public utility. There may very well have been a pragmatic component to this change in strategy: if he wanted to pursue astronomical studies and his patrons wanted work that was publicly useful, why not combine the two? Yet beyond this kind of strategizing, Kepler's own experiences with an increasingly discordant world, alongside his pursuit of the question of cosmic harmony, led him to embrace the idea that his work might be not only useful but essential to the public welfare. That his sole motivation was not simply continued employment is suggested by the fact that he began emphasizing this new kind of utility while his employment status was relatively secure, after his appointment as imperial mathematician to Rudolf II. More significantly, he began emphasizing it in particular to a man whose patronage would have been increasingly politically risky for Kepler—King James I of England.[68]

Kepler first reached out to James in 1607, when he presented a special copy of his *De Stella Nova*, a book on the new star of 1604, to the English king. He personally inscribed the copy of *De Stella Nova*, which he sent to James via his imperial ambassador, with the words "To a philosopher-king a servant philosopher / To a Plato a Diogenes / To the master of Britain one who at Prague begs for a small offering from Alexander / From his hired barrel sends and recommends this, his philosophical demonstration."[69] He also wrote a letter to James, where he included a prayer for James's success in seeking both personal and communal harmony: "May God bring it about that your Majesty govern Britain with such good fortune that you are never forced to bid philosophy farewell because of the heavy weight of business. May the thoughts and plans you undertake for the peace and improvement of the church be led to the safety of Christendom and the prosperity of the kingdom."[70] Kepler's conception of the "philosopher-king" had clearly changed from the days of the *Mysterium Cosmographicum*. There, it suggested the ultimate isolation and withdrawal of Charles V, but here, when applied to James, it meant something much closer to Plato's usage. Kepler indicated that James might be the ultimate philosopher-king if he applied his philosophical knowledge—knowledge provided by men like Kepler—to the real world and its problems.[71] Philosophy, hoped Kepler, would help James understand and implement the model that God had used in nature, a model that might be used to great effect in church and state.

In his *Harmony of the World* of 1619, Kepler stressed that this model was the model of harmony. When Kepler completed that book, he wrote to Christoph Besold of his plans to dedicate it formally to King James. Besold agreed with Kepler's vision of the monarch and heartily endorsed the planned dedication. He too believed, he wrote, that James was "special among rulers and among scholars, and alone in our generation is one who most happily joins together and harmonizes letters of every kind and expertise in ruling: thus, this new work of harmonies is owed to him."[72] James was, of course, a Calvinist and the son of a Catholic, yet the Lutheran Besold, like Kepler, clearly felt him to be an ally in the cause of religious peace. In Kepler's dedication to James, he declared that the harmony of nature he had explicated in the *Harmony of the World* could help provide James with a model for harmony on earth and equip him with the tools necessary to achieve his harmonic vision.

Why James? What was it about him that made both Kepler and Besold see him as the logical recipient of Kepler's harmonic vision? Kepler and Besold were, of course, not alone in their belief that James was uniquely positioned to help repair the breaches in Christendom, and they likely relied, in part, on reports of James's own speeches to Parliament. Already in his speech to the first Parliament of his reign in March 1604, he argued for religious unification and harmony. "I could wish from my heart," he said, "that it would please God to make me one of the members of such a generall Christian vnion in Religion, as laying wilfulnesse aside on both hands, wee might meete in the middest, which is the Center and perfection of all things."[73] Later, in his Premonition of 1609, the preface to a reissue of his Apologie for the Oath of Allegiance of 1609, James had argued more specifically that Christian princes should take it upon themselves to reunify their faith by calling for a general council, in which a certain baseline of common beliefs could be identified, and all points of division discussed and resolved.

It is a matter of some historical debate whether James I genuinely embraced the values of tolerance and ecumenicism, or whether his stances on these positions were themselves political strategies. While W. B. Patterson has argued forcefully for the former, others are less disposed to simply take James at his word. James did, after all, continue to burn heretics, and he persecuted the Separatist churches throughout his reign. His calls for compromise, for unity, and for a middle way allowed him to explicitly exclude "all incendiaries and Novelist firebrands on either side, as well Jesuits as Puritans."[74] In this way, his stated interest in harmony went hand in hand with persecution of those who placed themselves outside the religious settlement he had crafted. Moreover, in his role as a "new Constantine," James positioned himself as a powerful international arbiter and may well have viewed his attempts to cobble together a unified front in Europe as merely a prelude to the ultimate battle against the Turks.[75]

Yet despite all this, and even if we regard James's irenicism more skeptically than Patterson, Kepler was not alone in seeing James as a source of hope. Others believed in James's calls for churchly unity and religious peace, and many of them flocked to his court. Among them were Isaac Casaubon, Georg Calixtus, and Hugo Grotius, all of whom promoted some form of religious irenicism or conciliation. Casaubon stayed in England from 1610 to 1614, during which time Calixtus and Grotius also visited; like James's own words, the books of these three men "helped to make England a center of irenic activity . . . and kept the idea of a religious concord before a European audience."[76] Moreover, James did not confine his irenicism to words alone; the Jacobean church reached out to Protestant churches on the Continent and tried to establish contacts with the Eastern Orthodox Church as well. At the start of the Thirty Years' War, James sent emissaries to the empire to help the Protestants in Bohemia and the Austrian Hapsburgs arrive at a peaceful resolution to their conflict. These actions, and the words of both James and the men at his court, seem to have suggested to Kepler, at least, that James might be specifically receptive to his message.

In the dedication to the *Harmony of the World*, Kepler took care to clarify his reasons for linking James to the cause of world harmony. He noted that James obviously appreciated philosophical studies, as his own writings and the opinion of his subjects made clear. In particular, he had proven himself wise when it came to studies of the heavens—he had "deemed the astronomy of Tycho Brahe . . . worthy of the ornaments of his talent" and had visited Brahe's observatory in Denmark, and he had "marked the excesses of astrology with public censure" in a book he had authored. James would therefore be able to truly evaluate and appreciate Kepler's masterpiece, since he would have "complete understanding of the whole of this work and of all its parts" (certainly an exaggeration, but likely, Kepler believed, with a grain of truth to it). Yet he insisted that James's philosophical acumen, unusual for a ruler, was not the main reason for Kepler's dedication to him. Instead, Kepler's choice stemmed from "that manifold dissonance in human affairs." Though Kepler believed that "it was God who regulated all the melody of human life," and therefore that all dissonance served a purpose in the grand harmony that God had orchestrated, he still yearned for peace in his own time. He had turned to James, he explained, because "my longings prompted me to look for some basis for reconstructing consonance from your Davidic harp." In James's union of England and Scotland he had "produced one kingdom and one harmony (for what else is a kingdom but a harmony?)." This had proven, in Kepler's mind, that James was destined for even greater and grander undertakings in the service of world harmony. Accordingly, Kepler felt that it was his duty to reach out to James and provide him with the model whereby he could reproduce his domestic successes on an international scale, so that "at last this enduring dissonance . . . will end in pure and abiding harmony." He prayed that James would "look upon this work on harmony . . . [and] stir up in [himself] by the examples of the brilliance of concord in the visible works of God the zeal for concord and for peace in church and state." If James truly took the lessons of his book to heart, Kepler believed that he would be impelled to seek a peace on earth that followed the model of the harmony of the heavens.[77]

The Nature of Political Harmony: Kepler and Jean Bodin

Thus far, Kepler's claims seem rather vague. In what sense might cos-mic harmony serve as a political model? We've seen already, partic-ularly in this book's introduction, that political order had long been described in terms of musical harmony; a peaceful, well-ordered gov-ernment was a harmonious one. Is Kepler here simply suggesting that harmony ought to remind his patrons of the importance of peace? To get a better sense of the precise connection Kepler hoped to draw be-tween the cosmic and political orders, we need to look more closely at his debate, within the pages of his *Harmony of the World* itself, with another theorist who applied the harmonic model to politics in very specific ways—Jean Bodin.

Jean Bodin—humanist, politician, jurist, and political philosopher —is today most famous for his theory of absolute sovereignty. In the *Six Books of the Republic* of 1576, in which he first articulated this the-ory, Bodin hoped to help resolve the civil unrest in France that had begun with the start of the Wars of Religion in 1562. He would do so, he argued, by demonstrating the best way to organize the state, draw-ing on the teachings of both history and philosophy for guidance. His description of such a state hinged largely on the figure of the absolute monarch, answerable to no one but God. His arguments clearly rever-berated with sixteenth-century readers; by 1600 at least twenty-four editions of the book had been published, both in the original French and in Latin and other translations.[78]

Like so many other political writers of the sixteenth and seven-teenth centuries, Bodin was greatly influenced by Tacitus, whom he described as "wonderfully keen and full of prudence."[79] Like that of other Tacitists, Bodin's work was deeply historical; his *Method for the Easy Comprehension of History* formed the backdrop for his *Six Books of the Republic*, and his goal was to discover order amid the dense political and juridical details of the past. He emphasized prudence, which he called the "arbiter of human life,"[80] and argued that all norms needed to be situated contextually, adjusted to their places and times. And like Machiavelli before him, whose aim had been to use prudence and the knowledge of history to "induce universal laws of political power,"[81] Bodin hoped that his studies would reveal the laws that underpinned all political systems, past and present.

In this, Bodin was influenced as well by the new school of the French Ramists. Ramus taught that in all fields of study one needed to look at the undifferentiated data, organize it carefully and appropriately, and extract universals from particulars. In many ways, he saw mathematics as a model; though he did not distinguish himself as a mathematician, "the quantified world was the real spiritual home of this apostle of 'method.'"[82] Bodin may well have attended Ramus's lectures while in Paris;[83] he was certainly familiar with his works;[84] and the word "method" in the title of one of his books is further evidence of Bodin's debt to Ramus. Bodin's interest in examining past forms of government and justice to discover truths that applied to all peoples may likewise be linked as much to the Ramist tradition as to the political tradition that joined Machiavelli to the Tacitists who followed in his footsteps.

Ramus's emphasis on the foundational role of mathematics may have encouraged Bodin's decision to frame his own political laws in explicitly mathematical terms. To do so, Bodin reached back to the long tradition linking politics and the language of harmony, and to the idea of mathematical harmony in particular.[85] Harmony had long been used to link the realm of the natural and the social. In particular, the tradition of political harmony, the body politic, and the connections between macrocosm and microcosm had been used to support a political system with one supreme head, like the absolute sovereign Bodin endorsed. Bodin relied directly on much the same rhetoric of harmony and argued that the primacy of the monarch was apparent from daily experience of the natural world. "There is no need to insist further that monarchy is the best form," he wrote,

seeing that the family, which is the true image of the commonwealth, has only one head, as we have shown. All the laws of nature point towards monarchy, whether we regard the microcosm of the body, all of whose members are subject to a single head on which depend will, motion, and feeling, or whether we regard the macrocosm of the world, subject to the one Almighty God. If we look at the heavens we see only one sun. We see that gregarious animals never submit to many leaders, however good they may be. . . . The true monarchical state, like a strong and healthy body, can easily maintain itself. But the popular state and the aristocracy are weak and subject to many ills.[86]

Bodin did not stop with this general linkage, however. Reaching back to the Pythagorean tradition, Bodin argued that as harmony was a musicomathematical concept, one needed to consider the mathematics that underpinned it and apply that mathematics to the political sphere.

To do so, he invoked the Pythagorean notion of the three means or proportions, mathematical relationships that created different kinds of numerical series: arithmetic, geometric, and harmonic (also called sub-contrary). Archytas, who first mentioned the proportions, explained them as follows.[87] When comparing the three terms in the arithmetic proportion, "the first exceeds the second by the same amount as the second exceeds the third (e.g. 6, 4, and 2)." In the geometric propor-tion, "the first stands in the same relation to the second as the second to the third (e.g. 8, 4, and 2)." In the harmonic proportion, "the first term exceeds the second by the same fraction of itself as the fraction of the third by which the second term exceeds the third (e.g. 6, 4, and 3, where 6 − 4 = 2, i.e. 1/3 of 6, and 4 − 3 = 1, e.g. 1/3 of 3)."[88]

Following Archytas, several thinkers of the ancient world linked the first two of these mathematical proportions (though not the final, harmonic one) to different forms of justice and political constitutions. This was because they came to embody two different kinds of equality. Though in the arithmetic proportion the distance between each term is equal, the ratio between the terms is increasingly unequal the higher up the scale one goes. By contrast, in the geometric proportion the dis-tances between the terms are unequal, but the ratios remain the same no matter how high on the scale one ascends. If the terms are under-stood as people, and the numerical values of each term are understood to correspond to the values of the individuals (whether conceived in terms of virtue, wealth, or nobility of birth), then different political conclusions clearly followed.[89] Plato, for example, suggested in the *Gor-gias* that justice that corresponded to geometric proportion was the best kind. More specifically in the *Laws*, Plato referred to "two kinds of equalities (i.e. arithmetical and geometrical proportion)"; while the former, he wrote, was lauded more frequently, the latter—geometrical proportion—was "of a better and higher kind . . . for it gives to the greater more and to the inferior less, and in proportion to the nature of each . . . and this is justice, and is ever the true principle of states, at which we ought to aim."[90] Aristotle, too, took up the question of the means and argued in his *Politics* that—in contrast to Plato's preference for geometric proportion—some combination of the means offered the best model for government. According to Aristotle, the arithmetical series corresponded to democracy, while the geometrical series corre-sponded to oligarchy; neither, however, worked perfectly all the time. "It is a bad thing," he wrote, "for a constitution to be organized un-qualifiedly and entirely in accord with either sort of equality. . . . Hence numerical equality should be used in some cases, and equality accord-ing to merit in others."[91]

Like Plato, Bodin addressed the question of justice, and—when it came to the nature of the arithmetic and geometric means—he understood things quite similarly. The arithmetic proportion created sets in which the numbers, but not the proportions, differed by equal amounts. According to Bodin, this corresponded to commutative justice, or the principle of equality, in which justice is "regulated by fixed and invariable laws, not susceptible of any equitable interpretation nor admitting any privilege or exception of persons."[92] The geometric proportion, by contrast, created sets in which numbers differed by the same ratios but unequal amounts. According to Bodin, this corresponded to distributive justice, or the principle of similarity, in which different levels of society were subject to different laws, such that "it is agreed that the execution of the law ought to be adapted to the circumstances of each case."[93] Unlike Plato and Aristotle, however, Bodin invoked the harmonic proportion in his political discussion.[94] That proportion, in his view, united the other two, by directly combining elements of the arithmetic and geometric.[95] Bodin argued that this harmonic "blending" of the two series corresponded to a harmonic justice that avoided "the unmitigated rigidity of the commutative principle, and the variability and uncertainty of the distributive."[96]

More broadly, much as Plato had hinted and Aristotle had made clear, Bodin maintained that these proportions corresponded to different forms of government. According to Bodin, the arithmetic proportion corresponded to a democracy, where the nobility were given no special privileges, and equality under the law reigned supreme. The geometric proportion corresponded to an aristocracy, where some individuals were privileged over others, and the law was able to take those privileges into account. Finally, Bodin once again invoked the harmonic proportion, which he understood to combine the principles of the other two. This, according to Bodin, corresponded to a monarchy with an absolute sovereign. In such a system, the laws allowed for a measure of equality, but some distinctions between nobles and commoners remained, and the sovereign was given the power to supersede the law to achieve the best results for his people. Indeed, in Bodin's view the person of the sovereign guaranteed the harmonious blending of the commutative and distributive principles. According to Bodin, "the wise king ought therefore to govern his kingdom harmoniously, subtly combining nobles and commons, rich and poor. . . . In doing this the prince reconciles his subjects to one another, and all alike to the state."[97]

Bodin then moved beyond the proportions to number symbolism—

especially harmonic number symbolism—more generally in order to highlight the absolute sovereignty of the monarch. The orders of society in a harmonic government, he argued, corresponded to the numbers 1 to 4. The king "exalted above all his subjects, whose majesty does not admit of any division, represents the principle of unity, from which all the rest derive their force and cohesion." Below him, representing 2 to 4, were the three estates: the clergy, the military, and the people. Bodin maintained that to have a truly harmonious relationship between these groups, "the union of its members depends on unity under a single ruler, on whom the effectiveness of all the rest depends. A sovereign prince is therefore indispensable, for it is his power which informs all the members of the commonwealth." Since the king represented the number 1, his sovereignty and authority needed to be absolute and indivisible, just as he needed to be above the law to properly blend the different kinds of justice. Moreover, since all harmonic consonances could be produced with the ratios of the numbers 1 to 4, the monarchical system was complete—any change would "mar the harmony, and make an intolerable discord."[98]

By mathematizing politics, Bodin was not simply continuing the Pythagorean/Platonic tradition of old; indeed, the linkage between governments and mathematical proportions had rarely been invoked since the time of Boethius.[99] Prompted by the discord of his own time, Bodin endeavored to revive it and to combine it with the Ramist and Tacitean principles of his own day. Bodin hoped to harness the Ramist approach—which sought to locate simplicity and universality in a mass of disordered particulars and highlighted mathematics as a model—to the political sphere. Like other Tacitists, he wanted to decipher and organize the complexities of history and politics and in so doing to uncover whatever laws, maxims, or guidelines they revealed. He hoped in particular to use those universal laws to highlight the best forms of justice and government, in order to serve the power of the prince and strengthen the power of the state.[100] And he pinpointed the Pythagorean language of mathematical harmony as the best way to articulate those laws with clarity and certainty.

Kepler, as both a politician and a mathematician—and as someone for whom harmony figured centrally—would clearly have found Bodin's work of great interest. It should come as no surprise, therefore, that Kepler engaged directly with it. He did so within the *Harmony of the World* itself, at the very end of book 3 in a section entitled "Political Digression on the Three Means." This digression was a late addition to the text; Kepler opened it by explaining that when he had first prepared the text, he had included only a brief paragraph in the midst of book 3 on the idea of harmonic proportions in the state. There, he had written a marginal note pointing to "the splendid passage in Bodin on the state." However, this paragraph was accidentally omitted, because the pages of the original had been "carelessly distributed" during the printing. Upon recognizing the omission, Kepler wrote, he decided instead to add the paragraph to the very end of book 3 and expand it into a longer discussion of the politics of harmony. And instead of merely a brief reference to Bodin in the margins, Kepler decided "to transcribe from Bodin himself the main heads of this political dissertation, shaping the words and arrangement as far as possible from my understanding of that passage."[101] He focused on Bodin, he explained, to clarify and correct Bodin's own discussion of political harmony, which he felt was both obscure and mathematically faulty, and he focused on politics more generally in order "to lighten the tedium of dour mathematical demonstrations, of which the whole book consists, by the interpolation of some enjoyable popular material, and to display a foretaste of its considerable usefulness in understanding the State."[102]

By titling this section a "digression," Kepler invoked a classical rhetorical technique that is most closely associated with Cicero and is described at length by Quintilian in his *Institutio oratoria*. Quintilian had defined *digressio* as "the handling of some theme which must however have some bearing on the case, in a passage that involves digression from the logical order of our speech."[103] Rhetorical digressions were somewhat paradoxically titled: while departing from the main subject of the speech and seemingly irrelevant, they were actually designed to directly affect the reception of the speech as a whole. Quintilian enumerated several potential uses of the *digressio*, but most centered on manipulating the emotions of the audience to make them more receptive to the larger discussion. This was particularly the case when the topic being discussed was one to which the audience may have been unreceptive or unprepared to accept. The *digressio* usually involved a topic that was interesting or diverting; as Quintilian wrote, rhetoricians might digress "to some pleasant and attractive topic with a view to securing the utmost amount of favor from their audience."[104]

Kepler drew on this rhetorical tradition both in his use of the title *"digressio politica"* and in his explanation at the beginning of the section. He stated that the material he planned to discuss departed from his primary argument about harmony in nature in order to discuss Bodin's application of harmony to the political realm. He likewise noted that this topic was supposed to be a more generally pleasing topic to readers—indeed, unlike the rest of the material, he deemed it "popular" material, intended for the nonspecialist. This popularity should serve, he hoped, to convince his readers—some of whom, he implied, were not themselves overly concerned with purely philosophical debates—that his seemingly esoteric discussions of harmony were actually quite useful, particularly when it came to questions of politics. The use of the *digressio* technique allowed Kepler not only to try to sway the reader to agree with him but also to demonstrate how something seemingly off-topic—here political theory—was actually centrally relevant to the larger themes of the book.

Since he geared his political digression toward the general reader, Kepler began with an explanation of the three mathematical proportions. Though Bodin had argued that the harmonic proportion was a combination of the arithmetic and the geometric, Kepler disagreed—but not in order to return to the Pythagorean definition. Rather, he focused on harmony as something that was both mathematical and musical, something that could be *heard* rather than simply a series of numbers that could be calculated. He noted that there were many instances in which the arithmetic and geometric proportions could be combined in the manner described by Bodin but would not yield a harmonic series in the musical sense. And there were harmonic series that were also only geometric or only arithmetic, with no combination of the two—and some that were neither. Kepler referenced his own earlier discussion of musical harmony in *Harmony of the World*, which provided examples of such series, and he emphasized that harmony could not be arrived at simply via recourse to the mathematical theories of the ancients. Experience of the senses, too, needed to play a role in the determination of what constituted a harmony; Bodin, in his attempt to link harmony only to the combination of the geometric and the arithmetic, had "rebel[led] on the authority of the ancients against the sense of hearing."[105] Kepler, as we saw as well in the introduction, sought to frame harmony as a principle that was *both* mathematical *and* empirical; the Pythagoreans, he felt—and Bodin like them—had emphasized the former at the expense of the latter.

This criticism—that Bodin had sacrificed the complicated nature of true harmony for mathematical simplicity—was one that Kepler applied to Bodin's politics as well. Bodin had insisted that in justice and government, as in music, harmony was produced by the blending of the arithmetic and the geometric—that is, by the blending of the principles of equality and similarity. Kepler, by contrast, argued that oftentimes such blending simply destroyed both principles and did not arrive at true harmony. For true harmony to be achieved, Kepler insisted, it had to be thought of as a principle sui generis, one that superseded the ideas of both equality and similarity. When it came to issues of justice or morality, Kepler therefore maintained that the principle of harmony symbolized the common or public good: "the public good," he wrote, "has a certain correspondence with the way in which singing in harmonic parts is pleasing."[106] In politics, Kepler maintained that the principle of harmony stood for the good of the state. Like the good Tacitist that he was, Kepler explained that "this one supreme law, the mother of all laws—that anything on which the safety of the state depends is ordered to be sacred and lawful—is . . . consistent . . . with harmonic ratios . . . even if that law contains nothing further similar either to geometric or arithmetic proportions."[107] In both instances, if one argued for the harmonic proportion, one implied that the common good or public welfare stood above and beyond the needs of individuals, and beyond any formulaic maxims of justice or truth. There was, in other words, no universal law of harmony in politics that could be described in mathematical terms.

To elaborate these ideas, Kepler first turned to Bodin's conception of justice and focused on an example in which he agreed with Bodin's conclusion but not his reasoning. Bodin had retold a story from the childhood of Cyrus of Persia, in which Cyrus had observed a tall man wearing a short tunic standing near a dwarf with an overly long tunic.[108] Cyrus had argued that the two should exchange garments to obtain what would be best for them both. His master had instead ordered that each should keep his own garment. Cyrus, according to Bodin, had focused on the geometric proportion, where justice ought to take account of what was best for each individual. His master, by contrast, had focused on the arithmetic proportion, where each individual ought to keep what was rightfully his. Bodin suggested that the harmonic principle of justice could be achieved if the tall man had paid the dwarf money to exchange garments. For Bodin, this was harmonic because it combined the specific needs of each individual with the equality of their resources—that is, it was a combination of both earlier suggestions. Kepler agreed that this was a harmonic resolution, but only because it arrived at the greatest common benefit—"for the common benefit of both is compared with the pleasantness of singing in harmony."[109] The difference between these two positions is not great but comes down to this: for Bodin, harmonic justice can be achieved through a formula, by calculating the resources and needs of both parties. For Kepler, harmony is a question, not of mathematical exchange, but of something much more general and intuitive: the greater good of the whole. This need not be broken down into mathematical terms, and often it could not be.

Harmonic justice, according to Kepler, needed to stand on its own merits. To demonstrate, Kepler again offered an example in which he agreed with Bodin's result but not his rationale. Bodin had noted that punishments for murder tended to accord with harmonic justice, since "in the divine law all murderers are punished by death with arithmetic equality, but the kind of death to be inflicted is within the power of the judge in geometric correspondence with the dissimilar facts and variety of circumstances."[110] In particular, Bodin noted that the punishment for killing a head of state was far greater than for killing a peasant. Kepler agreed with Bodin that this was an instance of harmonic justice, but not because it sought to combine the arithmetic and the geometric. "This inequality in punishments," Kepler wrote, "is due not so much to the individual persons injured as to the safety of the whole republic."[111] Kepler argued that this was harmonic justice only because it sought to preserve the public good above all else, by preserving those who were charged with safeguarding it. Here again, Kepler emphasized harmony as something essentially social, something that needed to be applied to the community as a whole rather than the individuals within it.

Kepler focused not just on Bodin's approach to justice but also on the various forms of government. While Kepler did not follow Bodin in specifically endorsing monarchy as the ideal form of state, he approved of Bodin's use of the term "harmonic" in his description of monarchy and believed that here Bodin had followed Kepler's definition of the term inasmuch as he "relates all policies not so much to individual orders or men, as to the whole body of the state, and its safety and mutual love and peace."[112] Yet though he approved of this focus on the good of the state, Kepler argued against the mathematized politics of Bodin, in which a harmonic series could be broken down into its component geometric and arithmetic parts to yield a specific formula for governance. "If the harmonic proportions of numbers bring any light to bear on the understanding of politics," he wrote, "they do it on their own account, independently of any relation with geometric proportions."[113] Harmony was a sui generis principle, Kepler emphasized yet again, and the lessons it yielded for statecraft were also sui generis.

What, then, were those lessons? Kepler remained vague on the particulars and chose to focus only on the general concept of political harmony as public good. He demurred that he himself was not so politically experienced, nor did his book focus primarily on politics; but he suggested that even so, his political perspective was preferable to Bodin's. "Certainly if I had acquired knowledge of the state, and was dealing with politics in this book," he wrote, "Bodin would have learnt from this Harmony of mine . . . how to be a better political philosopher."[114] In particular, he offered one clear and central point of disagreement with Bodin. Bodin had sought to apply his mathematical ideas to both law and government and had argued that the harmonic principle (in his view, the combination of geometrical and arithmetical principles) was the best in both cases. Since Kepler, in contrast to Bodin, thought that harmony could *not* be reduced to clear laws and formulae, he emphasized that it should be applied specifically to government; justice might sometimes benefit from Kepler's loose conception of harmony, but it might sometimes benefit from the kind of clear application of the law that Bodin emphasized. "I should say," Kepler wrote, "that the condition of the state and the pattern of its government were one thing, and the administration of justice another, for they differ as part and whole; just as in mathematics geometric and arithmetic proportions in numbers are one thing, and musical harmonies expressed in numbers another."[115] Judges, that is, might at times justifiably administer the rule of law strictly according to either the geometric or the arithmetic proportion—that is, with careful consideration for the letter of the law or for fairness.[116] The ruler, by contrast, was "exercising a higher office, safeguarding the state and its individual limbs,"[117] and ought always to be guided by harmonic proportions; he should, that is, have the prerogative to depart from all considerations of legality or individual fairness at will and to focus solely on public harmony and the welfare of the state.

Here, Kepler was deliberately vague on what sort of ruler he had in mind—and, by extension, on what he felt was the ideal form of government. While Bodin had argued that mathematics had definitively pointed to the royal state as the ideal, Kepler instead described his ruler simply as "this regent, whether he be king, or the aristocracy, or the entire people"—for all these could, in principle, be guided by the notion of political harmony as Kepler had described it. Kepler further emphasized the fact that harmony, when it came to government, was not an objective mathematical blueprint for how to govern but rather an argument for the free and subjective judgments of the ruler. If those responsible for the welfare of the state were bound by detailed mathematical guidelines, by the intimate details of the law, or the demands of fairness, Kepler argued, "for God's sake, what a crop of arguments there would be!" Though it was appropriate for the decisions of judges to be subjected to intense scrutiny and to be "turned inside out and subjected to the most detailed examination by those learned in the law," the direction and condition of the state at large, "unfettered by such great compulsions, should be adapted to the general well-being at the will of the ruler, according to the circumstances, without a great commotion."[118] His own vagueness on the ideal form of the state and the appropriate decisions of the ruler was itself a political stance guided by harmony, as it left the ruler with the freedom and discretion to guide the state to safety and security.

That Kepler's idea of political harmony encompassed many different potential models was clear both from the content of the political digression and from the framing of the *Harmony of the World* as a whole. Kepler dedicated the book to James, the king of England, yet was explicit in that dedication about his own position as adviser to the Holy Roman Emperor. He wrote appreciatively of the political work of Bodin, who supported monarchy in France, while also referring within his discussion of Bodin to his own identity as a German and to at least one area where the German approach to justice ought to be preferred. In framing the book with reference to England, France, and the Holy Roman Empire, Kepler demonstrated that his conception of political harmony might, in theory, embrace all three, so long as their rulers heeded his lessons. If the end result was political harmony and peace, then Kepler deemed it a positive one, regardless of the specific form and constitution of the government. At the same time, in citing Bodin, a Catholic, and dedicating his book to James, a Calvinist, Kepler, as a Lutheran, sought to demonstrate that one specific confession was not essential to the establishment of political harmony.[119]

In his analysis of Kepler's political digression, August Nitschke has emphasized that though Kepler's appeal to political harmony seems to allow for the arbitrary and uncontrolled power of the ruler, in fact Kepler offers certain clear guidelines for rulers to follow.[120] Harmony for Kepler, Nitschke reminds us, is not only an analogy but also a cause. It has clear and visible effects throughout nature. Kepler had explained earlier in the *Harmony of the World* that the heavenly bodies moved as an expression of harmony, astrological influences were harmonic, and human souls "take joy in the harmonic proportions in musical notes which they perceive. . . . They move their bodies in dancing, their tongues in speaking, in accordance with the same laws. Workmen adjust the blows of their hammers to it, soldiers their pace. Everything is lively while the harmonies persist, and drowsy when they are disrupted."[121] In nature, Kepler argued, harmony led to motion, to action, and to joy. In politics, too, harmony for Kepler was a cause in its own right, not simply a rhetorical excuse for arbitrary rule. When Kepler wrote that the harmonic ruler was concerned with the welfare of the state above all else, Nitschke argues, he relied on the idea of harmony as a political cause—as that which fashioned men into a united political community. Consequently, in Kepler's view rulers did not have unlimited power to make whatever arbitrary decisions they chose; rather, the good ruler followed the principle of harmony only if all his actions contributed to political unity and the public good. The function of the ruler, that is, was both to follow the model of harmony as a political ideal and also to create harmony in the state.

Bodin and Kepler thus had much in common in their focus on harmony. Both saw it as a central principle that underpinned the natural world.[122] Both men believed that harmony entailed the embrace of diversity and an appreciation for the beauty of the whole. Similarly, both sought to link the harmony of nature to the harmony of church and state. Both lived in a world beset by uncertainty, violence, and strife, and both hoped to use their work to improve that world and pave the way for peace. Yet when it came to the relationship between natural and political harmony, Bodin and Kepler opted for two very different strategies. Bodin emphasized the mathematical foundations of his harmonic theory to highlight its certainty. As the foundations for political order seemed increasingly unstable, Bodin attempted to provide a new, secure basis for the French monarchy. He argued that absolute sovereignty was demonstrably rooted in the mathematics of harmony and, hence, that mathematics provided clear and direct rules, or laws, by which politics ought to be organized and conducted. By ending his *Six Books of the Republic* with a discussion of harmonic theory, Bodin sought to support his entire discussion of politics with a mathematical foundation, described via the language of music.

By highlighting Bodin's work in his political digression, Kepler associated his own ideas with those of Bodin; indeed, he noted that when it came to the general linkage between the harmony of nature and the harmony of the state, "I agree with his purpose as much as anyone."[123] Even in praising Bodin so highly Kepler highlighted the practical goal of worldly harmony, for few of his fellow Protestants would have had anything positive to say about the work of Bodin. Yet though Kepler linked his mathematical and musical arguments to his political discussion, he did so in a manner that differed dramatically from that of Bodin. Bodin had sought to demonstrate that harmony was a mathematical combination of the geometric and arithmetic proportions, while Kepler argued that harmony was a principle unto itself. Further, when it came to politics, harmony provided no specifics for governance. Harmony pointed only to the fact that the public good and the welfare of the state should be preeminent. It offered relatively free rein to rulers—royal, democratic, or aristocratic, as the case may be—so long as they sought to actively foster the good of the state above all else. In fact, establishing too specific a model for government would only hinder this goal, in Kepler's view, and lead to further disagreement.

Conclusion: Kepler versus Hobbes, and *Prudentia* versus *Scientia*

Bodin ultimately straddled a divide between prudential approaches to politics, which emphasized the empirical and the contextual, and scientific approaches to politics, which emphasized the universal and the mathematical. While the former approach was that of the Tacitists, the latter approach has been linked in particular to Hobbes, who positioned himself squarely against *prudentia* and in favor of *scientia* in politics. Though Kepler never explicitly engaged with the work of Hobbes, they both worked within a similar intellectual framework and dealt with a similar array of possibilities in constructing their political visions. Thus, despite the lack of a personal connection between them, by way of conclusion I want to briefly compare their perspectives, in order to add texture to the varying positions considered in this chapter and to illuminate more clearly Kepler's distinctive approach.

On the surface, Hobbes and Kepler, like Bodin and Kepler, appear to share certain core beliefs: both emphasized the role of geometry, and both argued that geometrical science could yield important political truths. Kepler, as we've already seen, wrote in his *Harmony of the World* that geometry was "coeternal with God, and . . . supplied patterns to God . . . for the furnishing of the world."[124] Hobbes, along similar lines, wrote in *Leviathan* that geometry "is the only science that it hath pleased God hitherto to bestow on mankind."[125] In fact, Hobbes linked the power of the sovereign directly to the power of geometry *and* the power of God in the frontispiece of the text, which juxtaposed the figure of the sovereign with a verse from the book of Job that alluded to God as geometer.[126] If the power of geometry was linked to the power of God, Hobbes argued that it likewise distinguished all modern progress: "whatever in short distinguishes the modern world from the barbarity of the past," he wrote, "is almost wholly the gift of Geometry."[127]

In what did the immense power of geometry reside, for Hobbes? Despite Hobbes's invocations of the divine, the answer was not quite as self-evident as it was for Kepler. For Kepler, to say that geometry was coeternal with God meant to suggest that somehow it was also "God himself (for what could there be in God which would not be God himself?)."[128] As such, it *must* embody certain truth. Hobbes, by contrast, denied that people could really know anything at all about God, since they had no image of him; Hobbes's idea of geometry as a divine gift seems to be no more than a rhetorical flourish.[129] For Hobbes, geometry's certainty was rooted in two factors. The first was that it made its definitions clear at the outset. Language, according to Hobbes, was ambiguous and unmoored from the reality of things. It could be *made* clear and precise, however, by clearly specifying the exact meanings of things at the start of every endeavor.[130] Definitions themselves were a matter of human construction; in Leibniz's summary of Hobbes's view, "the truth depends on the definition of terms, but the definition of terms depends on human will."[131]

It is this idea of *construction* that lay at the heart of the second reason for geometry's certainty: geometry was the ultimate art of construction, for we make both the definitions and the figures themselves. "Of arts, some are demonstrable, others indemonstrable," Hobbes wrote, "and demonstrable are those the construction of the subject whereof is in the power of the artist himself, who, in his demonstration, does no more but deduce the consequences of his own operation. . . . Geometry therefore is demonstrable, for the lines and figures from which we reason are drawn and described by ourselves."[132] Because we make it, Hobbes argued, we can know it with certainty, by working back from our construction to the original causes and definitions. This claim resonates with Kepler's own discussion of the "knowable" geometrical figures. For Kepler, a figure was knowable only if it was measurable or, crucially, constructible with a ruler and compass. Yet for Kepler, this too ultimately linked back to God, as unknowable figures were unknowable even to God; since they "have remained outside the Mind of the eternal Craftsman,"[133] they did not contribute to the harmonic archetype that Kepler described in his book. The rules for God's construction were much the same as the rules for human construction, though Kepler believed that human construction often introduced errors not present in the work of God. Only geometry itself, then, was *truly* certain. For Hobbes, by contrast, the certainty of geometry was linked not to God but strictly to its construction by people. Human construction, Hobbes insisted, was *itself* the hallmark of certainty. Hobbes then extended this claim to the realm of the political and argued that if one certain science was geometry, the other was politics; as geometry was demonstrable, Hobbes wrote, "civil philosophy is demonstrable, because we make the commonwealth ourselves."[134] He likewise insisted in *De Homine* that "politics and ethics . . . can be demonstrated *a priori*; because we ourselves make the principles— . . . namely laws and covenants."[135]

Like the Machiavellians, the Tacitists, and the reason-of-state adherents, Hobbes believed that politics was the art of construction and fabrication. Yet unlike Machiavelli, Lipsius, and Botero, Hobbes did not therefore emphasize the model of prudence *or* the model of empiricism and artisanal craft. If political construction was for many—including for Kepler himself—a matter of craft, trial and error, and persuasion, for Hobbes construction was a matter of *scientia*, the opposite of *prudentia*. Whereas both Kepler and Hobbes emphasized the certainty of geometry, and both hoped to illuminate some connection between geometry and the political realm, Kepler insisted on the *uncertainty* of politics and, like his fellow Tacitists, stressed the importance of *prudentia* and empirical knowledge to the political art. Geometry could illuminate politics, but politics would never share in geometry's certainty or its absolute clarity.

Hobbes, by contrast, disdained the prudential tradition in politics and argued in *Leviathan* that "signs of prudence are all uncertain; because to observe by experience, and remember all circumstances that may alter the success, is impossible."[136] He sought rather to offer a politics based on science, one that would be certain and absolute. His politics had therefore no need to look back at the histories of old or to rely on rhetoric. "The end of philosophy," he wrote in *De Motu*, "is not to influence, but to know certainly; therefore it does not consider rhetoric; and to know the necessity and truth of the consequences of a universal proposition; therefore it does not consider history, much less poetics, for these narrate singular givens, and furthermore deny the truth by profession."[137] The point of modeling politics on geometrical science rather than on prudence, history, and rhetoric was that it would therefore be so unequivocally and demonstrably true that it would brook no dissent. Hobbes insisted that science was the opposite of rhetoric: "the signs of this [i.e., rhetoric] being controversy, the sign of the former, no controversy."[138] This was not because science was grounded in natural truths; on the contrary, it was only because it was grounded in convention and rooted in the power of the absolute sovereign that its construction was clear and its definitions were explicit.

Shapin and Schaffer famously described another set of contrasts: between Hobbes, who bolstered absolutism in both science and politics and believed that "any working solution to the problem of knowledge was a solution to the problem of order," and Boyle, who sharply separated science and politics and supported the absolute sovereign in politics while emphasizing "managed dissent" in empirical science.[139] Kepler shared Hobbes's belief in geometrical certainty, though his reasons differed; by contrast, he shared Boyle's more empirical approach when it came to questions of politics and astrology. He did not insist on an absolute separation between natural philosophy and civic philosophy as did Boyle, and he thought, with Hobbes, that geometry could yield important insights for the state. Yet unlike Hobbes, the insights Kepler highlighted were not singular or absolute, nor did they point toward only one way to construct and manage the state.

For all of Hobbes's pessimism about human nature—a pessimism that lay behind his insistence on *enforcing* order and compelling assent—Hobbes was far more optimistic than Kepler about the ways in which the geometrical model foreclosed debate and about the possibilities of applying this to the political realm. At times Hobbes suggested that geometry's uncontroversial nature rested merely on the fact that it touched on no man's interest, and therefore its principles engendered no debate.[140] But he also suggested that it was the clarity of the principles themselves that made them impossible to dispute, even if one wanted to. "For who is so stupid," he asked in *Leviathan*, "as both to mistake in Geometry and also to persist in it, when another detects his error to him?"[141] The goal, then, was to model this kind of clarity in civil philosophy—to make the sovereign so strong and to establish his proclamations on such secure foundations that it would simply be impossible to disagree. Consequently, Hobbes wrote in *De Cive* that "if the patterns of human action were known with the same certainty as the relations of magnitude in figures, ambition and greed . . . would be disarmed, and the human race would enjoy such secure peace that . . . it seems unlikely that it would ever have to fight again."[142]

Kepler once had this kind of optimism. As we saw in chapter 1, he once believed that his geometrical demonstrations might definitively settle a debate just as fierce as the political disputes that Hobbes observed with horror: the debate over the nature of Christ's presence in the Eucharist. Yet he came to realize that no matter how strong the proof, the leap from geometry to theology was no simple matter—no one listened, and no one was convinced. He similarly chastised Bodin for arguing that politicians should follow particular mathematical guidelines; while leaving things to the subjective judgment of wise rulers might yield a fruitful, if arguable, resolution, pointing to mathematical principles would cause only more debates. "For God's sake," he wrote about such a scenario, "what a crop of arguments there would be!" Indeed, Hobbes's critics often disagreed first and foremost, not with his politics, but with his geometry—and they used that disagreement to discredit his entire program.[143] They, it seems, were not quite as convinced as he that geometry brooked no dissent.

While Hobbes broke with Aristotle's conception of politics, Kepler agreed with the Aristotelian position that politics was simply not a field that allowed for precision, because human affairs were so complex and variable.[144] Hobbes had argued for unity and uniformity by contrasting the idea of the unified "people" with the "multitude," characterized by its lack of unity. The goal, for Hobbes, was to eliminate the multitude and craft a people who might be considered a homogeneous unit and who could act decisively and in perfect agreement.[145] Kepler, who emphasized the diversity of human perspectives and ultimately embraced that diversity as a positive good—a mark of a harmony—would have embraced the multitude rather than consigning it to illegibility. If the multitude was monstrous to Hobbes, Kepler, as we saw in the introduction, argued that monsters themselves might serve as models for the harmonious community. He would have embraced the monsters in the titles of Hobbes's books rather than their banishment within those books' covers.

In the end, Hobbes serves as a complement to Bodin. Both were men who wanted to make politics certain and scientific, and both invoked mathematics in order to do so. Bodin began with prudence and particulars but ultimately sought to mathematize politics by creating a numerical proof of absolute sovereignty. Hobbes disdained prudence from the start, emphasized the certainty of geometry, and saw it as a model for politics and for absolute sovereignty in particular. Kepler, like Hobbes, saw geometry as a political model, but a model of an entirely different sort. Kepler's conception of politics, rooted in his own experiences and the ideas of Tacitism so popular at the time, rejected both the possibility of certainty that Hobbes embraced and the singular form of government at which he arrived. For although the rules of geometry were clear and unchanging, politics dealt with the changeable, the varied, the mistaken, and the confused. This didn't mean that politics should be avoided, despite Kepler's occasional proclamations to the contrary. It simply meant that it should be handled with care, like "a ship . . . shaken by dangerous storms." Geometry might light the way for that ship and guide it toward greater harmony. If in the end even that failed, one might at least take refuge in the geometrical studies themselves. "When the storms are raging, and the shipwreck of the state is frightening us," Kepler wrote, "let us let down the anchor of our peaceful studies in the ground of eternity."[146]

SIX

"The Christian Resolution of the Calendar": Kepler as Impartial Mathematician

Matthias Hafenreffer believed that Kepler should embrace his identity as a mathematician and avoid questions of theology. "Act as a strict mathematician," the theology professor urged Kepler in 1598.[1] Over the years, Kepler continued returning to the heated terrain of confessional disputes, while Hafenreffer continued reminding him to shut his eyes to everything but questions of mathematics. "Remember," Hafenreffer insisted yet again in 1619, "my Christian distinction between you the mathematician and you the theologian."[2] Kepler, of course, disputed the very distinctions that Hafenreffer urged him to recollect; he saw mathematics not as an abstract and instrumental device for calculation but as a real description of the cosmos, and he understood God himself as a geometer. Mathematical work, to Kepler, was *itself* theological, insofar as it offered insight into both God and the world he had created.

All this we've already seen, and all this made mathematics and the mixed mathematical sciences like astronomy focal points of study for Kepler. Yet beyond the question of what a mathematician could and should do, or what precisely the *objects* of mathematical study were, stand the mathematical *subject* and the question of what it meant to be a mathematician. Kepler referred to himself frequently as a *mathematicus* and argued not only that the mathematician had unique skills and abilities but also that he had particular virtues that might be profitably used or modeled in other domains. In particular, he posited impartiality as a primary epistemic virtue of mathematicians, one with distinct political and confessional implications. "Our studies are impartial," he wrote of mathematicians in a dialogue on calendar reform in 1604.[3] By focusing on Kepler's engagement with the disputes over the acceptance of the Gregorian calendar, this chapter seeks to explore the meaning of impartiality for Kepler and his world and the ways in which Kepler understood impartiality as a virtue that applied to mathematicians in particular.

It may seem obvious both what impartiality connotes and why mathematicians would be particularly impartial. That obviousness, however, is itself a result of specific historical developments that tend to be effaced in hindsight and that require careful excavation. Theodore Porter's account of the rise of quantification illuminates one part of this story. As Porter notes, "quantification is a way of making decisions without seeming to decide."[4] Porter's account, which focuses on the nineteenth and twentieth centuries, highlights the push for impersonality via quantification as a strategy marshaled in cases where trust was needed but lacking. Yet Porter's story is not really about impartiality itself, but rather about objectivity, understood here *as* impartiality. If "objectivity means fairness and impartiality" in this story, then impartiality itself is understood to imply "the exclusion of judgment, and the struggle against subjectivity."[5] Porter's narrative reveals how that particular kind of impartial objectivity came to be allied with trust in numbers.

Other accounts have focused more closely on how objectivity and impartiality themselves became conjoined values. In her work on objectivity, Lorraine Daston has highlighted objectivity's layered and multivalent meanings, which betray "a complicated and contingent history, much as the layering of potsherds, marble ruins, and rusted cars would bespeak the same in an archaeological site."[6] She has pointed to impartiality—understood primarily as disinterestedness—as one of the historical faces of objectivity, which she calls "aperspectival objectivity." In its fully articulated form, aperspectival objectivity "was the ethos of the interchangeable and therefore featureless observer."[7] This impartial kind of objectivity required "breach[ing] the boundaries of language, confession, nationality, and theoretical allegiance."[8] Daston highlights Thomas Nagel's "view from nowhere" as the motto for aperspectival objectivity—and by extension, as the meaning of impartiality itself.[9]

Though Porter and Daston shed much light on the history of objectivity, their accounts, which focus on the nineteenth and twentieth centuries, spotlight moments in which the meaning of impartiality seems itself settled and transparent. Objectivity doesn't always mean impartiality and isn't always achieved quantitatively, but impartiality always means the particular form of objectivity that is aperspectival, and numbers are always impartial. Yet like objectivity, impartiality has a complicated history, a history that has left it too laden with potentially contradictory meanings. Following the lead of Daston and those who have highlighted the historically contingent meanings of objectivity, scholars have therefore begun to turn to impartiality itself and have emphasized its multiple early and disparate invocations across divergent fields—"historiography, natural philosophy, moral philosophy, news publications, aesthetics, education, and religion among them."[10] Its meanings, they emphasize, have varied greatly; impartiality could "be interpreted, variously, both as a retreat from partisanship and a refusal to adhere to any party, or an exercise in judicious judgment: it could be both a quality of mind, and a characteristic of debate; it could be dissimulated, or paradoxically partisan; it could be criticized as disingenuous, inequitable, dangerous, or lazy."[11]

Kepler's reference to mathematicians as impartial—*unpartheylich*, also sometimes translated as "nonpartisan"—is therefore not as transparent as it might seem. What makes Kepler's invocation of impartiality particularly interesting, in fact, is that the word itself was a relatively new coinage in his day. While the word *partialis* had long been used to describe those who judged unfairly or with prejudice, its opposite was not impartial judgment but rather judgment conferred *aequabilis*, "equitably."[12] There is no classical Latin word for impartiality itself, which seems to enter vernacular usage—particularly in the German, as *unpartheylichkeit*—in the sixteenth and seventeenth centuries and then to later migrate into the Latin.[13] That the word itself was new suggests that its resonances, while drawing on the earlier tradition of *partialitas*, might be themselves new and indicative of concerns particular to Kepler's world. The emphasis on both partiality and impartiality in the postconfessional world, that is, may be very much a product of a time in which the language of *party*—as the confessions were often called— was newly important, and when the question of how the parties ought to relate to one another was suddenly and relentlessly pressing.

What, then, might early invocations of *unpartheylichkeit*, or impartiality, suggest? Contemporary dictionaries betray the variety of meanings noted earlier. In Hulsius's German-Italian/Italian-German dictionary of 1605, one of the first in which the term appears, *unpartheyisch* is understood as "neutral"; though there is no German entry for it, it is offered as a definition of the Italian *"neutrale, che no tiene piu da vno, che dall' altro."* Kaspar Stieler's *Teutscher Sprachschatz* of 1691, which lists it as an independent German entry, suggests that even at this later date the term was laden with several potentially contradictory meanings. The entry for *"Unparteylichkeit, die"* reads *"Neutralitas, sinceritas, candor, indifferentia,"* though "indifference" and "sincerity" seem to suggest opposing stances. In Adam Frideric Kirsch's eighteenth-century Latin-German dictionary, *"unpartheylichkeit"* is defined as *"neutrarum partium stadium,"* suggesting neutrality, like Hulsius's definition; by contrast, its opposite, *"partheylichkeit,"* is offered as a synonym for *"cupiditas,"* suggesting simply excess zeal rather than simple partiality. In English dictionaries, one sees a similar variety of meanings; Randal Cotgrave defined "partial" as "Solitarie, priuate, retired, vnsociable, all for himselfe; also vnequall, factious, more affected to one then another" in 1611, while John Wilkins defined "impartial" as "uncorrupt, sincere" in 1668.

The definitions of "impartiality" thus appear to move along a spectrum, in which they suggest either no stance at all, an open and sincere stance, or a stance held without excess zeal. All these may have been real possibilities in early modernity, yet not all would have been considered positive values. Indeed, though today our understanding of impartiality aligns most closely with the first definition—the view from nowhere—it would have made little sense to most men and women of the sixteenth and seventeenth centuries to regard having *no* opinion as something praiseworthy, particularly regarding issues that were deemed important. Though it might be necessary, according to some, for states to tolerate different confessional beliefs and practices, this did not mean that those in power were themselves neutral or disinterested, nor that they believed all sides to have equal claim to the truth.[14] Certainly when it came to the perspective of the individual (and even when it came to the polity, according to many), taking a stance mattered; neutrality was often described as "wretched" or "detestable."[15] Some believed that strong beliefs were to be celebrated and even imposed with force on others for their own good; others, who demurred, argued not against the merit of taking a side but rather against the zeal with which opposing sides were embraced and promulgated.

As Richard Scholar notes, Montaigne is perhaps the best sixteenth-

century representative of this latter stance.[16] Montaigne did not himself use the word "impartial," but he did invoke the language of parties,
and in so doing suggested a notion of impartiality that—like Kirsch's—
stood as a counterpart to *cupiditas* rather than a synonym for *neutralitas*. "When my will gives me over to one party," Montaigne wrote, "it
is not with so violent an obligation that my understanding is infected
by it. . . . My own interest has not made me blind to either the laud-
able qualities in our adversaries or those that are reproachable in the
men I have followed." He likewise insisted that "I adhere firmly to the
healthiest of the parties, but I do not seek to be noted as especially hostile to the others."[17] It was zeal that was dangerous, for Montaigne, not
political judgment itself. Without using the term, Montaigne suggested
that the impartial man might well have an opinion, so long as he did
not embrace it with such zeal that he dismissed outright all those who
disagreed.

We get further insight into the nature of impartiality by looking at
its early usage by Sebastian Franck, the sixteenth-century German humanist and reformer. Arguing against the tenor of the confessional disputes, Franck called in 1534 for an *"unparteyisch Christenthum."* "God
is not partial," he argued, "and we can know him as a god of heathens
and Christians, Turks and Germans."[18] If God himself could be understood as impartial, then clearly the word connoted something different from either "disinterested" or "indifferent." It suggested, instead,
great interest, but in the many rather than the few. It was this kind of
impartiality that spurred many of the irenical projects of the sixteenth
and seventeenth centuries, with their focus on churchly reconciliation;
strong adherence to only one confessional party stood in the way of
such unity, according to irenicists, while the impartial (but not disinterested or neutral) embrace of a plurality of confessional perspectives
might ultimately lead to a reunified church.

At the end of the seventeenth century, Lutheran Gottfried Arnold
invoked the word similarly when he wrote his *Unparteyische Kirchen-
und Ketzer-historie*.[19] Far from indifferent, the book took a strong stance
against the excessive zeal of the orthodox parties of Arnold's day; the
heterodox were more truly Christian, argued Arnold, because they
were the ones who embraced peace and unity. "We raise a powerful
protest," he wrote, "against all those who . . . would like to pick a fight
about this or that doctrinal point . . . considering that the entire history demonstrates how much malice, error, excess and destruction has
come out of such wars about words."[20] Arnold's book was impartial,
he claimed, not because it didn't see the parties at all, but because it

stood above them—because it encompassed them all, rather than just the very limited subset embraced by his contemporaries.

We can link this notion of impartiality to the one articulated by J. J. Berns in his seminal article that situates the emergence of impartiality alongside the spread of newspapers in the seventeenth century. Berns quotes from those newspapers to argue that their self-proclaimed impartiality did not reside in the fact that the individual articles themselves were thought to lack a point of view; rather, the newsman's impartiality stemmed from his ability to *balance* the partiality of the authorities he cited. Kaspar Stieler, for example—whose dictionary we turned to earlier—explicitly argued for *unpartheylichkeit* as a virtue of the newsman and noted that "fathoming the pure truth from so many *partheylichkeiten* . . . is much like seeking midday in the evening twilight."[21] Truth and impartiality, in this view, were quite different. Impartiality was specifically a question of perspective. But the impartial newsman did not have a view from *nowhere*; rather, he had a view from *everywhere*—or at least from as many places as possible. Impartiality was constituted by combining a plurality of perspective-dependent points of view.

Even when it came to judicial impartiality—probably the original domain in which *partialitas* was invoked—impartiality could be more personal and perspectival than impersonal and removed. One need only think of the story of Solomon, the paradigmatic impartial judge, who needed to decide between two women who both claimed a child as her own. As Kathryn Murphy and Anita Traninger note, Solomon's initial judgment—divide the child in two and give half to each—is the most impartial in the modern sense of the term. Yet this judgment was merely a feint: when the false mother betrayed *herself* as indifferent while the true mother protested, Solomon used the latter's deep emotion—her partiality, in modern parlance—to decide the case in her favor. If Solomon himself was impartial, then this version of impartiality clearly relied upon the partial perspectives of the claimants rather than dismissing them.[22] Sorana Comeanu likewise emphasizes that in the early modern period, impartiality—even of the judicial variety— did not include "the erasure of the person" that characterizes the objectivity that Daston describes in her work. Rather, she argues, "the key process geared toward 'universality' is not erasure but reorientation and transformation."[23]

In this chapter, we will see Kepler invoke impartiality as a virtue of mathematicians along similar lines—not as a lack of perspective but rather as a particularly capacious kind of perspective. We will see Kepler invoke it not in a mathematical treatise but in a dialogue between five characters: Catholic and Lutheran theologians and politicians and a mathematician. The dialogue genre is structured around an in-gathering of multiple perspectives; but in a work that strives for impartiality understood in today's terms, there is no place for a host of quarreling politicians and theologians. In a work arguing for the impartial mathematician, however, Kepler used not only the language of parties but also characters, both political and religious—boisterous, opinionated ones—who embodied those very parties.

More significantly, Kepler's dialogue was a debate about a particularly contentious and divisive political and confessional issue of the day: the possibility of accepting the Gregorian reform of the calendar throughout the Holy Roman Empire. Kepler used this dialogue, we will see, both to suggest the manner in which the fissures in Christendom could be repaired and also to offer a specific model for a kind of churchly unity, one that did not depend on absolute agreement on points of doctrine between the confessions. With regard to the former, Kepler highlighted impartiality as the opposite of *cupiditas*, along the lines of Montaigne's suggestions earlier. To be impartial was to be an open-minded communicator, to listen to other points of view—as all the characters in his dialogue, aided by the mathematician, ultimately do—despite the fact that one might disagree with them. In this manner Kepler suggested that both calendar reform and churchly re-unification could be achieved only if Catholics and Protestants would abandon their polemics in favor of impartial, honest, and respectful communication.

In his invocation of impartiality toward the actual resolution of the calendar debate, Kepler relied on the idea of impartiality as the embrace of multiple perspectives, the bringing together of viewpoints that were themselves partial but that might, when woven together, create an impartial whole. By the end of Kepler's dialogue, the Catholics and Protestants don't actually agree on any one solution to the question of the calendar. Yet the mathematician has used his skills to solve things anyway, for he has crafted an alternative mathematical method for the Lutherans by which they might celebrate Easter on the very same day as the Catholics, though using a different calendar. With this resolution, Kepler devised a method by which the two parties could differ in particulars while still coming together on issues of primary importance. In this way, Kepler suggested that in other matters of confessional dispute, there were surely similar points of commonality, where Catholics and Protestants might unite harmoniously while remaining committed to their own particular doctrines and practices. Like Franck and Arnold before him, Kepler's vision of the church was heterodox— but this didn't mean, as Arnold seemed to think, that it couldn't embrace orthodoxy too. Much as Franck did before him, Kepler emphasized the ultimate impartiality of God himself: "Christ the Lord," he later wrote in his 1623 *Confession of Faith*, "neither was nor is Lutheran, nor Calvinist, nor Papist."[24]

Impartiality, according to Kepler, was necessary for the reunion of the church, and the mathematician, who could use his skills to arrive at solutions via multiple pathways, had it in abundance. Kepler's dialogue on the reform of the calendar epitomized this in both its form and its content. As for others in the sixteenth and seventeenth centuries, for Kepler impartiality was essentially an irenic value. When Kepler invoked it, he did not mean objectivity, in any of its varied meanings, or disinterestedness or detachment or neutrality. He meant tolerance.

The Gregorian Calendar Reform and
Protestant Reactions: A Brief History

The solar Julian calendar was introduced by Julius Caesar in 46 BCE as the civil calendar of the Roman Empire, and it subsequently became the civil calendar of Christian Europe. Since it assumed the average year length to be 365¼ days, it divided the year into 365 days, with a leap day added to February every four years to account for the extra quarter days. The vernal equinox was assumed to be March 25. In addition to this civil calendar, the church needed a method by which to correlate lunar and solar time in order to properly set the dates for movable feasts like Easter, which were tied to particular seasons of the year. For this purpose, the church used the nineteen-year Metonic lunisolar cycle, which helped to determine on what solar dates the new moons of a given year would occur.

There were two problems with this overall arrangement. First, the periods of revolution of the lunar and solar years are incommensurable, so that the Julian year is somewhat longer than the year in the Metonic cycle. This introduced an error in the lunar cycle of approximately one day in 300 years. Second, the Julian length of the solar year is slightly longer than the actual tropical year (the time it takes the sun to return to the same position along the ecliptic), leading to an error of 11 minutes and 14 seconds each year, or one day in 129 years. Both of these errors were principally problematic because of the need to accurately calculate the date of Easter. According to the Council of Nicaea in 325 CE, Easter Sunday was established as the first Sunday after the first full moon occurring immediately after the vernal equinox. The council, recognizing that some slippage had already occurred since the time of Julius Caesar, fixed the date of the vernal equinox to March 21, rather than 25, but did not establish measures to correct for the faulty length of the Julian year; consequently, the astronomical vernal equinox continued to move forward in time. Likewise, nothing was done to better correlate the lunar and solar years and allow for a more accurate determination of the date of the new moon. By the mid-sixteenth century, the astronomical vernal equinox was occurring on March 11, ten days earlier than the set date of March 21, and astronomical new moons were occurring four days before ecclesiastical new moons.

The problems with the Julian calendar were apparent well before the sixteenth century, and attempts at reform were initiated, though never completed. Pope Clement VI briefly considered the question of calendar reform in 1344, as did the Council of Constance in 1417—both times without resolution. Nicholas of Cusa sought to reform the calendar at the Council of Basel shortly thereafter, also unsuccessfully, while in 1476 Regiomontanus was called to Rome to consult on the question of the calendar, but he died before the question could be seriously addressed. At the Lateran Council, Paul of Middelburg took up the reform of the calendar and recruited other expert opinions on the subject. Copernicus, asked for his views, declined to comment, arguing that the motions of the sun and moon were not yet well enough understood to allow for successful reform of the calendar, and the council ended with no conclusion reached. The Council of Trent finally jump-started the reform process in 1563, with the decision to amend the missal and breviary, which included the calendar on which they were based. And the reform process was finally completed in 1582, when Pope Gregory XIII created a commission, headed by the Jesuit mathematician Christoph Clavius, to reform the calendar on the basis of the recommendations of astronomer Aloysius Lilius. The commission wrote up their plan in the *Compendium Novae Rationis Restituendi Kalendarii*, which they sent out to mathematicians at Catholic universities throughout Europe for feedback. After some change based on the feedback, Gregory XIII enacted the reforms in his bull *Inter Gravissimas*, published along with the *Canones in Kalendarium Gregorianum Perpetuum*.[25]

The Gregorian reform encompassed several changes. First, in order to remedy the fact that there were approximately three days too many in each 400 years, the new calendar dropped three leap years in that amount of time by allowing the last year of a century to be a leap year only if it were divisible by 400. This left an error of one day in 3,333 years, which the reformers recognized but deemed small enough to be acceptable. The new calendar also reestablished March 21 as the true vernal equinox, in accordance with the decree of the Council of Nicaea, by dropping ten days from the calendar; the papal bull specified that October 4, 1582, would be followed by October 15. Finally, Lilius's primary achievement in reforming the calendar was his correction of the epact cycle. In brief, there were two methods by which the lunar phases could be tabulated in the Metonic cycle in order to easily determine on what solar date a new moon would occur and, by extension, to set the date of Easter—the epacts were one method, and the golden numbers another. Both methods were in error by the time of the Gregorian reforms. In Lilius's calendrical modifications, the golden numbers, then considered most important for the calculation of Easter, were replaced by an improved epact cycle, which adjusted both for the discrepancy between the ecclesiastical and astronomical new moons and for the changes to the solar calendar made in the Gregorian reforms.

The Gregorian calendar was accepted immediately in Catholic Europe, yet was received by Protestants with suspicion and hostility. This was due in large part to the manner in which the reform had been conducted, and in particular to its papal origins, rather than to any strong opposition to the idea of calendar reform in itself. After all, Luther himself had recognized the problems of the Julian calendar in his 1539 *Von den Konziliis und Kirchen.* Yet he had argued, first, that it was unnecessary to tie the solar date of Easter to the lunar calendar at all and recommended instead a fixed date for Easter, following the example of Christmas. Second, and more importantly for the matter at hand, he argued that reform of the civil calendar ought to be left to political, rather than religious, authorities. Following Luther's lead, German Protestants objected to the pope's meddling in what they saw as political business, suitable for the emperor alone.

In fact, the issue of the calendar came under discussion in the empire shortly after the introduction of the papal calendar in 1582, at the Reichstag in Augsburg of the same year. There, Rudolf requested reports on the calendar reform from all his electors. Protestant electors like the Landgrave of Hesse and Augustus of Saxony argued against the introduction of the new calendar, but Rudolf ultimately decided to introduce the reform throughout the empire. He issued a proclamation in September 1583 ordering the adoption of the Gregorian calendar, starting in January 1584; in it, he took care to avoid any reference to its religious character or its papal author, noting only that it was necessary for the empire to conform to the dating of other lands. Yet Protestants in the empire were well aware of the papal bull and—convinced that this was a religious issue—refused to adopt the new calendar.

The city of Augsburg presents one particularly noteworthy example of the calendar conflict and its effects.[26] At the time of calendar reform, Augsburg was a biconfessional city, ruled by Catholics but with Protestants forming the majority of the populace. Even before the emperor's general proclamation, the city council of Augsburg decided in January 1583 to introduce the new calendar, and maintained that it did so "for purely civic and political reasons . . . without the least intention, however, of obstructing or interfering in any way in the teaching, belief, order, or ceremonies of one or the other of the two religions."[27] Yet the Lutherans were not convinced; preachers argued against the new calendar from the pulpit, and the Lutheran residents continued to conspicuously observe holidays according to the Julian calendar. Butchers refused to slaughter according to the times of Gregorian calendar holidays, and Lutheran members of the city court refused to appear during Julian calendar holidays. The issue became violent when the Catholic city council attempted to banish Georg Müller, the leading Lutheran preacher in Augsburg and a strident opponent of the new calendar. Müller was freed by the city's Lutherans as he was being escorted out of town, and an armed riot ensued. Eventually, Lutheran clergymen ordered the people to lay down arms and return home, but the conflict itself was anything but resolved.

In addition to public sermons, Protestant theologians across the empire published short pamphlets and longer tomes arguing against the Gregorian calendar reform. One noteworthy example is Jacob Heerbrand, professor of theology at Tübingen, who published the 1584 *Disputatio de Adiaphoris, et Calendario Gregoriano*. In it, he referred to the new calendar as a product of the "the anti-Christ and the devil" and declared that if German Protestants accepted the calendar, they would ultimately be accepting the religious authority of the Catholic pope.[28] "We do not recognize this legislator, this calendar-maker," he wrote, "just as we do not hear the shepherd of the flock of the Lord, but a howling wolf."[29] Likewise, the theologians at the University of Tübingen issued an opinion on the calendar in 1583, in which they argued that the new calendar "has manifestly been devised for the furtherance of the idolatrous popish system. . . . If we adopt his calendar, we must go into the church when he rings for us. Shall we have fellowship with the Antichrist? . . . Satan is driven out of the Christian Church. We will not let him slip in again through his representative, the Pope." It was clear, they believed, that the new calendar served no agricultural or commercial purpose: "Summer will not come sooner or later if the vernal equinox should be set a few days farther back or forward in the calendar; no peasant will be so simple as, on account of the calendar, to send out his reapers at Whitsuntide, or the gatherers into his vineyard at St. James's Day."[30] They therefore claimed that the only reason for the pope to seek calendrical reform was to extend his own religious or political power, both of which were unacceptable.

Though most of the arguments against the calendar reforms were religious ones offered by theologians, the new calendar was criticized by astronomers and mathematicians as well. Some, like Joseph Justus Scaliger and Sethus Calvisius, accepted that calendar reform was necessary and timely but objected to the details of the Gregorian reform and offered alternative reform plans of their own. Others, like Michael Maestlin, argued both against the reforms of Pope Gregory in particular and against the very idea of calendar reform more generally—or, at least, against the present need for calendar reform.[31] In his 1583 *Ausführlicher und gründtlicher Bericht*, Maestlin wrote that "although in the old Roman calendar, still employed by us, some defects have crept in, they have not yet spread so far, nor by no means are they so important that they need a correction."[32] In fact, calendar reform would only confuse the common people, making life more, not less, difficult. Though the old calendar was not perfect, it would continue to suffice for a while yet. Indeed, argued Maestlin, the Last Judgment was swiftly approaching, making calendar reform pointless, for the errors in the calendar would not increase so much before the end of time that they required imminent correction. Maestlin also claimed, like Heerbrand and the Senate of Württemberg, that even if correction were warranted, the pope had no authority to institute what was essentially a political reform, and that he did so only to increase his power in Protestant regions.

Maestlin's mathematical and astronomical objections to the new calendar, though discussed briefly in the *Ausführlicher und gründtlicher Bericht*, were considered in more detail in his 1586 *Alterum Examen*. There, Maestlin noted that even in the new calendar the vernal equinox would not always fall on March 21 and that it was foolish to affix it permanently to that date. He argued that the current available astronomical tables—the Alfonsine and Prutenic—were imperfect, and therefore, it would have been sensible to wait for improved tables before establishing what Clavius and Gregory had vainly called a "perpetual" calendar. He likewise objected to the fact that the new calendar relied on the mean, rather than the true, motions of the sun—any good calendar, he felt, should stick as closely to the true astronomical facts as possible. He attacked the new epact cycle of Lilius as riddled with errors, often resulting in an incorrect date for Easter. He concluded that rather than calling it a perpetual calendar, "if you wish to find a worthy epithet to apply to the Gregorian Calendar, a very correct one would be A VILE MEDLEY OF ALL ERRORS."[33] Since the pope and his mathematicians could not have been unaware that their reform contained so many errors, Maestlin believed, they clearly had ulterior motives at heart.

Kepler and Calendar Reform, c. 1597

As we've seen, acceptance of the Gregorian calendar tended to fall along confessional lines; Catholics typically supported the new calendar, while Protestants tended to reject it because of its papal origins. Kepler is often cited as an exception to this rule. Conventional wisdom holds that Kepler was one of the few Protestants able to rise above his confessional allegiance and recognize the superiority of the Gregorian calendar; this is taken to reveal the extent to which Kepler placed more value on astronomical precision than on theological wrangling.[34] As this chapter demonstrates, such a claim is misleading; Kepler ultimately came to suggest multiple possible ways to resolve the calendar disputes, most of which did not involve the embrace of the Gregorian calendar without any alteration. Further, by positing theology and astronomy in opposition, such a claim ignores the centrality of theology to all of Kepler's life and, in particular, to his evolving approach to the question of the calendar. Yet the idea that Kepler strongly supported the Gregorian calendar against his own confessional allies does not come from nowhere; rather, it reflects the stance of Kepler as a young man, at the very start of his career.

The calendar debates still raged furiously in 1596, as Kepler, a young teacher and district mathematician in Graz, finished writing his first book, the *Mysterium Cosmographicum*. Kepler was eager to see the book in print, yet this was no simple proposition, for Graz was at some distance from the main centers of printing, and Kepler had not yet made the sort of contacts in Graz who could easily help him with the task. Instead, he relied on his old teacher and friend Michael Maestlin to see the work readied for publication at Tübingen. Maestlin undertook the task gladly; he was, after all, a strong supporter of Kepler and a great admirer of the book. Yet he did so at some cost to himself, for the efforts he expended on Kepler's behalf caused him to delay the completion of his own work; in particular, he noted in a letter to Kepler, "I couldn't finish my work against the new calendar before the book fair, on account of which I received very great criticism from the Senate."[35] The work in question was Maestlin's *Examina eorumdemque Apologia*, the latest of his polemical works against the Gregorian calendar reform, in which he responded to Clavius's 1588 rebuttal of some of his previous arguments.

Kepler and Maestlin had much in common—from their political and confessional allegiances to their belief that the Copernican system truly represented the state of the cosmos. Yet on the question of calendar reform, the twenty-six-year-old Kepler parted ways with his mentor—and with the majority of his fellow Protestants—and believed that the Gregorian calendar ought to be adopted by Protestant Germany.[36] Though Kepler had never publicly addressed the issue of calendar reform, he took it up in his reply to Maestlin, using Maestlin's offhand mention of his own work against the calendar to express his disagreement with Maestlin's approach. To ease his criticism of Maestlin's position, he first took pains to note those areas in which he and Maestlin were in agreement. The Gregorian calendar was not error free, he admitted, and many of Maestlin's own mathematical or astronomical arguments were sound—so much so, Kepler punned, that "neither a *Clavi*[37] nor a wedge nor, indeed, even the whole machinery of the heavens will resolve [them]." Kepler added that he himself was not fully satisfied with the details of the new calendar; in particular, he found the Gregorian calendar's removal of ten days of little advantage. Though he did not believe, like some, that the end of the world was immediately at hand, he knew it would come eventually, and hence, he was not overly worried by the "empty fear," as he mockingly put it, "that the world will endure so long that Easter will fall in autumn, and the heavens will fall into the earth." Moreover, he felt that from a political perspective the new calendar was as necessary "as a fifth wheel to a carriage." Finally, Kepler admitted that the fears of the Protestant theologians about this Catholic innovation were not totally groundless: the theologians were right, he averred, to "look around and make sure that there isn't a snake lying in the grass."[38]

Kepler clearly agreed with Maestlin in many respects. Yet "though I support all this," wrote Kepler, "as for the rest, my position is highly heretical." That is, despite his objections, he firmly believed that the new calendar ought to be embraced by its opponents—by individual Protestants and by Germany more broadly. In explaining why he felt this way, Kepler focused on the fact that the new calendar had already been widely accepted and was clearly there to stay. The objections he had raised at the start of the letter, he implied, may well have been sufficient reasons to reject the new calendar in 1582, but they were no longer relevant in 1597. With the new calendar already a fixture throughout most of Europe, Protestant Germany remained one of the few places to persist in refusing the reform, and doing so put it at a distinct disadvantage. "What should half of Germany do?" he asked. "How long will it separate from the rest of Europe?"[39] In contrast to Maestlin, Kepler argued that the reform of the calendar would not lead to widespread confusion; it should be relatively easy to change calendars at this late date, since the way had been paved by all those who had already done so.

In addition, Kepler argued, it was unclear what alternative the opponents to the Gregorian calendar really anticipated. Unless they waited to see "whether some deus ex machina comes and illuminates all those magistrates [who have accepted the new calendar] with the light of the gospel," there were really only two other options to accepting the Gregorian reforms. On the one hand, they could simply stick with the old Julian calendar. Yet he noted that all the great astronomers of the past 150 years, and even Luther himself, had argued that the calendar needed some correction, and it would be foolish to ignore this. On the other hand, the opponents of the Gregorian reforms could theoretically propose a new calendar of their own—and, indeed, many did just this—yet Kepler doubted that a better alternative would really surface. This was likely in part due to the astronomical difficulties involved in the issue of calendar reform, but also, as he explained, because once one alternative had been widely accepted, introducing another one would certainly lead to only greater confusion. Therefore, the only real solution at this point was to reform the calendar following the Gregorian model. Even if it was not the ideal solution, it was a workable solution for the present—and this was sufficient, for "we are not afraid for remote centuries."[40]

Kepler also responded directly to one of Maestlin's arguments: that astronomers should avoid assisting in the reform of the calendar, since it made no practical difference to them what calendar was widely accepted. Calendars were simply a convention for the ordering of time, Maestlin had asserted, and though astronomers needed to agree on what convention they used, the details were unimportant for their work. In response, Kepler pointed to one of the primary arguments of his *Mysterium Cosmographicum*, the book that Maestlin had delayed his calendar publication to support: astronomers "are concerned not only with utility but also with order and beauty." In the dedication of the *Mysterium Cosmographicum*, as we saw in the previous chapter, Kepler had argued at length against those who maintained that the book was not "useful." Kepler had claimed that beauty and intellectual delight were still higher standards. In a similar vein, he argued to Maestlin that even if the calendar had no bearing on the practical pursuit of astronomy, the Gregorian calendar was still more astronomically pleasing than the Julian, and for this reason it should be supported by astronomers. "For if it pleased God to decorate the world with perfect quantities," he reasoned, "why should not some perfection in the calendars also please astronomers?"[41]

Kepler likewise noted that though Protestants worried about the theological consequences of accepting the new calendar, they would be wise to acknowledge the theological consequences of rejecting it as well. The opponents of the reform had argued that adopting the pope's calendar would be tantamount to submitting to the theological authority of the pope. Yet Kepler claimed that this was no longer the case. Having rejected the calendar already for nearly twenty years, Protestants had made it clear that they did not need to obey the pope's decrees; if they willingly accepted the calendar now, without additional compulsion, it would be clear that they did so not because of the pope's authority but simply because they judged the reformed calendar to be preferable. By continuing to reject the Gregorian calendar, however, they ran the risk that in the future an emperor more hostile to Lutherans than Rudolf II would arise and would force them to accept the new calendar against their religious objections. "It would be preferable," he wrote, "to voluntarily accept it while there is still no constraint."[42]

Kepler ended his letter to Maestlin by noting that the turn of the century approached, which would be an ideal time to adopt the new calendar—1600 was, after all, a more appropriate year to inaugurate a new reckoning of time than 1582. He suggested that if the emperor framed the edict for the new calendar as a political matter alone, and did so in open consultation with his mathematicians, it would make clear that Protestant Germany had accepted, not Pope Gregory's bull, but rather the advice of their own mathematicians and the decree of their emperor. He sincerely hoped that this would happen soon, for the good of his country and the pride of his countrymen; "it is a disgrace to Germany," he wrote, "having restored the art of reformation, to alone lack reform."[43]

The damage to Germany's image caused by the rejection of the new calendar was particularly upsetting to Kepler because he believed that that image was specifically tied to the issue of mathematics—as he once wrote, "mathematics is the true pride of Germany alone."[44] And if Kepler upheld Germany as the pinnacle of mathematical aptitude and innovation, he often represented Italy as exactly the opposite. When Galileo wrote in 1597 of his hesitation to publish his Copernican ideas, Kepler noted that Germany presented fewer impediments than Italy to its mathematicians.[45] Likewise, when he heard that Galileo's telescopic observations were rejected in Italy in 1610, he noted that such foolishness was only to be expected in a country whose astronomers rejected parallax, a phenomenon almost universally accepted elsewhere.[46] When writing to congratulate Samuel Hafenreffer—the son of theology professor Matthias—on a mathematical disputation he had written, Kepler noted that he had done both the University of Tübingen and the German nation proud with his work. "Look at the nations," he wrote in his congratulatory letter. Unlike the Germans, "the Italians [only] dream." Kepler named only two mathematicians of merit in Italy, Federico Commandinus and Giambattista Benedetti; "for [even] Clavius," he noted, "is a German."[47] And Kepler cited the fact that he himself was "a German, in nationality and character,"[48] as a reason why he could not move to Bologna when offered the chair of mathematics there after the death of Giovanni Magini. Germany was where mathematics flourished, Kepler believed, while Italy was known for its backwardness with respect to mathematical pursuits. Yet when it came to the question of the calendar, Kepler found the situation suddenly reversed: Italy had pioneered a mathematically advanced new calendar, while Germany refused to adopt it and persisted with the old calendar despite its errors. That the country known for its mathematical expertise should refuse to accept an obvious mathematical improvement—and one originating from Italy, no less—was a disgrace, in Kepler's eyes.

Yet behind Kepler's arguments to Maestlin there is also a clear element of religious and political naïveté. He wrote to Maestlin that "the path to renovation is very easy for us,"[49] clearly not realizing quite how contentious an issue the adoption of the new calendar really was. Having only recently moved from Tübingen to Graz, and still only an unknown schoolteacher, Kepler had not yet felt the full effects of the raging confessional disputes, nor had he yet had any exposure to the complications of political debates in the empire. The following years brought some dramatic changes to Kepler's life. Expelled from Catholic Graz as a Lutheran, Kepler moved to the cosmopolitan city of Prague, to work for Tycho Brahe at the court of Rudolf II. Within only a few short years he found himself granted the position of imperial mathematician following Tycho's death. His personal exposure to the violent effects of the fiercely entrenched confessional disputes was quickly coupled with his immersion in the world of politics. Both these factors would affect his views on the question of the calendar.

Kepler's Dialogue on Calendar Reform, c. 1604

As imperial mathematician, Kepler was often asked by Rudolf and other members of the ruling elite both for astrological prognostications and for his opinion on issues of political importance that pertained to mathematics. The new Gregorian calendar was clearly such an issue, and between 1603 and 1613 Kepler received numerous requests for reports or opinions on the question of the calendar. Of course, Kepler was not at liberty to respond in any way he chose, for the emperor had his own views on the calendar, particularly after his early attempt to introduce the Gregorian calendar failed in the Protestant areas of the empire. Yet Kepler did address the question of the calendar on his own terms, rather than the emperor's, in at least one document: an unpublished dialogue that Kepler wrote in 1604 and modified in successive drafts over a number of years.

Written during his tenure as imperial mathematician, the dialogue reflects Kepler's growing awareness of the difficult political and religious issues at stake in the calendar debates. In fact, the immediate impetus for the work, which Kepler titled "Ein Gespräch von der Reformation des alten Calenders," was likely the 1603 Reichstag convened by Matthias, then Archduke of Austria (and later to become emperor after Rudolf II), during which he invited discussion on the question of calendar reform.[50] Kepler may have drafted the work in preparation for some of his official reports for the emperor and his electors, though it differs in some key respects from those reports—or he may have used it simply as a way to think through the issues on his own. In either case, the fact that it does differ from those official documents, and that Kepler never published it or even showed it to the emperor[51]—though he felt it important enough to return to it multiple times over the years and mentioned it in letters to his correspondents—suggests that, like his earlier letter to Maestlin, it offers us some insight into his personal perspective on the question of calendar reform.

As the title of the work indicates, Kepler intended it not simply as a discussion of the already reformed Gregorian calendar but rather as a more general "dialogue on the reformation of the old calendar"—that is, a discussion about how to reform the *Julian* calendar, one that does not take the Gregorian reforms as a given. Kepler clarified in subtitles that the dialogue would actually consider three separate options: whether the states of the Holy Roman Empire should adopt the Gregorian calendar, remain with the Julian calendar, or adopt a third, newly reformed Julian calendar. Kepler crafted the work as a literary dialogue, with each position represented by speakers who debated them at length over the course of the work. It might have made sense, Kepler noted in the preface, to name the characters in the dialogue after the principal figures in the calendar reform debates—Clavius, who supported the Gregorian reforms on behalf of the pope, and Maestlin, who argued against them on behalf of his fellow Protestants. Yet Kepler worried that because the tenor of the debate between the two had become so hostile, doing so would awaken old grudges, which would "agitate my friendly dialogue."[52] Therefore, he decided to create two fictional characters on each side of the debate, one political and one religious. In support of the Gregorian calendar reforms (representing the Catholic side, that is) he introduced the political Cancellarius and the religious Confessarius, and in opposition to the Gregorian reforms (representing the Protestant side) he introduced the political Syndicus and the religious Ecclesiastes. To these four characters he added a fifth, the young Mathematicus—who, it soon becomes clear, was intended to represent Kepler himself.

The dialogue begins with each character making one of the stock claims often cited either in support of or against the Gregorian calendar reform. Cancellarius first asserts the need for the Holy Roman Empire to finally adopt the reformed calendar, since the maintenance of two different calendars in the empire has led to "unrest in political affairs."[53] Confessarius supports him by noting that only by adopting the new calendar can its opponents return to obedience to the church. By contrast, Syndicus argues that Protestant mathematicians have demonstrated a great many defects in the Gregorian calendar, and Ecclesiastes maintains that only if the emperor, and not the pope, reforms the calendar could Protestants possibly accept it, "since we differ from one another in doctrine."[54] The four characters then debate whether

or not the calendar is a religious or political artifact, with each side using the examples of antiquity—particularly Julius Caesar and Emperor
Constantine—to support its claims. The Catholics, who argue that it
is religious, maintain that it is a matter for papal correction, while the
Protestants, who argue that it is political, deem it a matter of imperial
concern. Amid this debate, Cancellarius raises another question, which
he proposes they not discuss immediately but keep in mind for consideration: "If a similar reformation would be done anew, could it be done
better than the one done already [by Pope Gregory]?"[55]

It is at this point that Mathematicus first enters the discussion. Syndicus calls on Mathematicus to speak his mind about the questions
they've been discussing, because "you are younger . . . [and] will be
impartial" (Kepler himself was just over thirty at the time), and also
because "today's astronomy is better."[56] Rather than take up the debate,
Mathematicus denies that the entire discussion has any relevance to
him. He asserts that

if clever and overly hot-tempered and idle individuals in Germany would spend
enough time on their studies that they would learn to understand these things
themselves . . . not only would they forgo the heat and haze of quarrel on this account, but also they would finally see that they have no reason to debate the matter
so much. Our studies are impartial and are devoted to the utility of mankind, to
quiet, peace, and unity. You theologians have confused so much—decide things
without the mathematicians and don't mix us into it, and thereby make our studies hated. It doesn't matter to us at all whether we calculate according to Egyptian,
Chaldean, Greek, or Julian years, or according to Gregorian years, when our work is
already piled high for us and needs careful attention.[57]

Mathematicus here argues two things. First, he claims that the *arguments* about the new calendar are not rooted in mathematics at all.
Rather, calendar debates have become debates about theology. For this
reason, mathematicians should not be asked for advice or embroiled
in the contentious debates at all; instead, the question of the calendar
should be settled by the quarreling parties themselves. Second, he argues that the calendar *question* itself is not one that pertains to mathematicians, because it is irrelevant to their studies. For astronomical
purposes, the calendar could be reckoned according to any number of
systems, with little practical difference.

Mathematicus's claims here are surprising on a number of levels. For
one, Maestlin had made exactly the latter claim in his rejection of the
Gregorian calendar—a claim that Kepler himself had discounted in his

1597 letter, where he argued that even if there were no practical as-tronomical difference between calendars, astronomers should still sup-port the one that was the most accurate and beautiful. Further, Kepler's invocation of the "impartial" mathematician in this context seems to differ from the one outlined in the introduction to this chapter. His comments rather seem to suggest that mathematicians are impartial because they are detached, uninvolved, and disinterested in questions of theology and politics. Yet this cannot be the case, for Kepler, himself a mathematician, was already very clearly "mixed into" the debate on calendar reform; he had earlier taken a clear position in his letter to Maestlin and was obviously still engaged with the debate in his politi-cal role as imperial mathematician. Indeed, the writing of the dialogue itself represents an attempt, by a mathematician, to settle the issue, in ways that seem far from detached. This might simply suggest that Mathematicus does not represent Kepler's view, yet the character, of an age and profession with Kepler, is clearly intended as a medium for the author's own voice. Mathematicus's initial refusal to participate in the dialogue must therefore be seen as a rhetorical strategy; in start-ing with an exaggerated denial of the calendar debate's relevance to mathematicians, Mathematicus can slowly begin to articulate his ideal view of the exact role that mathematicians should serve in the calen-dar debates—and of the true nature of the mathematician's virtue of impartiality.

The process of slowly defining the role of the mathematician begins as all the parties in the debate immediately attack Mathematicus's re-fusal to participate. "How can you push the matter so far away from yourselves?" wonders Cancellarius. "Who else first brought up the mat-ter two hundred years ago, but mathematicians?"[58] Confessarius like-wise notes that all the Catholic mathematicians of note were consulted in the drafting of the Gregorian calendar, while Ecclesiastes points out that Protestant mathematicians are the very ones who have said that the Gregorian calendar is riddled with error and therefore should be rejected. Clearly, both Catholics and Protestants argue, the debate is inextricably tied to mathematics. Mathematicus does concede that mathematicians have been closely involved with the correction of the calendar. But he asks that the others concede to him that "mathemati-cians are not to blame in the calendar debate and conflict, for they did not state their opinion so that people could quarrel on account of it—just as God cannot be assigned guilt for men who kill one an-other, as though he should have not let wine come to be."[59] He insists, that is, that mathematics itself is pure and unassailable, as are its true practitioners—but like any good thing, it can be wrongly used.

Yet even this argument is challenged by Syndicus, who maintains that "you mathematicians first began this controversy yourselves."[60] Mathematicus rejects this claim by defining what it means to act as a mathematician: "I call someone a mathematician who speaks, writes, debates, and argues on the basis of his art—that is, so long as he stays within his boundaries. However, when he disputes outside his art and gets by with political or theological arguments or is affected by his appointed position in government, then I consider him a theologian or a politician, and in this matter I have nothing in common with him."[61] Kepler is not here arguing that a mathematician cannot *hold* theological or political positions. He is also not suggesting that a mathematician cannot be at the same time a politician or a theologian—indeed, he cannot be so arguing, since, as we saw earlier, he claimed both the latter titles for himself over the course of his life. He had already said that mathematics and theology were inextricably intertwined, much as astrology (a mathematical art) and politics were. What, then, are we to make of the claim that mathematicians—to qualify as such— must maintain boundaries that differ from those of theologians and politicians?

With this claim, Kepler is in fact continuing the process of fashioning he has already begun in this dialogue, by creating an ideal "mathematician" and then slowly painting a picture of the space he occupies and the methods he uses to pursue his craft. This ideal mathematician, it seems, occupies a realm distinguished primarily by its *difference* from politics and theology, a difference that revolves around the notion of impartiality. Recall that Mathematicus has just defined mathematics as impartial in that it was "devoted to the utility of mankind, to quiet, peace, and unity." Politicians and theologians, as Kepler frames them here (via the arguments of Syndicus, to whom Kepler responds), were devoted to something quite different—not peace and unity but strife and division. Had not the calendar debate proven precisely this point? Those who devoted their energies to controversy and blame were not mathematicians, even if they claimed to use mathematical arguments. They were, rather, theologians and politicians, occasionally armed with mathematical tools. Mathematicians, Kepler argues, could not be grouped together with these divisive figures, for mathematics was *by nature* a peaceful art. Mathematicians who contravened the purposes of peace were not mathematicians at all.

Though Kepler initially created the characters of his dialogue to avoid invoking the animosity of the Clavius-Maestlin debates, Clavius and Maestlin enter the picture now, as Mathematicus is further pressed to explain whether he views their involvement in the calendar debates to be "mathematical." Those mathematicians like Clavius who were involved in the reformation of the calendar, responds Mathematicus, were acting not only as mathematicians but "either had theological or political thoughts on their own, and then supplied them to the authorities, or were summoned by the pope and empowered by him."[62] Likewise, Maestlin "was also impelled by his political authorities and theologians, so that he wrote about it not only mathematically but also theologically and politically."[63] Again, the key issue is not simply the political or religious ties of Maestlin or Clavius but the fact that they used those ties both to bolster their mathematical authority and to further political and theological divisions. True mathematics, by contrast, ought to rely on its own authority and seeks its own ends—the ends of peace. "In mathematics," Kepler insisted, "one disputes on the basis of neither the authority of the pope nor the authority of the emperor."[64]

As the debate in the dialogue continues, Mathematicus is pressed about mathematical involvement in the initial Gregorian reform of the calendar. He allows that this involvement was warranted, since all mathematicians agreed that the Julian calendar was mathematically problematic, and there was a clear need for some mathematical resolution. Though Syndicus points out that Maestlin had observed similar errors in the Gregorian calendar, Mathematicus argues that such errors were far fewer in number than in the Julian calendar. More to the point, he highlights the fact that the reformers of the calendar were aware that such errors existed and "knowingly allowed them."[65] From this point on, the question had ceased to be mathematical. As he explains, "What then can a pure mathematician bring against that from his art? It is not appropriate for an astronomer to command church and state, as though they ought to orient themselves along with the times of the year according to the motion of the heavens. Rather, one ought to leave the matter to the higher faculties [of the university], so that then, when a conflict arises relating to the reform, they will prefer thereafter to leave the mathematicians to themselves."[66] Mathematics can aid in the reform of the calendar, emphasizes Mathematicus, but only when mathematical questions are at issue. When the debate has clearly turned away from mathematics and mathematicians knowingly take up the new political or religious terms of the debate, they have crossed the boundary to politics or religion. And in attempting to "command church and state," mathematics will ultimately be commanded by them—for in involving themselves in issues beyond their purview at the outset, mathematicians will ultimately be called upon by those in power to promote specific religious or political stances regardless of their merits, leading to further conflict. By sticking only with mathematics from the start, however, mathematicians can make it clear that theirs is an art of peace, not of zeal, and an art that embraces multiple possible outcomes rather than only one. If they do this, others "will prefer . . . to leave the mathematicians to themselves"—not isolated from the world but isolated from the conflict that characterizes it. The power, that is, will lie with the mathematicians themselves rather than with the rulers whose divisive aims have brought the world to the brink of ruin already.

The Catholics and Protestants press on, with each side claiming, for various political or religious reasons, that they are right. Mathematicus then interjects once more. "Thank you, my lords," he exclaims. "With these words you have demonstrated yourselves that a mathematician has nothing to do with the calendar debate." He explains: "For both initially and now you could argue with one another without my mathematical instruction. If only you would spare other [mathematicians] also and not say, our mathematicians have found things to be so, while ours have found it otherwise. Stay closer to the truth of the matter and say, instead, that whatever each of us wants, he can use his mathematicians to achieve it. For they are our servants."[67] As before, Mathematicus claims that not only have mathematicians improperly employed political or theological arguments, but they have become the servants of church and state, who produce arguments at the command of their political or religious superiors. Likewise, he makes clear that his point is not that mathematicians should remain in an ivory tower, far removed from worldly conflicts; mathematics can help to resolve those conflicts, but only when it stays within its own boundaries and recognizes no master beyond its own principles.

By now Mathematicus has finally convinced the various parties about the place of mathematics in the reform of the calendar. "You fill me with such wonder," exclaims Cancellarius, "that I must believe that you are right. Please tell me the right and pure reasons for establishing or reforming the calendar."[68] Now that Mathematicus is on firm ground with regard to his proper role, the dialogue unfolds in an extensive discussion of the reform of the calendar, finally with his full participation. Throughout the remainder of the dialogue, Mathematicus acts as a referee of sorts, clarifying what requires clarification and taking sides on specific mathematical or astronomical questions. He makes clear that a new calendar is necessary for neither political nor agricultural reasons, but only in order to properly calculate the date of Easter according to the rules established by the Council of Nicaea. There is some disagreement, however, about how best to comply with those rules, and the parties begin to debate the merits of the different calculation cycles. Amid this debate, Syndicus interjects, "I would like to propose a plan to my most gracious lords that . . . would be useful for a long time and would serve the interests of obtaining peace between us":

My thought is this. A communal decree should be instituted at a Reichstag that Easter be celebrated in the future according to the decree of the Council of Nicaea, and the mathematicians should be charged with diligently computing it from the foundations of astronomy, and not from the old and faded *computus*, according to which we find ourselves celebrating Easter several days away from the intention of the council. I would urge the Protestants most strongly to accept this decree . . . and both parties will have it so that [the Protestant and Catholic dates for] Easter will fall almost always together in our lifetimes, but by different decrees.[69]

Cancellarius immediately objects that such a process would not be admissible under Pope Gregory's bull, and the debate continues. Yet here Syndicus has outlined the basic operating principle for Kepler's dialogue, one that the Protestants eventually work out in some detail, though the particulars differ from those suggested here by Syndicus. The way to resolve the conflict, Syndicus suggests and Kepler ultimately clarifies, is to devise a means by which Catholics and Protestants can disagree about the details of calendrical reform while still arriving at synchronized dates for Easter.

The two sides move on to discuss the correct length of the solar year and the need to improve the solar calendar, as well as the related issue of the moving equinox and the reasons for or against removing ten days from the calendar. The Protestants argue against the omission of ten days from the calendar; Mathematicus agrees with the Protestants that since the actual equinox would take place in the coming years on March 10, the Gregorian commission should rightly either have removed eleven days from the calendar or designated the equinox as occurring on March 20. He notes, however, that this may not be a mathematical issue, for it depends on the rationale used in the determination of specific calendar dates. If those establishing the calendar "maintain that today the equinox falls on March 11 or 21, I can contradict them from my art."[70] But if, knowing when it actually happens, they chose to designate the equinox on the twenty-first for the purpose of calculating the date of Easter, he admits that they have the full right to do so. "What does it matter to an astronomer," he asks, "what day is pleasing to them for the *terminus lunae quartae decimae* of Easter? Easter is a holiday, not a star."[71] Here he emphasizes that there is an important difference between mathematics and astronomy, a difference that resides in their degrees of flexibility. Astronomy was not flexible, for Kepler was invested in a noninstrumental view of astronomy in which astronomical theories described what really happened.[72] When it came to the calendar, however, things were more complicated. The length of the solar year and the specifics of the lunar cycle certainly had their source in astronomy, and calendars should stick reasonably closely to that astronomical source—but the decision about the specific date of Easter was ultimately a religious one. And for this reason, Kepler claimed, it was flexible. In particular, it was mathematically flexible. The stars moved in one particular way, but mathematicians could imagine other ways and could calculate accordingly if those who crafted the calendar so desired. This, yet again, was an instance of their impartiality.

Over the course of further discussions, the Protestants, with the aid of Mathematicus, eventually determine that it might be possible to produce a corrected golden-number cycle, in contrast to the corrected epact cycle of the Catholics. The Protestants, again with the aid of Mathematicus, likewise determine not to remove ten days from their calendar and to still take March 10 as the equinox. This results, as Mathematicus explains, in the Protestants celebrating Easter one month earlier than the Catholics in certain years—specifically, when the full moon occurs on March 10. Yet this would happen only five times in one hundred years, according to Mathematicus—and more importantly, for the remaining years, the date of Easter calculated according to the corrected golden-number cycle would almost always be consistent with the date of Easter calculated in the Gregorian calendar on the basis of the epacts. The participants have thus arrived, essentially, at the solution articulated earlier in the dialogue by both Syndicus and Mathematicus: a method of calculation that differs from that of the Catholics but that enables regular agreement when it comes to the celebration of Easter.

This is not to say that the Catholics and Protestants then embrace with open arms. "If only," Syndicus remarks, "Pope Gregory had instituted this method of correcting Easter. For then we would have followed his correction without contradiction."[73] "I don't believe it at all," counters Cancellarius. "For you have too many quarrelsome people among you."[74] The two sides continue to argue for their own version of the calendar, though without excessive hostility. Finally, as the dialogue winds down, Mathematicus extends his "blessing" and reminds everybody that "mathematicians are not at fault in the calendar debate, but rather each party uses its mathematicians to achieve what it wants."[75] With this in mind, he graciously offers the others "my services for the Christian resolution of the calendar, in part or in whole, as much as is possible according to my art, so that with this matter settled a beginning is made to settle the other impending Christian conflicts and henceforth to live with one another peacefully and amicably."[76] Here he emphasizes once again that though mathematicians ought not to employ theological or political arguments or to operate on the basis of specific theological or political motivations, their arguments can still have important theological or political *effects*. For mathematics—a "peaceful art," as he noted earlier—can be employed to help settle theological or political disputes like the issue of calendar reform, presenting a solution that would be acceptable to all parties and allowing Christians to live "peacefully and amicably." And with a final "Amen" on the part of all, the dialogue comes to a close.

The calendar dialogue, though whimsical in character, clearly occupied Kepler seriously, as he returned to it and made corrections over a number of years. His position in the dialogue is also quite different from the one he articulated ten years earlier in his letter to Maestlin. On the one hand, this change was tied to Kepler's growing frustration with the calendar's main explicator, Christoph Clavius. In 1603 Clavius had published his *Romani Calendarii a Gregorio XIII Restituti Explicatio*, a massive text in which he sought both to better explain the new calendar and to defend it from attacks by its Protestant opponents. Kepler read the book and wrote to Maestlin in 1607 that Clavius's approach had "filled [him] with anger." Though he had once urged Maestlin not to overturn the Gregorian calendar, he wrote, he had since himself recommended a separate calendar, "for the sake of a defense against the errors of Clavius." Though he had no intention of "overthrow[ing] the established customs of nations," he also felt that it would be wrong to "endure the corrupting mathematics of Clavius." In particular, he believed that it was necessary to clarify the errors in the Gregorian calendar and to offer suggestions for alternatives so that the Gregorian calendar could eventually be updated and improved, though this would likely take place at some time in the future. Though Pope Gregory had called the new calendar "perpetual," Kepler believed that with regard to particular questions, like the relationship between the lunar and solar calendars or the omission of the leap days, "the pope himself reserved the judgment to posterity."[77]

He elaborated on his problems with Clavius's approach to the calendar in a short, unpublished document, labeled "Explicationis Calendarij a Clavio Scripta Emaculatio." These relate in particular to the manner in which the date of Easter was calculated, the unclear designation of the equinox, the wrongly assumed length of the synodic month and the tropical year, and the lack of sufficient reasons given for the rejection of the golden number. To these Kepler added two additional objections. The first relates to the problematic labeling of the calendar as "perpetual." Even supposing that in future years the astronomical knowledge of Kepler's time was still accepted as true, he wrote, there were still many things that the bull did not—and indeed could not—anticipate, making the "perpetual" label clearly problematic. Clavius certainly knew this, and yet he deliberately wrote the bull "very confusedly" so that it would appear flattering to the pope, honored as the founder of a calendar so good it would last for all time. Clavius, that is, "was obliged to remain silent [about the problems with the calendar] because of his office and in this way garnered a reward."[78] Similarly, Kepler complained that in relating the reasons why the new calendar did not follow the true motions, Clavius had omitted the real reason: the authority of the church. Clavius pretended, that is, that the new calendar was established on the basis of mathematics alone, when it was clear that some of the key decisions at play were influenced by extramathematical considerations.

Much as he had argued in the calendar dialogue, then, Kepler found Clavius's involvement in the calendar problematic not merely because of specific mathematical and astronomical errors he had introduced but also because Clavius was dishonest about his true motivations and, more importantly, because those motivations had pushed him toward one side alone and made him hostile toward the other. And his dissatisfaction with Clavius's approach seems to be at least partially responsible for the change in his attitude between 1597 and 1604. At the same time, Kepler's new attitude toward the calendar reform was likely also closely linked with the changes in his life that accompanied it: his expulsion from Graz, his growing consciousness of the intensity of the confessional dispute, his new position as imperial mathematician, and the political awareness that came with it as he watched the emperor struggle to rule a divided empire. Unlike the younger, more naïve Kepler of 1597, the older Kepler of 1604 had come to realize that the calendar dispute could not be solved simply by noting the superiority of the Gregorian calendar; the debate was too entrenched and too tied to the larger confessional conflict for this tactic to be successful. With time, Kepler had come to realize that a more creative approach to the problem was warranted, one that took seriously the hostility and inflexibility on both sides of the issue.

Yet though Kepler's approach to the calendar was influenced by his political connections, it still differed in important ways from those of Emperors Rudolf and Matthias, as is apparent from the official (nondialogic) proposals he eventually drafted and promulgated on their behalf. Specifically, both emperors saw the removal of ten days from the calendar as essential for the preservation of unity in the empire, whereas Kepler retained the ten days in his dialogue, as had been his preference since his 1597 correspondence with Maestlin. The difference between Kepler's own solution to the calendar and the desires of Rudolf, his employer at the time, may explain Kepler's decision not to publish his calendar dialogue—and, indeed, not to show it to Rudolf at all. Similarly, Kepler's decision to keep the dialogue private may have been tied to his insistence, in the dialogue, that mathematicians not act as "servants" to church or state—for the emperor likely saw him, the imperial mathematician, as just that. Kepler's own view of the purity of mathematics was at odds, to some extent, with the political demands of his official position. This is not to say that Kepler felt compelled to betray the principles he articulated in the dialogue in his official position, but only that he may have recognized that it was not politically expedient to declare publicly that he was a servant to no one.

In many ways, ultimately, Kepler's dialogue was a private exercise in the fashioning of both self and community (one that he repeated in other formats and contexts more publicly, as we've seen throughout this book). When it came to self-fashioning, Kepler used his dialogue to focus on his identity as a mathematician. In so doing, he joined an ongoing conversation about the status of astronomers both within and without the university. Astronomers in the sixteenth and seventeenth centuries had begun to posit a new relationship between mathematics and natural philosophy, one that granted the astronomer the right to both calculate celestial motions *and* demonstrate the causes of those motions.[79] Both Kepler and Copernicus before him were central figures in the attempt to argue for a physical astronomy that could make claims about the real world and penetrate into the true causes of things. In so doing, they argued that the astronomer was a philosopher—indeed, because the astronomer mathematically explicated the perfect works of God, he was the philosopher most able to speak with certainty about the world around him. The move from the university to the court was one way in which some astronomers bolstered their status, legitimated their authority, and reformulated their disciplinary roles. Though Kepler was wary of too close a linkage between politics and mathematics, he used his 1604 dialogue to promote a similar goal. In it, he emphasized the centrality of his Mathematicus to the resolution of the politically charged calendar dispute. By this means, Kepler portrayed the mathematician as a figure with unique insight into the world around him, one whose skills were a crucial resource for all those struggling to resolve seemingly intractable conflicts. He thereby recast mathematical astronomy, portraying it not as a technical discipline of calculation and prediction but, rather, as Copernicus had once claimed, as "the summit of the liberal arts," a discipline "most worthy of a free man."[80]

Alongside this reformulation of the astronomer's role, Copernicus had argued that the distinctive expertise of mathematicians and the high status of their discipline meant that only other mathematicians were qualified to evaluate each other's claims. "Mathematics is written for mathematicians," he famously declared. Copernicus therefore discounted the relevance of any critiques of astronomical claims coming from those without mathematical expertise—in particular, from theologians. He wrote: "Perhaps there will be babblers who claim to be judges of mathematics although completely ignorant of the subject and, badly distorting some passage of Scripture to their purpose, will dare to find fault with my understanding and censure it. I disregard them even to the extent of despising their criticism as unfounded."[81] Kepler's arguments in the 1604 dialogue echoed those of Copernicus. In emphasizing the fact that mathematicians ought not to employ theological or political arguments, Kepler argued at the same time that theologians and politicians ought not to attempt to comment on mathematical claims and should instead "leave the mathematicians to themselves." Much of the authority of the mathematician as mediator and conciliator stemmed, that is, not only from the mathematician's unique insight into the world around him but also from his privileged position as expert—a position that served to shield him from the criticism of all those but other mathematicians, who alone would be able to properly evaluate his claims.

This links Kepler's dialogue to another, larger narrative within the history of science: the rise of the expert as an apolitical actor. In examining the emergence of the early modern notion of expertise, Eric Ash notes that one criterion by which the expert came to be identified was his perceived objectivity and independence: "an expert was ideally one with a better understanding and a superior position from which to give advice in part because he had no personal stake in the work itself, being above it all."[82] While Ash applies this idea to a variety of disciplines, Theodore Porter's *Trust in Numbers*, as we saw at the start of this chapter, tells a later story of the process by which objectivity came to be specifically invested in numbers and quantification. The reliance on impersonal numbers, Porter argues, created a sense of objectivity that bridged gaps of distance and distrust in the nineteenth and twentieth centuries; in this way, Porter posits quantification in part as a political solution to a political problem. In Porter's narrative, mathematics is able to function as a political solution only by claiming to occupy an apolitical zone; the supposed neutrality of numbers allows them to be mobilized as a political tool. Kepler, as I've argued here, was not claiming that the mathematician was *neutral* but rather that his perspective was broad, fair, and tolerant. Yet like Porter's actors, it was numbers that granted their users this unique perspective, one that steered clear of religious or political hostility. The mathematician was impartial in the older, multiperspectival sense; his mathematical skills allowed him to see multiple possible solutions to the pressing problems of the day. Crucially, though, those solutions were *mathematical* ones— through the flexibility of numbers alone, Kepler's mathematician arrives at a solution with political and religious significance. Kepler's Mathematicus, like Porter's actors, is simultaneously apolitical and political; he is clearly himself a political actor who solves a very political problem, yet he does so by emphasizing that mathematicians can answer only mathematical questions and can be evaluated only by other mathematicians.

Alongside Kepler's self-fashioning, I've argued that the calendar dialogue shows us Kepler's fashioning of the community, by offering a model for the kind of communication that might successfully resolve the calendar debates. More broadly, it offered a model for interconfessional communication and collaboration and, indeed, for the possibility of large-scale interconfessional reconciliation. Through the voice of Mathematicus, Kepler emphasized that mathematicians could operate without hostility or unwarranted bias and could use their skills to support any number of ends. He likewise highlighted the fact that mathematics was a discipline devoted to harmony and peace and opposed to conflict of any sort. Both these claims allow the two parties in the dialogue to accept Mathematicus as a mediator, and ultimately, it is only with his assistance that they come to any resolution at all. In this way, Kepler suggested that if mathematicians on both the Catholic and Protestant sides could cultivate a similar persona, they could act as crucial mediators in the confessional disputes more broadly and, through their art, provide models of harmony for others to follow.[83]

Though Kepler emphasized in the calendar dialogue that mathematicians needed to steer clear of religious or political motivations and arguments, he clearly believed that pure mathematics was still politically and religiously relevant and could be marshaled to settle political or religious disputes. Yet for mathematicians to successfully function as models and mediators, and to pave the way for a new reformation of the church, they needed to make it clear that they stood above the theological conflicts surrounding them and were motivated by the desire to unite rather than to divide. He later cited Proclus, who likewise argued that "those features which to the uninitiated in the truth of divine matters seem difficult to grasp and lofty are by mathematical reasoning shown to be trustworthy, manifest, and uncontroversial."[84] God was a geometer, after all, and the harmony he had instilled throughout the cosmos was a geometric one. Mathematicians were therefore ideally suited to broadcast the message of harmony to a war-torn world and to provide models of ordered mathematical reasoning that could be applied to other, more disorderly spheres.

Despite Kepler's best efforts, the calendar debate in the empire remained unresolved in his lifetime. In fact, it was only in 1699 that the Protestant half of the empire adopted a new calendar—though not the Gregorian one. Popularly known as the *Verbesserte* (improved) calendar, the calendar eventually endorsed at the Regensburg Reichstag in September 1699 followed the spirit of the model Kepler had advanced on behalf of Rudolf and Matthias years earlier. Ten days were removed following February 18, 1700, and the rule for leap days followed the Gregorian calendar; however, the method for the calculation of Easter differed from the Gregorian calendar, as it relied directly on astronomical observations and calculations for both the equinox and the full moon.[85] Additionally, it relied on Kepler's *Rudolfine Tables* rather than the older *Alfonsine* or *Prutenic Tables*. In this way, Kepler was able to play a central role, though not the one he had hoped, in Germany's adoption of an updated calendar.

Conclusion: Dialogue and Community

In closing, I want to return to dialogue, as both genre and practice. At the start of this chapter, I emphasized that impartiality for Kepler did not suggest detachment, understood as neutrality. Here, I want to emphasize that it likewise did not suggest detachment understood as isolation. If, as I've also suggested, Kepler's Mathematicus might stand at the start of a trajectory that ends with the modern notion of expertise, he does not do so unproblematically, for he does not embody the kind of expertise that defines itself *against* the larger public. Though Kepler may have gestured toward Copernicus's dictum that mathematicians were beholden only to other mathematicians, in the end the knowledge of Kepler's Mathematicus was produced socially, in conversation with nonexperts, rather than alone in hermetic seclusion or in consultation with other mathematical experts. The social production of knowledge and the centrality of conversation, debate, and dialogue were important elements of Kepler's larger message; he did not select the dialogue genre for his 1604 work on the calendar merely on a whim.

Dialogues have always been, after all, enactments of both conversation and community. They have been used to imagine shared social and cultural spaces that might create a kind of solidarity, if not equality, among mutual participants in a conversation. And, as Jean-François Vallée has noted, they have long been linked to the topos of friendship itself. Vallée points out that many of the central works in antiquity on friendship—Plato's *Lysis*, Cicero's *De amicitia*, Lucian of Samosata's *Toxaris*—were written as dialogues, and through the interactions of their characters, they "illustrate performatively . . . the intellectual and/or ethical theses developed on friendship within the utterances of the characters."[86] The dialogue genre, so popular in antiquity, rose to new prominence once again in the Renaissance, and the linkage between dialogue, friendship, and community remained strong. Humanists produced a great profusion of dialogic writings, and at the same time they envisioned their own community—the epistolary community of the Republic of Letters—as a dialogue of another sort. In his discussion of the community of Christians in the postclassical age, Richard Lim has described letters as "post-diasporic dialogues,"[87] and this is very much how the members of the Republic of Letters—spread across Europe, and further still—understood their own epistolary networks and communications. Erasmus, for example, described the letter as "a kind of mutual exchange of speech between absent friends."[88] Letters and dialogues were ultimately *both* the signature forms of the Republic of Letters, forms that embodied friendship and polite discourse. The letters that Kepler and his correspondents shared, along with the dialogues that they created, gestured to a communal, public space; the former created that space, and the latter modeled it.

In this way, when Kepler used his calendar dialogue to model a solution to the problems facing Christendom, that solution ultimately looked a great deal like the Republic of Letters itself, a community in which everybody was presumed equal, and which, though it encompassed many allegiances and perspectives, was still fundamentally a harmonious unity.[89] Of course, this characterization of the Republic was often an ideal rather than a reality; the scholarly communities of early modernity were not immune to bitter wrangling and dispute.[90] But there is value in the articulation of an ideal, even if it is not always realized in practice. In this sense, Kepler's dialogue aligns quite closely with another genre that had long been allied with that of the dialogue—the utopia. Nina Chordas and others have emphasized this connection between dialogues and utopias and argued that it rests, in part, on the "problematic standing of both dialogue and utopia in relation to fiction."[91] Both dialogues and utopias created alternative worlds and modeled social orders; they commented on reality while offering an alternate reality in its place. As in Kepler's dialogue, the worlds and orders they imagined might be at once criticisms of existing social relations and suggestions for newer, better ways of organizing the community.

Both dialogues and utopias were artistic inventions, yet they offered the *pretense* of reality; they were fictionally real. As Chordas notes, Plato himself played on this tension between the fictive and the real by using the very medium of the dialogue to banish poets from his imagined republic.[92] Kepler—like Plato, like More, like Campanella—allowed us to overhear a conversation in which a new kind of world was imagined. If, as we saw in the previous chapter, Kepler thought about creating a utopia of his own with the *Somnium*, he may not have ultimately abandoned that dream. If he did not create it with a journey to the moon, he may rather have done so in a more mundane setting: at a table around which some theologians, politicians, and a mathematician solve one problem, the problem of the calendar, and suggest a way in which the world of Christendom might be imagined anew and in which the wounds of the body politic and the body of Christendom might be healed.

Some have argued that the appeal of the dialogue genre may lie in its ambiguity, an ambiguity that could enable an evasion of responsibility and a lack of commitment.[93] The multiplicity of voices, in this view, might allow authors to suggest controversial ideas without standing behind them; just because one character in a dialogue represented a particular viewpoint, it did not mean that the author of the dialogue *him-*

self agreed with it. For Kepler, this was not the point at all. Rather than evading responsibility, Kepler used his dialogue to embrace a different kind of responsibility, and rather than refusing to commit to an ideal, Kepler used his dialogue to suggest a different kind of ideal. The point was not to hide one's particular viewpoint behind an imagined character and then to put other perspectives in other mouths, as tricks or distractions. The point was rather to show that *many* perspectives were important, and that inclusivity itself was an ideal worth embracing.

Montaigne understood this; he wrote that when Plato used the dialogue genre, he did so "to put more fittingly into diverse mouths the diversity and variation of his own ideas."[94] Montaigne's *Essays* may themselves be read as a kind of imagined dialogue, one that Montaigne used to replace the very real dialogue he had enjoyed with his lost friend, La Boétie. That dialogue is at once with himself, with his contemporaries, and with the ancient authors whose voices fill the pages of his book, and it is one that rejects dogmatism and inflexibility in favor of multiplicity and changeability.[95] In writing about Dostoevsky, Mikhail Bakhtin likewise offered a new synonym for "dialogue," or "dialogism": "polyphony." "The polyphonic novel is dialogic through and through," he wrote.[96] Dialogic novels, according to Bakhtin, included "a plurality of independent and unmerged voices and consciousness, a genuine polyphony of fully valid voices."[97] Though Bakhtin focused on the literature of modernity, his words have remarkable resonance when applied to Kepler. Bakhtin, in fact, argued that Dostoevsky carried out "a small Copernican revolution."[98] In his view, the monologic novel was different from the dialogic in the same way that the geocentric cosmos was different from the heliocentric; the latter two were decentered, in constant motion, and multivoiced. Kepler's dialogue, his cosmology, and his quest for harmony were all characterized by this polyphony, by multivoicedness. The polyphonic author, like the polyphonic God, creates, according to Bakhtin, "not voiceless slaves (as does Zeus) but *free* people capable of standing *alongside* their creator, capable of not agreeing with him and even of rebelling against him."[99] Kepler's dialogue was polyphonic *and* impartial in this sense—it offered the view from everywhere, the true God's-eye view.

Perspective, Perception, and Pluralism

Kepler's vision of a better world, I've argued, rested on harmony. He believed that heavenly harmony should serve as a blueprint for earthly harmony, particularly given the increasing confessional and political dissonance around him. He understood harmony to be active and changeable, as were the planets in their cosmic symphony; like their harmonies, earthly harmony too should be multivoiced and required difference—and even discord—to give it life. Many of the themes I've explored in the previous chapters—polyphony, tolerance, accommodation, diversity, and dialogue among them—were all, in some sense, coextensive with harmony as Kepler understood it. A harmonious community, like the harmonious cosmos, was one that embraced many perspectives rather than just one.

It is with this idea of perspective and point of view that I'd like to conclude. In Kepler's engagement with optics and perspective via the camera obscura and in his broader considerations of cosmological and historical perspective, he explored the difficult relationship between the true and the deceptive and between the particular and the universal. My forays here will be snapshots rather than sustained and detailed analyses, but they will help to answer some questions that linger behind Kepler's discussions of earthly harmony. In the ideal harmonious world that Kepler envisioned, what is the balance between tolerance and conviction and between pluralism and unity? Must we sacrifice truth on the altar of harmony? Is discord simply inevitable, or is it desirable?

Kepler and the Camera Obscura

Essentially a pinhole camera outfitted with lenses and mirrors, the camera obscura, or "dark chamber," projected a reversed image onto a flat screen. Used across a wide array of disciplines and domains in early modernity (astronomy, natural magic, painting, mapmaking, etc.), it was also a potent metaphor and model for human vision and human personhood—indeed, Jonathan Crary argues that during the seventeenth and eighteenth centuries it was "without question the most widely used model."[1]

There were various ways that a camera obscura could be constructed and used; Kepler's own camera obscura, for example, was a tent camera that projected an image onto a piece of paper, which allowed him to create panoramic images and to view astronomical events like sunspots and solar eclipses.[2] More importantly, though, as an object that projected a reversed image, in his optical studies Kepler used the camera obscura as an analogue to the eye itself. Here, he followed Della Porta, who had made just this analogy in his *Natural Magic* and had hoped that "ingenious people will be much delighted in this."[3] Kepler acknowledged himself "very pleased"[4] and noted the similarity between the various components of the eye and the camera: "The pupil

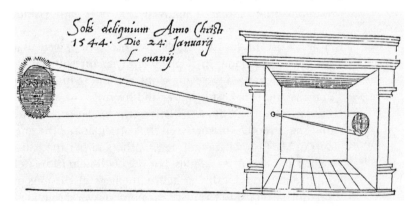

C.1   Picture of a camera obscura from Gemma Frisius's *De Radio Astronomica et Geometrica* (1545)

plays the role of a window and the crystalline behind plays the role of a screen," he wrote.[5] Kepler's use of the camera obscura metaphor has played a central role in debates over his theory of vision and the extent to which he represented the culmination of the medieval perspectivist tradition or a break from that tradition.[6] More interesting for our discussion is another question, however: just what did this analogy suggest about the nature of the knowledge obtained by both apparatuses, the natural eye and the artificial camera?

Crary has argued that the camera obscura functioned as a metaphor for objectivity in its early modern invocations.[7] Indeed, this is how many have understood Kepler's own linkage between the camera and the eye, operating on the assumption that to mechanize means to objectify and to purify. T. Kaori Kitao thus credits the "Keplerian eye" with the birth of the myth of objectivity, of vision "purged of the subjectivity inherent in the physiological and psychological process of perception."[8] According to Kitao, Kepler's theory of vision, centered on the camera obscura, "carried on the isolation of science from human concerns when he eliminated perception from vision in his optics."[9] Yet such arguments fundamentally misunderstand the linkage Kepler highlighted between the camera and the eye. The fact that the eye was analogous to the mechanical camera obscura meant, not that its observations were therefore *less* problematic, but rather that they were *more* so. Kepler's examination of the camera obscura led him to recognize the significant errors produced by the device, errors that he suggested were present in the natural process of vision as well. In her analysis of the relationship between art and vision in the seventeenth century, Svetlana Alpers understood this; for her, Kepler represented someone who recognized that any lens, whether natural or artificial, necessarily distorts.[10]

As we saw in chapter 4, the early telescopic discoveries were received with ambivalence, and at times even with hostility; Kepler, in his role as astronomer, was thus invested in justifying the images obtained by artificial devices like the telescope and the camera obscura. As Raz Chen-Morris and Ofer Gal argue, his justification ran precisely contrary to that of some earlier thinkers, who justified observations gleaned via the camera obscura by arguing that it was a neutral object and that, as such, it did not interrupt the natural process of vision, the "seamless flow of forms through the eye to the intellect."[11] The instrument was safe, in other words, so long as it did not interfere with the trustworthy eye. Kepler, by contrast, argued that the instrument did change things, but so did the eye; both worked the same way, and the observations of the telescope or the camera obscura were as reliable as those of the eye, for both were equally mediated. As Chen-Morris and Gal note, this equivalence—and this manner of bolstering the authority of the instrument—came "at a steep epistemological price: if the instrument is not prone to error more than the eye, it is immediately implied that the eye is as vulnerable to error as the instrument."[12] Kepler thus asserted in his *Optics* that "it has been demonstrated most clearly, from the very structure of vision, that it frequently happens that an error befalls the sense of vision."[13] Mechanizing the eye led, not to its heightened trustworthiness or reliability, but rather to a renewed sense of the deceptiveness and unreliability of sense perceptions.[14]

Did this mean that such perceptions were to be rejected? No, for perfect loyalty to the things themselves was not the goal for Kepler. It was the archetypal, harmonic structures *beneath* the things that were most important; the senses were important for the hints they provided of what really mattered. Further, understanding the process of vision, the nature of light, and the distortions inherent in the lens allowed the viewer to accept observations for what they were—constructs, yes, but constructs that were collective, and distortions that joined all viewers together, along with their instruments. Sharing those observations, combining visual reports and telescopic discoveries, reproducing experiments, dissecting the visual process itself—all this led to a perceptual space that was reliable precisely in its commonality.[15]

The very process by which the camera obscura worked—and by extension the eye as well—modeled this construction of a collective, shared, and thereby reliable space. Jean Pelerin, in the very first printed account of linear perspectival construction, had linked perspective to the idea of harmony when he explained that it made multiple points of view "concordable."[16] In his description of the use of the camera obscura in the *Optics*, Kepler likewise linked perspective to harmony, understood as the successful combination of difference and of the creation of unity in diversity. To create an image with the camera obscura, Kepler recognized, he needed to set up a procedure and then repeat it over and over, from different positions and points of view.[17] Only the overlaying of each figure from each position created the ultimate, reliable image. The image consisted "of shapes that are potentially infinite, . . . mutually overlapping."[18] Observation, whether by the eye or by the camera obscura, was situated, flawed, and perspectival, in Kepler's view; it was a construct, but one that was still subject to geometrical analysis. Any observation was a *pictura*, painted either on the eye, the canvas, or the camera's screen.[19] It needed to be interpreted and understood both in the context of its own production and in relation to other forms of knowledge.

Erwin Panofsky famously wrote that perspective "subjects the artistic phenomenon to stable and even mathematically exact rules, but on the other hand, makes that phenomenon contingent upon human beings, indeed upon the individual."[20] To Kepler, likewise, perspective seemed to straddle an important divide; the camera revealed that artificial observations and natural observations were very much the same. Observation was objective and subjective, universal and relative, trustworthy and deceptive at the very same time. Our best hope of emphasizing what was right, rather than what was wrong, lay in technologies that overlaid points of view one on top of the other and created a shared, common space, and in communal structures that did the same.

Kepler and Cosmological Perspective

If the sense of vision was prone to error, this was particularly so in cosmological contexts, when motion was involved and the observer was confined to his position on earth. Because "the sense of vision is in error about the movable," Kepler recognized, it was incredibly difficult to judge between the heliocentric and geocentric theories on the basis of observations alone, particularly as there was no way "to carry us across to the moon or to another of the wandering stars."[21] Kepler, of course, did provide many of his own justifications for the Copernican system, which he viewed as the real and true description of the cosmic order; he did not, like many astronomers of his day, believe that both systems were equally true or, rather, that neither system made any physical truth claims at all.

Yet things are not quite as simple as this would imply, even for Kepler. If the Copernican cosmos was the true description of the motions of the planets, this was not the only kind of truth that mattered. Perception mattered too, and perceptions were based on the earth, and on the humans who inhabited it. Kepler argued that if God had seen fit to create human beings in his image on this particular planet—indeed, to give his own Son flesh on this particular planet—it must mean that the perspective afforded from this point in the cosmos was particularly relevant, and particularly true. It was that perspective that God had used in order to adjust the proportions of the celestial bodies to one another; as Kepler wrote in *The Epitome of Copernican Astronomy*, since the earth was "the home of the contemplative creature," and since "contemplation originates from the vision of the stars," then God determined "the beginning in proportioning the bodies of the world on the basis of the vision of the sun from the earth."[22]

Kepler further claimed not merely that the planets circled the sun but also that they produced a harmonic symphony in their musical ratios. Since harmony made no sense without someone to perceive it, the perceptual focus of the cosmos was actually the earth rather than the sun. In fact, the harmonies could not exist without the earth or, more specifically, without human beings who could appreciate them. As Kepler wrote: "for some sensible harmony to exist, and for its essence to be possible, there must be in addition to two sensible terms a soul as well which compares them. For if that is taken away, there will indeed be two terms, which are sensible things, but they will not be a single harmony, which is a thing of reason."[23] To provide for this rational soul, and thus to allow the harmonies to exist, "the Creator deemed it sufficient to shape souls, which control earthly creatures, in such a way that they expected and observed and noticed the harmonies, all round the circle. . . . That is the case with the apparent motions of the planets beneath the zodiac, as seen from the earth."[24] Additionally, without our earth-centered perspective, not only the harmonies but also all of astronomy would not exist: "If the earth, our home, did not measure out its annual circuit in the midst of the other spheres, changing place for place, position for position," Kepler argued, "human reasoning would never struggle to the absolutely true distances of the planets, and to the other things which depend on them, and would never establish astronomy."[25] The preeminence Kepler afforded to perception is evidenced by his rejection of the possibility of an infinite cosmos, a thought that filled Kepler with "a hidden horror; one finds oneself wandering in this immensity, in which any boundaries, any center, and therefore any fixed places are negated."[26] As Judith Field notes, he opposed the idea not merely because it undermined his articulation of the structured, archetypal, and harmonic cosmos but also because of "the limitations it would place upon the value of observation."[27] An unobservable universe was inconceivable to Kepler; the universe, like the harmonies, needed to be *perceived* to truly exist.

In his discussion of the relationship between knowledge and imagination in Kepler's thought, Guy Claessens thus argues that while there is a clear geometric or cosmological truth for Kepler, there are also other kinds of truths that are equally relevant. The appearance can matter as much as the reality behind that appearance. "For Kepler," Claessens writes, "the legitimacy of perception—and of the imaginary construction based on it—is not subject to discussion. . . . If the sun appears to revolve around the earth, this perception is relevant as well."[28] Kepler himself made precisely this point in his discussion of the biblical story wherein Joshua stops the sun in its passage across the sky. Though this was clearly not a description of the astronomical truth, it did not mean the story was false, for, in Kepler's words, "the perception of the eyes has its truth."[29] The question was how to interpret that truth and how to situate it alongside other, different kinds of truths.

To elucidate the possibility of different truths, and different kinds of centers, Kepler turned to the metaphor of the human body. The body was a unified system, yet its parts operated independently, with different functions geared toward different ends. The physicians knew this: "However much they may uphold the unity of the soul in man, they identify its diverse faculties according to the diversity of organs, the heart, the liver, and the brain, and even their faculties present in the particular organs. This is clearly also the case in the world, as the sun is a sort of heart and the earth thus plays the part of the liver or spleen."[30] The sun might be the absolute center from an astronomical perspective, in other words, but it was not the *only* center. In *The Human Condition*, Hannah Arendt associates both the Ptolemaic and the Copernican cosmological stances with the idea that there is only one center to a universal system, the central question being which it was. For Arendt, what distinguishes modernity is the move beyond this, to a point where "we no longer feel bound even to the sun, that we move freely in the universe, choosing our point of reference wherever it may be convenient for a specific purpose."[31] For Arendt, then, the modern observer is unbound; there is no center but the one freely chosen. For Kepler, neither this nor the earlier position Arendt outlines are quite accurate; there is neither one universal center nor none at all, but rather two, equally valid depending on the question asked. We are not unbound but twice bound.

Yet this, too, is not the end of the story. For as Robert Westman points out, one of Kepler's favorite moves was "the reversal of perspective, the logic of which was intended to break down the necessity of one and only one point of view."[32] Kepler makes this move most clearly in his *Somnium*. In his own footnotes to the book, Kepler described what he deemed its thesis: "Those things which are to us particular aspects of the whole world—the twelve celestial signs, solstices, equinoxes, tropical years, sidereal years, equator, colures, tropics, arctic circles, and celestial poles—are all tied to the very small earthly sphere and depend only on the imagination of earth's inhabitants. Thus, if we transfer the imagination to another sphere, everything must be understood to be changed."[33] The main thesis of the *Somnium*, in other words, was that much of what we take for granted as natural and necessary was actually only happenstance, or a result of our personal and limited perspectives. By moving from the earth to the moon, Kepler forced his narrator, and his readers, to see the world differently, to challenge their preconceptions, and to recognize the inherent subjectivity of much of their situation.[34] To do this, he structured the entire story as an exercise in distancing and perspective shifting. The story was framed via a dream narrative which allowed Kepler to adopt the perspective of a young Icelandic boy. Kepler explained, once again in his footnotes, why he believed this framing was necessary: "In this remote island," he wrote, "I looked about for a place to sleep and dream, in order to imitate the philosophers in this style of writing. For Cicero, about to dream, crossed into Africa, and Plato, in the same Western Ocean, fashioned Atlantis. . . . Finally, Plutarch, in his little book *The Face in the Moon*, after much discussion wanders into the American Ocean and describes to us the sort of island that a modern geographer would probably apply to the Azores, Greenland, and the territory of Labrador, regions around Iceland."[35] Distance of all kinds was crucial to the perspective shifting that Kepler's larger narrative embraced.

The reason for this perspective shifting was in large part cosmological: Kepler hoped to demonstrate that the seeming centrality of the earth was merely a question of our particular terrestrial perspective. To those who cried that the earth couldn't possibly be moving because it didn't *appear* to be moving, Kepler insisted that "to those who are on the moon, it does not appear to revolve but is considered motionless, just as our earth seems motionless to us."[36] Kepler could—and did—argue for the motion of the earth on other grounds as well, but here he relied on the importance of perception, perspective, and imagination. If it was obvious on the basis of sense perception that the heavenly bodies revolved around the earth, to the Lunarians it was equally obvious that the earth itself moved. "If others assert that the Lunatic senses of my Lunarian people are deceived, with equal right I retort that the terrestrial senses of our earthly inhabitants are lacking in reason."[37] To make this point all the clearer, Kepler painted a picture of the moon that was, in all its particulars, an inverted picture of the earth. As Chen-Morris incisively argues, the *Somnium* functions as a kind of camera obscura, "where the daemon's fantastic lunar astronomy is an inverted picture of the terrestrial world view," and the notes below the text "supply the reader with the inversion principles, thus revealing the real structure of the physical reality of the heavens."[38]

Timothy Reiss likewise urges us not to forget the literary nature of the text. The dream is not just a dream but a book; Kepler annotated that book, and the dream only begins when Kepler, as narrator, falls asleep while reading and is transported to the world of Duracotus. The narrator emphasizes the Frankfurt Book Fair at both the beginning and the end of the dream, and the differences between the text and its footnotes, and the story within the story, allow us to form multiple possible readings and interpretations of the whole.[39] Reiss likewise emphasizes that this is not the only place where Kepler's name was linked to the moon; when Kepler is mentioned in Galileo's *Dialogue on the Two World Systems*, it is to disparage his attribution of the tides to the power of the moon. As Salviati proclaims in the text, "I am more astonished at Kepler, than at any other. Despite his open and acute mind, and though he has at his fingertips the motions attributed to the earth, he has nevertheless lent his ear and his assent to the moon's dominion over the water, to occult properties, and to such puerilities."[40] Reiss notes that, of course, Galileo was wrong, and that though the linkage between the oceans and the moon might have been popular superstition, it was also quite true. He links this to the fantastical claims of the *Somnium* and concludes that "the perhaps not altogether untimely lesson of the *Somnium* is that although we may refer to 'fact' and 'superstition' as to two mutually exclusive classes of discourse, they are not so much 'opposites' as complementary, different from one another in their constructs of the same."[41]

In other words, the real lesson is that perspective matters, that perspective shifting can yield tremendous insight, and that different perspectives can yield different kinds of truths. Imagination was both a scientific and a political tool; it could foster astronomical insight and allow those limited, terrestrial inhabitants to imagine alternative worlds, both cosmic and much closer to home. Arendt wonders about the possibility that "this mathematically preconceived world may be a dream world, where every dreamed vision man himself produces has the character of reality only as long as the dream lasts."[42] Kepler embraced the dream, offered up another one, and encouraged his readers to dream as broadly as possible in order to realize harmony in all its possible permutations.

Kepler and Architectural/Historical Perspective

Perspective and perspectival decentering were thus equally relevant to Kepler's portrayal of vision and his conception of cosmology. They were relevant as well to his engagement with the astronomical realm by route of an entirely different discipline—architecture. In his very first book, Kepler had invoked architecture as a way of understanding God's relationship to the cosmos. He had written there that "just like a human architect, God has approached the foundation of the world according to order and rule and so measured out everything that one might suppose that architecture did not take Nature as a model but rather that God had looked upon the manner of building of the coming human."[43] In linking the cosmic to the architectural, Kepler was following in the ancient tradition of harmony, understood (as we saw in the introduction to this book) in both musical and organic terms. Both musical and organic harmony had long been understood to apply not merely to the works of God but also to the architectural works of man: on the one hand, through the ancient Vitruvian linkage between architecture and bodies and, on the other, through the Pythagorean linkage between architecture and music. Leon Battista Alberti, in the fifteenth century, brought all these tropes together when he invoked Pythagoras to argue that architects needed to derive their harmonic methods "from musicians, who have already examined such numbers thoroughly," and likewise invoked "the great experts of antiquity" to argue that "a building is very like an animal, and that Nature must be imitated when we delineate it."[44]

Alberti elaborated at length upon the specific implications of architectural harmony. One lesson he emphasized was that as there were different kinds of voices in music, and different kinds of bodies in nature, so too there should be different kinds of buildings. He explained that

I would not wish all the members to have the same shape and size, so that there is no difference between them. . . . Variety is always a most pleasing spice. . . . Just as in music, where deep voices answer high ones, and intermediate ones are pitched between them, so they ring out a harmony. . . . [S]o it happens in everything else that serves to enchant and move the mind. . . . [Just as] some [bodies] are slender, some fat, and others in between . . . [the ancients] concluded that by the same token each [building] should be treated differently.[45]

Alberti invoked the three different architectural orders, in particular, as analogues to the different kinds of bodies represented in nature.[46] Harmony implied variety, in Alberti's view, and this was as true of architectural harmony as of musical or organic harmony.

This did not mean, however, that all variety was harmonious; in fact, quite the opposite was true. The architect needed to be sure that "the mistake is avoided of making the building appear like a monster with uneven shoulders and sides."[47] Variety, according to Alberti, "is a most pleasing spice, where distant objects agree and confirm with one another, but when it causes discord and difference between them, it is extremely disagreeable."[48] Variety could as easily lead to a monster as a harmony, and the architect needed to be careful not to let the latter become the former. Here, too, Alberti followed Vitruvius, who had warned his readers not to create "monsters rather than definite representations taken from definite things."[49] While Alberti allowed for more variety than Vitruvius had, he agreed with his predecessor that monsters needed to be avoided at all costs. "Look at Nature's own works," he wrote. "For if a puppy has an ass's ear on its forehead, or if someone had one huge foot, or one hand vast and the other tiny, he would look deformed."[50]

Still further, Alberti understood the idea of harmony to apply temporally as well. Harmony described not only the way the different parts of a building fit together or related to other buildings but also the extent to which a building reflected the unified product of a single mind, created at a specific moment in time.[51] Medieval buildings, Alberti believed, were not harmonious, because they were too heterogeneous, the product of many creators over too long a time and thus irregular and discontinuous.[52] "Reasoned harmony," according to Alberti, was achieved in a building when "nothing may be added, taken away, or altered but for the worse."[53] This was a comprehensive standard and applied "even to the minutest elements," which needed to be arranged "in their level, alignment, number, shape and appearance, [so] that right matches left, top matches bottom, adjacent matches adjacent, and equal matches equal."[54] Once this was achieved, so that nothing could be added except to the detriment of the building, the harmony was complete.

In an ideal world, this harmonic unity would be achieved by a single architect in one fell swoop. As this was rarely the case, Alberti stressed that architects should act *as though* it was, de-emphasizing temporality in favor of architectural unity. He wrote that "the brevity of human life and the scale of the work ensure that scarcely any large building is ever completed by whoever begins it. While we, the innovative architects who follow, strive by all means to make some alteration, and take pride in it, as a result, something begun well by another is perverted and finished incorrectly. I feel that the original intentions of the author, the product of mature reflection, must be upheld."[55] In this view, the original architect had absolute control over the nature of the building, and departures from his intention were departures from the harmony of the ultimate product. As Marvin Trachtenberg argues, the goal of architectural harmony, according to Alberti, was "to collapse time to an unmoving point, thereby to build *outside time*."[56] Additions and historical accretions that departed from the original plan of the structure would only create a monster.

If this idea of architectural harmony, and the consequent rejection of the monstrous, began in conceptions of natural and cosmic harmony, it migrated back there as well in the writings of Copernicus. Copernicus's rejection of the Ptolemaic cosmos resonated strongly with Alberti's rejection of medieval architecture, for both were rooted in the same Vitruvian ideal.[57] Copernicus, in his preface to the *De Revolutionibus*, described the Ptolemaic cosmos as a monster, for it was "just like someone taking from various places hands, feet, a head, and other pieces, very well depicted, it may be, but not for the representation of a single person; since these fragments would not belong to one another at all, a monster rather than a man would be put together from them."[58] The problem, according to Copernicus, was that astronomers of old had not understood the true "structure of the universe and the true symmetry of its parts."[59] Like a building, the cosmos—designed by "the best and most systematic Artisan of all"[60]—needed to be harmonious, which suggested a kind of unity, or an obvious linkage between all its parts. Copernicus suggested that the cosmic picture of his day was filled with so many historical accretions and complications that it no longer had any unity of form and thus could not be representative of the true world system. His system was clearly to be preferred, for "in this arrangement . . . we discover a marvelous symmetry of the universe and an established harmonious linkage between the motion of the spheres and their size, such as can be found in no other way."[61]

Kepler, as we've already seen, embraced Copernicus's argument for the harmonious simplicity and unity of his system. "These hypotheses of Copernicus's not only do not sin against the Nature of things but greatly assist it," he wrote in the *Mysterium Cosmographicum*. "It loves simplicity; it loves unity. Nothing exists in it that is idle or superfluous, but more often one cause is designed for many effects."[62] Kepler praised Copernicus for having "freed nature from that oppressive and useless paraphernalia of so many immense orbs."[63] Yet if Kepler's description of the cosmos rejected the monstrous and embraced the kind of unity that Alberti had highlighted architecturally, there is another place where Kepler did the opposite and created just the kind of monster that both Copernicus and Alberti had decried: and that, in fact, was architecture. More specifically, it was the architectural frontispiece to the *Rudolfine Tables*.

In the frontispiece, which Kepler himself designed, he depicted the history of astronomy in architectural form. Astronomy is a monopteros, an open temple built on a platform with columns and a dome. It is dedicated to Urania, who sits on top of the dome holding a laurel. The floor of the temple is the foundation of astronomy, a map of the heavens. The dome is supported by ten visible pillars, which progress in sophistication from background to foreground, mimicking the historical progression of astronomical systems. The backmost pillars are bare tree trunks, which do not even reach the ceiling; these represent the ancient, Babylonian history of astronomy and are paired with a Babylonian astronomer without instruments. The next two pillars are crafted from roughly hewn stone, while the next four are brick pillars, cracked and patched. These are paired with early instruments like the armillary sphere and the astrolabe, as well as Hipparchus, who stands next to one of the brick pillars, and Ptolemy, who stands next to another. Finally, in the foreground are two sophisticated marble pillars, accompanied by Copernicus and Tycho Brahe; Copernicus's pillar is Doric, while Tycho's is Corinthian.

Viewed in light of Alberti's descriptions of architecture, Kepler's temple of astronomy is nothing other than a monster. Its parts do not fit with one another; it is the product not of one mind but of many; and it represents the unfolding of history rather than a frozen moment in time. And if it was important for Kepler to argue that the cosmos themselves, created by the divine artisan, were harmonious in the Vitruvian sense, it was equally important for him to emphasize that this was not the case when it came to the very human history of astronomy. Astronomy moved slowly and sometimes laboriously; its progress was made by building on the past, sometimes messily and haphazardly. Astronomical progress did not erase past achievements but acknowledged them; astronomical harmony, understood through the lens of history, was heterogeneous rather than homogeneous. It was monstrous; but this, yet again, was a monster that Kepler could embrace.

C.2   The temple of astronomy from the frontispiece of Kepler's *Rudolfine Tables* (1627)

Indeed, like optical or cosmological perspectives, historical perspective was a way of seeing beyond oneself and of recognizing that no one vantage point offered a privileged and absolute view of the truth. The truth needed to be built, piecemeal and over time, with wood, then stone, then bricks, then marble. Kepler viewed the search for true knowledge as a work in progress, built in large part on the work of the past and dependent on mistakes and imperfections as well. Kepler praised Copernicus not just *despite* his shortcomings but also precisely *because* of them. Copernicus, he wrote, had "no scruple in now and then disregarding or changing hours in time, quarter of degrees in angles, or more . . . and he would seem to deservedly incur censure, if he had not done it deliberately, believing that it was better to have an astronomy that was somewhat imperfect than none at all. . . . To aspire to establish knowledge with the least possible damage, as Copernicus dared, is characteristic of a brave man."[64] Kepler similarly did not represent his own work as perfect or fully complete. "I challenge as many of you as will chance to read this book," he wrote in the *Harmony of the World*: "come, be vigorous and either tear up one of the harmonies which have been everywhere related to each other, change it for another one, and test whether you will come as close to the astronomy laid down in Chapter IV; or else argue rationally whether you can build something better and more appropriate on to the heavenly motions, and overthrow either partly or wholly the arrangement which I have applied. Whatever contributes to the glory of Our Founder and Lord is equally to be permitted to you throughout this my book."[65] It was this give-and-take between past and present—this dialogue, in other words—that Kepler embraced as the truly human harmony, the temple of astronomy in all its glory. If "the path to the truth is long and circuitous,"[66] then the straight path would be the wrong path; multiple perspectives, rather than one, were necessary for the journey to proceed.

Kepler emphasized the importance of perspective not just through the temple of astronomy as a whole but also through the specific images within it. He depicted Copernicus and Tycho at the center in dialogue; Tycho points at the ceiling of the temple, decorated with a depiction of his world system, and asks Copernicus, "Quid si sic?" (What if it were thus?). Copernicus, however, is not convinced and—as Johann Hebenstreit explained in the poem accompanying the image—shakes his head in disagreement.[67] As Stephen Gattei notes, Kepler's image mimics an earlier one by Tycho himself at Uraniborg.[68] In that image, Tycho also asks a group of fellow astronomers, "Quid si sic?" but there it is clear that he is not actually soliciting opinions; he is, rather,

C.3 Detail from the frontispiece of the *Rudolfine Tables*: Copernicus and Tycho debate their theories

in Gattei's words, soliciting compliments.[69] By having Tycho address Copernicus directly and by signaling Copernicus's disagreement, Kepler effectively turned the earlier monologue into a dialogue.

Kepler also highlighted the importance of perspective in other ways in the image. Though the temple has twelve pillars, only ten of them are visible because of the limitations of the visual field. Similarly, though the image depicts six goddesses perched on the dome of the temple, each representing a discipline and instrument that Kepler needed for his reform of astronomy and production of the *Rudolfine Tables*, the poem accompanying the frontispiece imagined—and detailed for the reader—another six, hidden from view. The poem thus functioned as a means to illuminate what lay beyond the reach of the visible. There is always, it suggested, an image lurking in the shadows, and to ignore it is to miss a part of the picture. Finally, Kepler played with the ideas of perspective, history, and truth in his depiction of the base of the temple. He depicted himself on one of the panels there, sitting at a table having just completed his work—a candle is burning, an inkpot rests in front of him, and he has just finished writing some numbers on the tablecloth. At first glance, he seems almost ancillary to the primary scene within the temple; he is not one of the pillars of astronomy but rather its humble servant. Yet as Gattei emphasizes, at second glance it is clear that his role is a far more important one, for in front of him on the table is a replica of the dome of the temple of astronomy itself, the very temple in whose base Kepler's panel rests. Gattei views this as a sign that Kepler sees his own work as the temple's crowning achievement. In his interpretation, "the great astronomers of the past are important supporting actors, but the structure they support is Kepler's own work. His is the dome crowning the edifice. His is the final touch to the picture of our universe and the key contribution to the understanding of its structure."[70] Gattei is certainly right to emphasize Kepler's belief that he had uncovered something fundamental, and something unknown to those who came before him. Yet though Kepler depicted himself as constructing the dome of the building in whose base he toils, he did not believe that the temple of astronomy was finished or that the achievement of its construction was primarily his own. He understood his own place in history, as we've seen, and saw himself as part of a process that had long been unfolding and would continue to do so long after he had exited the stage. Through his self-representation Kepler may well have wanted to highlight the extent to which his system was superior to those that had come before it, but this was not the only message. His image also suggests that human construction and imagination play a role in any act of scientific knowledge, and that we at one and the same time discover and produce the world in which we live. As he emphasized with optics and again with cosmology, one way to transcend the limitations of human perspective was to recognize and embrace them—to realize that since we see through a glass darkly, we must rely on visual and imaginative aids, or the combined perspectives of others.

C.4 Detail from the frontispiece of the *Rudolfine Tables*: Kepler sits at the base of the temple of astronomy

In his various emphases on perspective, Kepler shared much with Gottfried Wilhelm Leibniz; though Kepler looked with foreboding at a world on the brink of war, and Leibniz looked back at the destruction it had inflicted, both highlighted perspective as a proof of and a corrective to the distortions of human knowledge. Both turned to optics to demonstrate, on the one hand, the incompleteness of the single perspective and, on the other, the harmony that could be found in the combination of multiple perspectives. Both turned likewise to architecture and spatial analogies, and both compared the complete God's-eye view with the incomplete human one.[71] One can hear the resonances of Kepler's own optical, cosmological, and architectural examples in Leibniz's claim that "God, by the creation of many minds, willed to bring about with respect to the universe what is willed with respect to a large town by a painter, who wants to display delineations of its various aspects or projections. The painter does on canvas what God does in the mind."[72] As Matthew Jones argues, to Leibniz this was not simply a fact to be understood but also a call to arms. Leibniz, like Kepler, believed that humans, "constrained to a point of view, should mimic God's production of views."[73] Humans should strive to broaden their perspective, to create what Leibniz called "diversity compensated by unity," or what Leibniz and Kepler both called harmony. This applied not just to studies of nature but also to the practical dilemmas of political and religious communities. Leibniz wrote that "the place of the other is the true point of perspective in politics as well as in morality."[74] Kepler's mission, to "collect the bits [of truth] wherever I can find them and put them together again,"[75] was similarly motivated. Perspective thus leads to a kind of pluralism, and even cosmopolitanism. It is to this idea—the nature and meaning of the cosmopolitan worldview—that I turn now in closing.

Kepler and Cosmopolitanism

In a 1610 letter to an anonymous recipient, Kepler offered a vision of his work and his hopes for the future: "If it has pleased the mind to contemplate what God makes, it should also please it to do what that same God instructs. As soon as this is achieved by all, then there will be nothing more for humankind to desire than for everyone throughout the whole world to inhabit one city, and in turn to delight in one another, far from every strife, as we hope will be the case in the future."[76] This vision—of a study of nature that might yield lessons for church and state—is one that Kepler repeated throughout his life, as we've seen over the course of this book. Kepler suggested here, as he did in his books and other letters, that if people truly recognized the patterns of harmony that God had imprinted throughout the cosmos, they might finally resolve their differences and live together in peace, far from every strife. The divisions of confession and of political allegiance would give way, in the utopian image of this letter, to one large world community whose shared values stemmed directly from the divine archetypal harmony.

This image, however, is beset by tension. What kind of city is Kepler describing here? Is this a city in which there *are* no differences because everyone agrees, or is it one in which differences are allowed to flourish under some larger, all-encompassing umbrella? Over the course of this book, I've suggested the latter and argued that while Kepler initially hoped to resolve all differences, he ultimately came to embrace many of them as not only irreconcilable but also beneficial to the body politic and the *corpus Christianum*. His resulting vision of harmony, which embraced both the particular and the universal, was not uncomplicated or even always internally consistent, but this is no coincidence, for these tensions—between the particular and the universal, the local and the global, the many and the one—lie at the heart of the very first articulations of cosmopolitanism and have dogged discussions of the cosmopolitan ideal ever since. In closing, I want to look more closely at the nature of the cosmopolitan vision and at the question that lies at its heart: does the cosmopolitan deny all ties or embrace many ties? And does the cosmopolitan seek one universal community or many different, overlapping communities?

Diogenes the Cynic first used the word *kosmopolites* in the fourth century BCE to describe himself as, literally, a citizen of the cosmos. As Kwame Anthony Appiah notes, "this formulation was meant to be paradoxical," for the very idea of the citizen suggested belonging to a particular polis, while the cosmos meant the entire universe, rather than any one place within it.[77] What could it mean, then, to be a *citizen* of the universe at large? To Diogenes, this was likely a negative, rather than a positive, ideal. Famous for sleeping in the marketplace and deliberately overturning popular custom, Diogenes did not mean to suggest that the cosmopolitan man affirmed his fellowship with others, but rather that he rejected the rules and standards of the city, and of civilization itself.[78] The cosmopolitan did not embrace all cities and men; he was, rather, indifferent to them all. The cosmopolitan was the quintessential stranger.

The Stoics later embraced the cosmopolitan ideal, but in doing so they transformed it. Zeno of Citium most famously represented the cosmopolitan perspective in his injunction to his followers to regard all men as "fellow citizens" and in his hope "that there should be one life and order as of a single flock feeding together on a common pasture."[79] Cicero, likewise, described the whole world as the "common home for gods and men, a city for both of them,"[80] and praised Socrates for his own cosmopolitanism. "When Socrates was asked which country he belonged to, he replied 'the world,'" wrote Cicero, "for he regarded himself as an inhabitant and citizen of every part of it."[81] Marcus Aurelius similarly described all people as "fellow citizens; and if that be so, the world is a kind of state. For in what other common constitution can we claim that the whole human race participates?"[82] To him, it was not the person who detached himself from the local body politic but rather the one who cut himself off from the larger world community who became an "amputated limb."[83]

In these more positive formulations of the cosmopolitan ideal, to be cosmopolitan was precisely to embrace *all* one's fellow men and to view them as part of a larger community. This was not because of the value of plurality itself but rather because of something understood to unite all men and to make them fundamentally alike (or capable of becoming so). In many cases, that something was empire. As Anthony Pagden remarks, "insofar as it had any content at all, the common law for all humanity . . . was a Greek law, not the happy multicultural amalgam which has so often been made of it. Similarly for the Romans, the world . . . was the Roman *civitas* extended to non-Romans."[84] If cosmopolitanism emphasized universalism, it was a specifically imperial universalism, a making of the larger world into the image of its Greek or Roman center. The early Christian embrace of these ideals worked on similar principles; when Saint Paul argued that "there is neither Jew nor Greek, there is neither bond nor free, there is neither male nor female: for ye are all one in Christ Jesus,"[85] he implied that Christianity *itself* was universal, the unifier that would erase all previous differences and create a true world community cast in its specific image.

These imperial or early Christian arguments for cosmopolitan universalism often overlapped with another justification for the world city: reason itself. The world was a common home to gods and men, for Cicero, because "they are the only reasonable beings: they alone live by justice and the law."[86] In this view, although the individual polis was often fallible and irrational, since it was governed by fallible men, the world in its totality was ruled by the gods and hence embodied the universal laws of reason and justice; it, then, was the true polis, home of all rational men hoping to follow more fully in the image of the gods.[87] Those who used their reason were fellow citizens, to be embraced as such; those who did not were not encompassed by the cosmopolitan ideal. Diogenes Laertius describes the Stoic position that "friendship . . . exists only between the wise and the good, by reason of their likeness to one another"; those who are neither wise nor good, by contrast, deserved the "opprobrious epithets of foemen, enemies, slaves, and aliens."[88]

Cosmopolitanism was thus neither simple nor simply pluralistic in its origins. Yet it was embraced, over the years, by those who sought to articulate a space for difference, perspective, and tolerance; this was particularly so for those Renaissance humanists who looked back at the ancient world and forward at their own and who attempted, in the increasingly divisive world of early modernity, to resolve theological disputes without resorting to violence. Erasmus implored "the promoters of bloodshed between nations" to remember that "this world, the whole of the planet called earth, is the common country of all who live and breathe upon it," and that "all men, however distinguished by political or accidental causes, are sprung from the same parents."[89] Though his concern was with Christian reconciliation, he extended these claims even to non-Christians: "Is not the Turk a man—a brother?" he asked.[90] Montaigne used the language of cosmopolitanism to suggest that pluralism was itself a valuable ideal. "I consider all men my compatriots," he wrote, "and embrace a Pole as I do a Frenchman, setting this national bond after the universal and common one. . . . Nature has put us into the world free and unbound; we imprison ourselves in certain narrow districts."[91] Lipsius used similar language when speaking of foreigners. "Are they not men sprung out of the same stock with thee, living under the same globe of heaven?" he asked. "Thinks thou that this little plot of ground . . . is thy country? Thou art deceived. The whole world is our country, wheresoever is the race of mankind sprung of that celestial seed."[92] These men, and their contemporaries, are the basis for Stephen Toulmin's linkage between the sixteenth century and the cosmopolitan vision, which he understands as the "readiness to live with uncertainty, ambiguity, and differences of opinion," in contrast to the rational certainties of seventeenth- and eighteenth-century philosophers.[93] Though his contrast may be too sharply drawn,[94] it highlights the extent to which the cosmopolitan vision, during the religious turmoil of the post-Reformation era, often embraced pluralism at its core.

Kepler's vision of world harmony, I've argued, aligned with this kind of cosmopolitan pluralism. Harmony without variety, he believed, "ceases to be pleasing altogether."[95] He called for a Christian community that included Catholics, Lutherans, and Calvinists, even with their differences unresolved. He emphasized the possibility of multiple different political configurations that might equally serve the public good. He portrayed mathematics as a tool of toleration, and the mathematician as someone who might craft a community that allowed for difference and disagreement in its midst. He did not embrace *all* differences, of course. He quoted Jews on questions of astronomy and chronology while at the same time describing them as "characterized by a virulent spirit and language, and an eagerness for sophistry," and hoping that with their conversion or destruction, God might "snatch away from the life of the Christians the greed, the inhumanity, the disregard for the neighbor, the scorn, the pomposity in clothing, [and] the lusts."[96] He remarked on the progress of the Turks, who had "cast off barbarity and learned civility" (not the highest of praise) and hoped for the eventual downfall of their empire and the "ruin of their religion."[97] He prayed not only for the conversion of the Jews and Muslims but also for the conversion of the pagans in the Americas.[98] Like the Stoics before him, he often spoke dismissively of the *vulgus*, the unlearned public, who, he believed, did not deserve the same respect as his learned peers.

Yet I've tried to demonstrate that, though narrowly circumscribed to include only the Christian world, his vision of harmony was, for its time, inclusive and plural; let us not forget that the rival confessions of his day typically saw each other as having succumbed to the Antichrist and, therefore, as deserving of death rather than dialogue. To Kepler, by contrast, they remained part of the *corpus Christianum*, following the model of the cosmic harmonies, which were polyphonic, plural, and often discordant. His conception of the nature of religious and political communities, of the possibilities of human vision and knowledge, and of the type of progress possible over time, all inclined him toward a (limited) pluralist form of cosmopolitanism far more often than toward either particularism or universalism. If, as Appiah comments, "there's a sense in which cosmopolitanism is not the name of the solution but of the challenge,"[99] then it was a challenge that Kepler consistently took up, and a conversation in which he constantly participated. Pagden ultimately argues that it is an error "to hope that we can ever achieve a truly cosmopolitan vision of the cosmopolis."[100] This may be true, but that doesn't mean that such a vision has not been articulated alongside the universalist one. Kepler, albeit imperfectly, offered us a model of what that alternative vision looks like.

At times, Kepler's cosmopolitanism led him to feel much like Diogenes, the quintessential stranger. "I am a guest, a foreigner, almost unknown," he wrote to Matthias Bernegger from Sagan. "I am all but deaf to the local language; I am in turn considered a barbarian. . . . There is no house of my own here, and I have no place in the church, from whose gate I have departed."[101] Yet even then, he turned to those friendships he continued to sustain, mostly through letters, as his anchors, both particular and universal. "Friendships are given life by harmonic tempering," he wrote.

For what concord is to proportion, that love, which is the foundation of friendship, is to the whole compass of human life. . . . Although friendship cannot survive frequent injustices, yet it rejects laws, and refers everything to the sound and sober judgment of love, dispensing now equality, now proportionality, and when neither, always dispensing what seems in the immediate situation to make for the preservation of love, which is also goaded on, as harmony is by discords, and as fire is by an iron poker, by a few injustices and renews its strength by free forgiveness of them.[102]

The harmony of personal friendship was to Kepler a microcosm of world harmony, and the occasional discord made it all the stronger. Erasmus had claimed, in his adages, that "friends hold all things in common" and had cited Pythagoras and Plato as proof.[103] To Erasmus, this applied in particular to the classical and Christian traditions, which he took to be in harmonious dialogue rather than in opposition. Kepler would likewise have extended his notion of both friendship and harmony to his ancient interlocutors as much as to his contemporary ones; his cosmopolitanism was diachronic as well as synchronic in nature.

Notably, Kepler's conception of cosmopolitan harmony did not require the rejection of particular ties in favor only of the universal. Kepler insisted to his dying day that he was a sincere Lutheran, while at the same time arguing for a reunified church that embraced other confessional stances. He identified strongly as a German and, at the same time, as a member of the international Republic of Letters; he wrote in both the German vernacular and the Latin language of the scholarly international elite, depending on audience and subject matter, and often moved back and forth in the same letter, and even the same sentence. His local, particular ties did not prevent him from asserting larger, more universal ones—though they often made it more difficult.

In her 1994 article endorsing the cosmopolitan vision, Martha Nussbaum argues against particularism (there patriotism, in particular) as something that "substitutes a colorful idol for the substantive universal values of justice and right."[104] Citing Diogenes, the Stoics, and ultimately Kant, she notes the loneliness that this kind of universalist cosmopolitanism can engender. "Becoming a citizen of the world is . . . a kind of exile—from the comfort of local truths, from the warm nestling feeling of patriotism, from the absorbing drama of pride in oneself and one's own."[105] Cosmopolitanism, in this view, offers "only reason and the love of humanity, which may seem at times less colorful than other sources of belonging."[106] Though she notes that world citizenship, even according to the Stoics, did not mean a rejection of local identifications, the picture she draws—of a series of concentric circles, moving from the particular to the universal—ends with our task "'to draw the circles somehow toward the center' . . . making all human beings more like our fellow city dwellers."[107] This remaking of the world in our own image sounds not that far removed from the kind of "civilizing," imperial vision that Pagden identifies at the core of the cosmopolitan ideal.

In a later article, Nussbaum draws back from this universalist articulation of cosmopolitanism, noting that "the denial of particular attachments leaves life empty of meaning for most of us, with the human psychology and the developmental history that we have."[108] Instead, she urges something more akin to the pluralist form of the cosmopolitan vision: "an uneven dialectical oscillation within ourselves, as we accept the constraints of some strong duties to humanity, and then ask ourselves how far we are entitled to devote ourselves to the particular people and places whom we love."[109] The pluralist cosmopolitan embraces the perspective of the local and the particular, so that life can be both personally meaningful and widely and peacefully shared.

This articulation shares much with Hannah Arendt's understanding of the "common world" necessary for democratic deliberation. According to Arendt, people who share a common world share, not a single dominant worldview, but rather a collective communal space that contains within it a plurality of worldviews. The common world is built out of "the simultaneous presence of innumerable perspectives and aspects . . . for which no common measurement or denominator can ever be devised."[110] Much as Kepler generated an image of a book with his camera obscura by looking at it from "the infinite points of the edges," Arendt's common world is constituted only by the very different locations of those who demarcate its boundaries. The community itself is shared by all; the perspectives that make it up, however, can and must diverge.

Kepler hoped that in his lifetime he might see a world "far from every strife." He saw, instead, the opposite: Europe torn apart by the most brutal war it had ever seen. His vision of world harmony often seems, these days, to be even more of an unrealizable dream for us than it was for him. Is the dream even worth recalling, or must it be shelved like some dusty and outdated relic, history having answered Kepler's call for harmony with a resounding no from which we have yet to recover?

Though unrealized and perhaps ultimately unrealizable, Kepler's vision is, I believe, worth embracing—in its hopefulness, its inclusiveness, and its recognition that we can be both many and one, both devout members of local communities and devoted members of a larger, more encompassing world community. Kepler drew hope from his conception of the origins of harmony. "It turns out," he noted, "that boys, that primitives, peasants, and barbarians, and the very wild beasts, perceive harmonies in notes, though they know nothing about the theory of harmony."[111] Harmony, in other words, was buried deep within every one of us. To Kepler, it was built into the very fabric of our being, for God had "breathed this particle of His own image into all souls absolutely."[112] But as Kepler himself would argue, we can hold a vision in common without agreeing to the particulars that support it in each case; the pursuit of harmony need not rely even on Kepler's God to sustain it.

It is hard to imagine how we might create a more perfect world, or what that might even mean. The cosmopolitan cosmopolis may seem as far off as the Lunarians of Kepler's *Somnium* and may require as dramatic a shift in perspective to envision, much less to create. Yet no temple is built in a day, as Kepler would remind us, and "the path to the truth is long and circuitous."[113] In the end, the pursuit of harmony may matter more than its ultimate attainment.

Acknowledgments

Writing this book while raising two young children, I have often thought about the similarities between the two endeavors: both take a great deal of time and effort; are alternately glorious and torturous; are difficult to send out into the wider world, to be judged on their own merits; and take an enormous amount of help from others to accomplish. There are so many people without whom I could not have completed this book, and certainly without whom it would have been far less than it is. Five, in particular, deserve special mention. Matthew Jones first introduced me to the history of science in my sophomore year of college and has continued to offer support, advice, and penetrating comments on my work in the years since. The late Michael Mahoney helped me immerse myself in the world of early modern science and encouraged this project in its early stages; I deeply regret that he was unable to see its completion. Anthony Grafton helped me set the story of early modern science in a much broader context, read countless drafts, continually offered wise advice, and encouraged me to do things my own way; his support and mentorship are truly unparalleled. Eileen Reeves's and Yair Mintzker's thorough and insightful comments on an early version of the manuscript helped me to reorient it and strengthen its arguments, particularly its overall framing. I would also like to thank the members of Princeton's History of Science Program Seminar and of the University of Chicago's Society of Fellows, many of whom have read drafts of some of these chapters and have enriched my perspective with their own work and ideas. For not including the individual names of so many people to whom I owe gratitude, and for any errors that remain herein, I beg forgiveness. Finally, I would like to thank my husband, Elliot Gardner—a true partner of heart and mind—for everything we have shared, especially our two boys. I dedicate this book to you.

Portions of chapters 2 and 3 were previously published in "From Cosmos to Confession: Kepler and the Connection between Astronomical and Religious Truth," in *Change and Continuity in Early Modern Cosmology*, edited by Patrick J. Boner, Archimedes Series 27 (Dordrecht: Springer, 2011). A portion of chapter 4 was previously published in "Forms of Persuasion: Kepler, Galileo, and the Dissemination of Copernicanism," *Journal for the History of Astronomy* 40, no. 4 (2009): 403–19.

Notes

1   *KGW* 17:747.
2   *KGW* 18:909.
3   *KGW* 18:941.
4   On the conflict and struggles that Kepler faced in his personal life, see Rublack, *The Astronomer and the Witch.*
5   *KGW* 15:357.
6   I am here positioning my own approach against that of John Hollander (*Untuning of the Sky*, 19), who argues that by the sixteenth and seventeenth centuries theories of cosmic harmony had become trivialized and served only as "decorative metaphor and mere turns of wit." I follow instead the advice of James Haar ("*Musica Mundana*," 444), who reminds us to "be cautious about labelling as metaphor only what was for so long believed in so earnestly."
7   In speaking of the fertile conjunction of the theory of cosmic harmony with theories of magic, Gary Tomlinson (*Music in Renaissance Magic*, 99–100) makes this point and argues that we consider the cultural changes of early modernity as "the reciprocal transformation of categories and the contexts in which they are reenacted." Those who invoked the language of harmony, from this view, "retold the old stories . . . and made them new."
8   The story of the origins of modern science has itself been "reconfigured" in recent days, with new emphasis placed on the role of religion in the science of early modernity even in introductory accounts of the period. See, e.g., Osler, *Reconfiguring the World*; Coudert, *Religion, Magic, and Science*; among others.

9   These narratives, too, have been challenged and pushed back of late. See esp. Laursen and Nederman, *Beyond the Persecuting Society*; and Laursen and Nederman, *Difference and Dissent*.

10   *Harmony of the World*, 2.

11   Ibid., 3.

12   Ibid., 4.

13   Ibid., 5.

14   Ibid.

15   The earliest known version of this story comes to us from the *Manual of Harmonics* of Nicomachus of Geresa (c. 60–c. 120).

16   Galileo's father, Vincenzo Galilei, reported in his 1589 *Discorso intorno all'opere messer Gioseffo Zarlino da Chioggia* that he had repeated Pythagoras's supposed experiment only to prove it false: the mathematical relationships produced, he argued, did not correspond to Pythagoras's musical scale.

17   In book 2 of *De caelo*, Aristotle attributes the idea that the motions of the planets produce a musical harmony to the Pythagoreans.

18   See Prins, "Harmony."

19   Gouk, *Music, Science, and Natural Magic*, 81.

20   Creese, *The Monochord in Ancient Greek Harmonic Science*, 7.

21   Van Orden, *Music, Discipline, and Arms in Early Modern France*, 56.

22   See Ziolkowski, "The Bow and the Lyre."

23   See L. Spitzer, *Classical and Christian Ideas of World Harmony*, 9. My discussion of harmony in this section owes much to this book.

24   See Lipmann, "Hellenic Conceptions of Harmony."

25   In book 4 of the *Republic* (424c), Socrates asserts that "the guardians must beware of changing to a new form of music, since it threatens the whole system. As Damon says, and I am convinced, the musical modes are never changed without change in the most important of a city's laws" (trans. Grube, 99).

26   Stahl, *Commentary on the "Dream of Scipio,"* 73.

27   Cicero, *De republica* 2.42, "The Republic and the Laws" (*Treatises of M. T. Cicero*, 42).

28   Daly, *Cosmic Harmony and Political Thinking in Early Stuart England*; and van Orden, *Music, Discipline, and Arms in Early Modern France*.

29   Shakespeare, *Henry V*, 1.2.180–83.

30   Le Roy, *Aristotle's Politiques or Discourses of Government*, 263–64 (originally published as *Les politiques d'Aristotle* [1568]). Quoted in Lloyd, "Constitutionalism," 277.

31   Bodin, *The Six Books of a Common-weale*, 456.

32   See, e.g., L. Spitzer, *Classical and Christian Ideas of World Harmony*, 64.

33   "A Song for Saint Cecelia's Day," in *Works of John Dryden*, 167–68.

34   John of Salisbury, *Policraticus*, 51.

35   Ibid., 67.

36 *De regimine principium*, in *Aquinas: Political Writings*, 11.

37 "Speech of 1609," in *Political Works of James I*, 307.

38 Bodin, *The Six Books of a Common-weale*, 790.

39 Hobbes, *Leviathan*, 3.

40 Ibid., 185.

41 Ibid., 257.

42 Ibid., 264.

43 See Cook, "Body and Passions."

44 1 Corinthians 12:12–13, 27.

45 Quoted in Honeygosky, *Milton's House of God*, 31.

46 Quoted in Eden, *Friends Hold All Things in Common*, 135.

47 Quoted in ibid., 26.

48 See in particular Duffy, *Stripping of the Altars*; and Muir, *Ritual in Early Modern Europe*.

49 Duffy, *Stripping of the Altars*, 92.

50 Ibid.

51 Ibid., 93.

52 Ibid., 44.

53 As Duffy (ibid., 6) writes, "Corpus Christi . . . was conceived and presented . . . as a celebration of the corporate life of the body social, created and ordered by the presence of the body of Christ among them."

54 Kaplan, *Divided by Faith*, 68–69.

55 Ibid., 69–71.

56 Muir, *Ritual in Early Modern Europe*, 170.

57 Quoted in Elwood, *The Body Broken*, 152.

58 Ibid., 151.

59 Ibid., 152.

60 "An Anatomy of the World," in *John Donne: The Complete English Poems*, 276, lines 213–18.

61 Quoted in L. Spitzer, *Classical and Christian Ideas of World Harmony*, 135.

62 See Pesic, *Music and the Making of Modern Science*, 52. See also Westman, "Proof, Poetics, and Patronage."

63 McColley, *Poetry and Music in Seventeenth-Century England*, 10.

64 In my description of Kepler's approach to music in the *Harmony of the World*, I rely heavily on Walker, "Kepler's Celestial Music."

65 *Harmony of the World*, 137.

66 Ibid., 139.

67 *KGW* 13:72.

68 *Harmony of the World*, 446–47.

69 Ibid., 446–48.

70 See Tomlinson, *Music in Renaissance Magic*, 68–71.

71 As he had written to his teacher, Michael Maestlin, as early as 1595, "in the sphere is the Trinity, the surface, center, and capacity. Thus, in the orderly world: the fixed stars, the sun, the air, or the intermedi-

ate ether; and in the Trinity, the son, the father, and the spirit" (*KGW* 13:23).

72 *Harmony of the World*, 305.

73 It may seem that despite his emphasis on the way harmony *sounded* and on the judgment of the ear, Kepler has just substituted one mathematical and a priori method of determining the harmonies—albeit a geometrical one—for the earlier Pythagorean method. Yet it is worth noting, as Walker points out, that Kepler's geometrical method is actually a very bad way to determine harmonies that are based on the system of just intonation, for it does not actually generate thirds and sixths. In order to include them, Kepler follows some convoluted (and not always convincing) pathways that take him farther afield from his original rule than he probably preferred. It is, however, easy to generate *Pythagorean* harmonies using Kepler's geometrical method—and this Kepler refused to do. Walker ("Kepler's Celestial Music," 239) writes: "If Kepler had accepted the still current, very ancient Pythagorean and Platonic system of intonation, involving only the consonances 1:2, 2:3, and 3:4, he would have had no difficulties at all: but he did not accept it, and that on purely empirical grounds—because, before he set out on his investigation into causes, he had already established by ear that just thirds and sixths are consonant."

74 *Harmony of the World*, 462.

75 Quoted in Cardano, *Writings on Music*, 212. See also Prins, "Harmony."

76 *Harmony of the World*, 250.

77 Ibid., 442.

78 Ibid., 281.

79 Bodin, *The Six Books of a Common-weale*, 718.

80 *Harmony of the World*, 268.

81 Ibid., 276.

82 Ibid., 268.

83 Hobbes, *Leviathan*, 257.

84 Pufendorf, *Present State of Germany*, 152–53, quoted in Berns, "*Partheylichkeit* and the Periodical Press," 106.

85 See, e.g., Knoppers and Landes, *Monstrous Bodies / Political Monstrosities*; and Daston and Park, *Wonders and the Order of Nature*.

86 See K. Long, "Odd Bodies."

87 "Of a Monstrous Child," in *Complete Essays of Montaigne*, 539.

88 *Harmony of the World*, 252.

89 Fernand Hallyn (*Poetic Structure of the World*) and Nicholas Jardine ("Places of Astronomy in Early-Modern Culture") both link Kepler's harmonic vision, with its complexity and union of differences, to his Mannerist aesthetics. In describing Kepler's embrace of the ellipse, Hallyn describes the ellipse as "a union of contraries," much like the harmonies of the world themselves (216).

90 *KGW* 12:29.17–19.

91 See Scribner, *For the Sake of Simple Folk*; and Hsia, "Monstrous Births,
    Propaganda, and the German Reformation."
92 Daston and Park, *Wonders and the Order of Nature*, 175.
93 *STRANGE NEWES of a prodigious Monster.*
94 *KGW* 12:31.4–13.
95 *Harmony of the World*, 462.
96 *KGW* 18:1072.
97 Antognazza, *Leibniz on the Trinity and the Incarnation*, xxi.
98 Calvin, *Institutes of the Christian Religion*, 2.36, 402. See also Edwards,
    "Calvin and Hobbes."
99 Antognazza, "Leibniz and the Post-Copernican Universe"; and Anto-
    gnazza, *Leibniz on the Trinity and the Incarnation*.
100 *KGW* 1:9.35–10.1.
101 "Windsor Forest" (1704), in *Works of Alexander Pope*, 48.
102 *Harmony of the World*, 498.

CHAPTER 1

1   Much of what we know of Kepler's perception of himself and his child-
    hood comes from his self-horoscope, essentially a self-analysis he wrote in
    1597 at the age of twenty-six. See *KGW* 19:328–37.
2   *KGW* 13:91.
3   "Vom Abendmahl Christi Bekenntnis," in *D. Martin Luthers Werke*,
    26:332b.
4   Quoted in Elert, *Structure of Lutheranism*, 231.
5   Quoted in Elwood, *The Body Broken*, 116.
6   Quoted in Elert, *Structure of Lutheranism*, 245.
7   Luther, "Disputatio contra scholasticam theologiam" (1517): "Nulla forma
    syllogistica tenet in terminis divinis."
8   See Kusukawa, *Transformation of Natural Philosophy*.
9   See I. Hunter, *Rival Enlightenments*. The following summary of the relation-
    ship between theology and metaphysics in the seventeenth century is
    heavily indebted to Hunter's analysis.
10 See Sparn, *Wiederkehr der Metaphysik*.
11 Literally, "forerunner of a cosmographic discussion containing the secret
    of the universe." Kepler often refers to the text as his *Prodromus*, but I
    will henceforth refer to it as it is commonly known today, the *Mysterium
    Cosmographicum*.
12 *KGW* 13:23.
13 For a discussion on the trope of "God's two books" in early modern sci-
    ence, with a particular focus on the issue of Copernicanism, see Howell,
    God's Two Books, esp. chap. 3.
14 See Field, *Kepler's Geometrical Cosmology*.
15 *KGW* 1:4.2–5.

16 See Aristotle, *Metaphysics* 1.5.985b–986a.

17 See *Harmony of the World*, 302.

18 *KGW* 1:23.7–9.

19 *KGW* 1:23.11–12.

20 Kepler elaborates on this distinction between *numeri numerantes* (counting numbers) and *numeri numerati* (counted numbers) in the appendix to book 5 of the *Harmony of the World*. There, in contrast to the position of Ptolemy, he argues that "there is no force in the numbers as counting numbers, and in their place [I] establish as the basic principles of the harmonies the counted numbers, that is, the things themselves which are subject to the numbers, in other words the plane regular figures and the divisions of the circle which are to be controlled by them" (500).

21 *KGW* 13:23.

22 See *KGW* 1:23.13–14.

23 *KGW* 1:23.20–22.

24 Cusa, *Of Learned Ignorance*, 73.

25 *KGW* 1:24.4–12.

26 *KGW* 8:30.6–8.

27 *Harmony of the World*, 304.

28 See, e.g., Carabine, *Unknown God*.

29 John 1:8.

30 Plato, *Timaeus* 28c, trans. Lee, 41.

31 See B. Smith, *Indescribable God*, esp. chap. 2.

32 Quoted in ibid., 26.

33 Ibid., 36.

34 *KGW* 13:64.

35 See *KGW* 7:267.15–16: "the reasons of geometry are coeternal with God."

36 Jean-Luc Marion makes this very point in his essay "The Idea of God," where he writes that "Kepler formulates nothing less than a new definition of God . . . ; indeed, it goes so far as to guarantee for the human mind a knowledge that is 'of the same nature as that of God.' . . . This epistemological univocity in fact implies that we understand God's divinity to the same extent that we understand mathematical possibilities" (269).

37 *KGW* 13:309.

38 Marion, "Idea of God," 274.

39 See Barker and Goldstein, "Realism and Instrumentalism in Sixteenth-Century Astronomy."

40 Copernicus, *On the Revolutions*, in *Complete Works, Nikolaus Kopernikus*, xvi.

41 See Westman, "Melanchthon Circle."

42 See N. Jardine, *Birth of the History and Philosophy of Science*, chap. 6.

43 *Complete Works, Nikolaus Kopernikus*, xvi.

44 For more on the contours of Kepler's reform of astronomy in physical terms, see Stephenson, *Kepler's Physical Astronomy*.

45 In *Birth of the History and Philosophy of Science*, Nicholas Jardine notes, too, but only tangentially, the role that geometry, and its linkage to God, plays in Kepler's radically realist approach to his astronomical and cosmological claims. According to Jardine, "it is because the human mind is formed in the image of the creator's mind that we are able to understand part at least of the plan according to which he created the world" (144).

46 *KGW* 1:15.5–7.

47 *KGW* 1:15.24–16.10.

48 See Westman, "Proof, Poetics, and Patronage."

49 *KGW* 1:16.19–23.

50 *KGW* 1:16.27–29.

51 *KGW* 1:16.33–34.

52 See his *Dissertatio cum Nuncio Sidereo*, where he contrasted his own "poor vision" with Galileo's "keen sight" provided by the telescope (*KGW* 4:290.21–22). See also the conclusion to this book.

53 *Harmony of the World*, 303–4.

54 See Claessens, "Imagination as Self-Knowledge."

55 See Cary, *Augustine's Invention of the Inner Self*, chap. 1.

56 *Harmony of the World*, 493.

57 See Barker, "Kepler's Epistemology."

58 Translated in ibid., 360. See *KGW* 21.1:446.38–45.

59 *KGW* 1:11.32–33.

60 *KGW* 8:9.16–22.

61 *KGW* 13:43.

62 Ibid.

63 Ibid.

64 Ibid.

65 See Caspar, *Kepler*, 49–50.

66 See Reich and Knobloch, "Die Kreisquadratur Matthias Hafenreffers."

67 *KGW* 13:44.

68 Ibid.

69 *KGW* 13:58.

70 *KGW* 13:48.

71 See Westman, "Melanchthon Circle," 181–86.

72 *KGW* 13:23.

73 See Blackwell, *Galileo, Bellarmine, and the Bible*.

74 *KGW* 13:64.

75 *KGW* 13:80. This story is recounted as well in Wilhelmi, *Die griechischen Handschriften der Universitätsbibliothek Tübingen*, along with the text of Hafenreffer's sermon and Martin Crusius's Greek summary of it.

76 *KGW* 13:92.

77 Ibid.

78 Hafenreffer's description of these disciplinary distinctions was, of course, an overly simplified one, as a growing number of astronomers (Maestlin

and Rheticus in the Lutheran camp, and Copernicus and Clavius in the Jesuit camp, among others) had already begun to dispute astronomy's place in the traditional hierarchy. For a more nuanced account of the place of astronomy and its evolution during the early modern period, see Donahue, "Astronomy."

79 *KGW* 13:93.

80 See Methuen, *Kepler's Tübingen*.

81 See Green, *Formula of Concord*.

82 In summarizing the situation for Kepler and his environs, I've simplified a story that was neither this uncomplicated nor this uniform. First, while the Formula of Concord did take seriously some concerns of both Philippists and Gnesio-Lutherans (or Flacians), it largely followed the more uncompromising Gnesio-Lutheran approach. Moreover, while in some areas, like Württemberg, it received overwhelming support, there were others, like Sweden and Denmark, where Philippism continued to flourish. Finally, there were specific issues in the Formula of Concord that continued to be disputed over time—one of which (the doctrine of ubiquity) concerned Kepler directly, as we will see below.

83 The start of this chapter includes a brief summary of the Lutheran position. For a more complete explanation of the various debates about the Eucharist, see Wandel, *The Eucharist in the Reformation*.

84 *KGW* 13:64.

85 *KGW* 12:29.17–19.

86 *KGW* 13:99.

87 Ibid.

88 *KGW* 13:85.

89 Kepler's belief in the utter persuasiveness of his argument and in the clarity with which he had made his points stands in marked contrast to his more cautious approach, and his attentiveness to the possibility of disagreement, in the *New Astronomy*. See Voelkel, *Composition of Kepler's "Astronomia Nova,"* 2.

90 *KGW* 13:99.

91 *KGW* 13:85.

92 *KGW* 13:99.

93 *KGW* 13:102.

94 *KGW* 1:14.21–27.

95 Years later, in 1621, Kepler released the second edition of the *Mysterium Cosmographicum*, complete with an extensive set of newly added footnotes. This was after the publication of the *New Astronomy* (1609), in which Kepler had published an extensive introduction on the relationship between Copernicanism and scripture. As a footnote to this contention about the primacy of scripture over nature, Kepler clarified that nature and scripture necessarily agreed with one another, and that therefore there could never actually be a disagreement between the theories of Copernicus and the

words of scripture. He explained that "it is clear that God has a tongue, but he also has a finger." The finger of God was clearly evident in nature, which Kepler believed he had proven to operate according to the theories of Copernicus. By contrast, God had adjusted scripture on the basis of the "common tongue of men"—a reference to the hermeneutical principle of accommodation. "Therefore," Kepler concluded, "in matters that are highly evident, anyone who is most religious should be very careful not to twist the tongue of God so that it refutes his finger in nature." *KGW* 8:39.32–41, 40.1–4.

96 *KOO* 1:26.

97 Funkenstein, *Theology and the Scientific Imagination*, 72.

98 See, e.g., Nadler, "Arnauld, Descartes, and Transubstantiation"; Watson, "Transubstantiation among the Cartesians"; and Ariew, *Descartes and the Last Scholastics*, chap. 7.

99 Ariew, *Descartes and the Last Scholastics*, 140.

100 Ibid., 147.

101 Ibid., 144.

102 Nadler, "Arnauld, Descartes, and Transubstantiation," 238.

103 Mercer, *Leibniz's Metaphysics*, 82.

104 Ibid., 83.

105 See Mercer's discussion "The Metaphysics of Substance" in ibid., 61–170, particularly the discussion on transubstantiation and the Eucharist at 82–93.

106 See Antognazza, *Leibniz on the Trinity and the Incarnation*.

107 Ibid., 87.

108 See McDonough, "Leibniz's Conciliatory Account of Substance."

109 Mercer, *Leibniz's Metaphysics*, 291.

110 Ibid., 338n70.

CHAPTER 2

1 *KGW* 13:107.

2 *KGW* 17:847.

3 *KGW* 17:750.

4 See Schinkel, *Conscience and Conscientious Objections*, esp. "The Religious Conscience before and after the Reformation," 171–202. The discussion of conscience below owes much to his account.

5 Ibid., 183.

6 See Bainton, "Luther's Attitudes on Religious Liberty," 20.

7 Schinkel, *Conscience and Conscientious Objections*, 195.

8 As Oberman (*Luther*, 204) notes in his biography of Luther, "Luther liberated the Christian conscience, liberated it from papal decree and canon law. But he also took it captive through the Word of God."

9 Quoted in Schinkel, *Conscience and Conscientious Objections*, 44.

10   Quoted in Bainton, "Development and Consistency of Luther's Attitude to Religious Liberty," 119–20.
11   Bainton, *Travails of Religious Liberty*, 114.
12   See his essay "Of Repentance" in particular, in *Complete Essays of Montaigne*.
13   Hobbes, *Leviathan*, 48.
14   Locke, *Essay concerning Human Understanding*, 71.
15   "Of Freedom of Conscience," in *Complete Essays of Montaigne*, 509.
16   Locke, *Letter concerning Toleration*, 33.
17   Hobbes, *Leviathan*, 424.
18   *KGW* 17:808.
19   *KGW* 17:750.
20   *KGW* 17:808.
21   See, e.g., Hotson, "Irenicism in the Confessional Age."
22   See Rummel, "Erasmus and the Restoration of Unity in the Church."
23   Ibid., 70.
24   This may, in part, have been because he was by then out of their ambit of authority, already ensconced in Prague as imperial mathematician.
25   For a more detailed summary of the Lutheran stance on ubiquity, see Schaff, *Bibliotheca Symbolica Ecclesiae Universalis*, vol. 1, from which I've excerpted the below summary here. See particularly the section "The Formula of Concord, AD 1577," 258–340, and, more specifically, "Part VII: The Christological or Ubiquitarian Controversy," 285–96.
26   Cited in ibid., 282.
27   The controversy eventually fizzled out over the course of the Thirty Years' War, and the milder form of ubiquity, following Chemnitz, predominated.
28   *KGW* 14:137.
29   *KGW* 12:7.3
30   *KGW* 12:7.9.
31   *KGW* 16:586.
32   Ibid.
33   *KGW* 17:750.
34   Ibid.
35   *KGW* 16:528.
36   Because of Kepler's insistence that he would sign the Formula of Concord "only conditionally," scholars have assumed that he never signed the Formula of Concord at all. This assumption has recently been challenged. In 2009 a manuscript was discovered that contained Kepler's signature and his subscription to the Formula of Concord in Graz in 1594 (see Schäfer, *Blätter für Württembergische Kirchengeschichte*). Though this has only begun to receive analysis, it should not dramatically change our perception of Kepler's stance. In 1594 Kepler was twenty-three and had only just left Tübingen for a new job in Graz. He was still three years away from publishing his first book and had only begun to work out his

own understanding of his place vis-à-vis the Lutheran Church. As Kepler himself never mentioned any subscription to the Formula of Concord in his letters, as his refusal to subscribe led to his excommunication, and as he strove to regain the right to take Communion while still refusing to subscribe, we can regard this signature from 1594 as at most the indiscretion of a youth still unsure of himself or his beliefs.

37 *KGW* 17:750.

38 Ibid.

39 *Nova Kepleriana*, 6:13.33–14.3.

40 *KGW* 16:610.

41 See Andreae, *Selbstbiographie*, 52.

42 Wagner, *Memoria Rediviva . . . Dn. Danielis Hitzleri*. Interestingly, the only mention the document makes of Kepler is in reference to Hitzler's astronomical pursuits, with a brief nod to Kepler's discussion of Galileo's *Starry Messenger*. Wagner either is unaware of or chooses not to mention Hitzler's role in Kepler's excommunication.

43 *KGW* 17:638.

44 See *Melanchthon Reader*, 217.

45 See Preus, *Inspiration of Scripture*, 9–10.

46 Quoted in ibid., 10.

47 Ibid., 130.

48 *KGW* 17:808.

49 Ibid.

50 Ibid.

51 Ibid.

52 See Whitman, *Origins of Reasonable Doubt*, 106.

53 *KGW* 17:750.

54 *KGW* 17:808.

55 On De Dominis, see Malcolm, *De Dominis*.

56 See Novosel, "From Periphery to Centre," 61.

57 See the chapter "Marco Antonio De Dominis," in Patterson, *King James VI and I and the Reunion of Christendom*, 220–59.

58 See *KGW* 12:30–35.

59 *Harmony of the World*, 252.

60 *KGW* 17:808.

61 *KGW* 17:829.

62 Ibid.

63 *KGW* 17:835.

64 Ibid.

65 On this point, Hafenreffer noted in the margins of Kepler's letter that "there is a great difference" between Kepler and the ordinary layman. In some ways, he was right: Kepler had worked as a Lutheran teacher and official in the past, he had tried to obtain a post at the university, and his role as imperial mathematician made him a far cry from your average

layman. Yet Kepler's claim was not without merit; had he not brought up his objections with Pastor Hitzler, he would likely not have been expected to sign the Formula of Concord at all in order to receive Communion at his church in Linz.

66 *KGW* 17:835.
67 Ibid.
68 See Ziebart, *Nicolaus Cusanus on Faith and the Intellect*, 98–117.
69 Cited in Kraemer, *Theology of the Laity*, 25.
70 See, e.g., the selection of essays in Kolb, *Luther's Heirs Define His Legacy*.
71 *KGW* 17:835.
72 Ibid.
73 Ibid.
74 Ibid.
75 Ibid.
76 *KGW* 17:750.
77 *Harmony of the World*, 304.
78 *KGW* 17:750.
79 *KGW* 17:835.
80 *KGW* 16:451.
81 *KGW* 17:643.
82 *Harmony of the World*, 129.
83 *KGW* 17:835.
84 Ibid.
85 *KGW* 17:843.
86 *KGW* 17:847.
87 Ibid.
88 Howell, *God's Two Books*, 114.
89 Bono, *The Word of God and the Languages of Man*, 69.
90 Harrison, *The Bible, Protestantism, and the Rise of Natural Science*, 57.
91 *Valerius Terminus*, in *Works of Francis Bacon*, ed. Spedding, Ellis, and Heath, 217.
92 Bono, *The Word of God and the Languages of Man*, 218.
93 Harrison, *The Bible, Protestantism, and the Rise of Natural Science*, 56.
94 *KGW* 4:204.33–35.
95 Quoted in Falcom, *Aristotle and the Science of Nature*, 49n36.
96 See Coudert and Corse, *Franciscus Mercurius van Helmont*, xi–xlvi.
97 Hobbes, *Leviathan*, 48.
98 Locke, *Essay concerning Human Understanding*, 37.
99 See Salmon, *Study of Language in 17th-Century England*, 131–34.
100 See Coudert and Corse, *Franciscus Mercurius van Helmont*, xxxiv; and Ishiguro, *Leibniz's Philosophy of Logic and Language*, chap. 3.
101 *KGW* 13:91.
102 *Assayer* (1623), in *Discoveries and Opinions of Galileo*, 238.
103 *KGW* 12:33.10–13.

104 See Ryle, "Letters and Syllables in Plato."
105 See, e.g., *Harmony of the World*, 209.

CHAPTER 3

1   *KGW* 15:333.
2   *KOO* 8.2:672.
3   Ibid.
4   See Dickson, "Johann Valentin Andreae's Utopian Brotherhoods." Dickson notes that Andreae referred to Besold's conversion to Catholicism as his "spiritual death" (771).
5   On Kepler's Jesuit friends in Graz, see Andritsch, "Gelehrtenkreise um Johannes Kepler in Graz."
6   *KGW* 18:1053.
7   See Grafton, *Worlds Made by Words*, chap. 6.
8   On the various theological forms of accommodation, see in particular Huijgen, *Divine Accommodation in John Calvin's Theology*. Much of my summary below is indebted to Huijgen's analysis.
9   Amos Funkenstein focuses much of his analysis on hermeneutical and pedagogical accommodation; see Funkenstein, *Theology and the Scientific Imagination*, 214, for an elaboration of this hermeneutical principle.
10 Huijgen, *Divine Accommodation in John Calvin's Theology*, 67.
11 Ibid., 96.
12 Funkenstein, *Theology and the Scientific Imagination*, 214–22.
13 See Benin, "The 'Cunning of God' and Divine Accommodation," 184.
14 See Graves, "Milton and the Theory of Accommodation," 253.
15 Ibid., 254.
16 Huijgen, *Divine Accommodation in John Calvin's Theology*, 82.
17 *New Astronomy*, 60.
18 See also the discussion of Kepler's appeal to hermeneutical accommodation in Moss, *Novelties in the Heavens*, 132–35. On Galileo's reliance on the same argument, see also McMullin, "Galileo on Science and Scripture."
19 See Howell, *God's Two Books*, 130–31.
20 Ibid., 131; Howell cites Kepler's claim that each planet's speed "est consilio accommodata ad harmonicas rationes."
21 Ibid.
22 See Johnson, "Stone Gods and Counter-Reformation Knowledges."
23 Funkenstein, *Theology and the Scientific Imagination*, 223.
24 Ibid., 224.
25 See Rummel, "Erasmus and the Restoration of Unity in the Church," 65.
26 For a detailed discussion of Kepler's approach in *De Stella Nova*, see Boner, *Kepler's Cosmological Synthesis*, chap. 3.
27 *KGW* 1:346.38–39.
28 See Boner, "Kepler v. the Epicureans."

29 See Boner, "Statesman and Scholar."
30 *KGW* 15:412.
31 See Baldini, "Roman Inquisition's Condemnation of Astrology," 87. For a broader survey of the relationship between astrology and the three major confessions, see Barnes, *Astrology and Reformation*.
32 Barnes, *Astrology and Reformation*, 91. See also *Bullarum, Diplomatum et Privilegiorum Sanctorum Romanorum Pontificum Taurinensis Editio* (1863), 646–47.
33 See Leitão, "Entering Dangerous Ground."
34 Giovanni Antonio Magini, for instance, served as judicial astrologer in the court of Mantua beginning in 1599.
35 In *The History of Magic and Experimental Science*, 6:171, Lynn Thorndike notes the publication of an astrological discourse by Gioanni Bartolini addressed to a cardinal at the papal court in 1618.
36 When it came to the Lutheran world, there were even fewer barriers to the practice of astrology: Melanchthon had specifically promoted astrology as a means to reveal God's providential governance, and astrology was crucial to the formation of the Lutheran confessional culture of the sixteenth century. See Kusukawa, *Transformation of Natural Philosophy*; Barnes, *Astrology and Reformation*; and Dixon, "Popular Astrology and Lutheran Propaganda."
37 *KGW* 15:412.
38 *KGW* 15:424.
39 Ibid.
40 Quoted in Burke-Gaffney, *Kepler and the Jesuits*, 31.
41 See Granada, "After the Nova of 1604."
42 *KGW* 15:424.
43 Ibid.
44 On Kepler's criticism of astrology more generally, see Rabin, "Kepler's Attitude toward Pico and the Anti-astrology Polemic."
45 See, e.g., *KGW* 1:292.1–6.
46 This is not the only instance of Kepler's playfulness, for play itself—as an imitation of God's own creative play—promised to offer insight into the cosmos. See Gerlach, "Humor und Witz in Schriften von Johannes Kepler"; and N. Jardine, "God's 'Ideal Reader.'"
47 Jürgen Hübner, for example, cites this interchange with Herwart as an example of the fact that Kepler embraced certain Catholic teachings openly and actively. See Hübner, *Die Theologie Johannes Keplers*, 155.
48 On Kepler's belief in the a priori nature and the certainty of cosmology, see Martens, *Kepler's Philosophy and the New Astronomy*, 48–56.
49 On the astrologer as a political figure, see chapter 5. See also Grafton, *Cardano's Cosmos*, chap. 7.
50 *KGW* 1:354.25–28.

51 *KGW* 11.2:368.5–7.

52 *KGW* 18:1072.

53 Ibid.

54 Ibid.

55 See Cameron, *Interpreting Christian History*, chap. 3.

56 *KGW* 18:1072.

57 Ibid.

58 *KGW* 18:1080.

59 Ibid.

60 *KGW* 18:1083.

61 Ibid.

62 *KGW* 4:295.13.

63 *KGW* 15:431.

64 Burke-Gaffney, *Kepler and the Jesuits*, 38.

65 Caspar, *Kepler*, 163.

66 Ibid., 164.

67 *KGW* 15:431.

68 *KGW* 15:433.

69 Quoted in Janssen, *History of the German People at the Close of the Middle Ages*, 238–39.

70 See Nelson, "Jesuit Legend," 97.

71 Quoted in ibid., 100.

72 *Life and Letters of Henry Wotton*, vol. 2, letter 372 to Sir George Calvert, July 29, 1622.

73 Quoted in Conley, "Vituperation in Early Seventeenth-Century Historical Studies," 171–72.

74 See Feingold, "Jesuits: Savants," 17.

75 Conley, "Vituperation in Early Seventeenth-Century Historical Studies," 172, 173.

76 See Goldgar, *Impolite Learning*.

77 Ibid., 197.

78 Ibid.

79 Feingold, "Jesuits: Savants," 3.

80 Baldwin, "Pious Ambition," 320.

81 M. Hunter, *Establishing the New Science*, 10.

82 See Loyola, "Constitutions of the Society of Jesus," chap. 12 in *Ignatius of Loyola*, 297.

83 Feingold, "Jesuits: Savants," 2.

84 *KGW* 15:398.

85 *KGW* 13:23.

86 *KGW* 13:91.

87 *KGW* 13:89.

88 *KOO* 4:446.

89 See, e.g., Verkamp, "Limits upon Adiaphoristic Freedom"; Jaquette, *Discerning What Counts;* and Manschreck, "Role of Melanchthon in the Adiaphora Controversy."

90 See Kolb and Nestingen, *Sources and Contexts of the Book of Concord,* 184.

91 See Manschreck, *Melanchthon,* 286.

92 See Cusa, *Nicholas of Cusa's "De Pace Fidei,"* 37.

93 See De Dominis, *De Republica Ecclesiastica,* 4.8, 676.

94 *KGW* 18:1083.

95 *Nova Kepleriana,* 6:13.33–14.7.

96 See Mitchell, *Paul and the Rhetoric of Reconciliation,* 130–37.

97 Quoted in ibid., 134.

98 *KGW* 12:23.30–38.

99 *KGW* 12:27.14–18.

100 *KGW* 12:27.22–25.

CHAPTER 4

1   *KOO* 8.2:672.

2   *KGW* 4:344.15–16.

3   See Skinner, *Reason and Rhetoric in the Philosophy of Hobbes.*

4   Quoted in ibid., 83.

5   See S. Jones, *Calvin and the Rhetoric of Piety,* 20.

6   Quoted in ibid. (See Cicero, *Orator* 8.24.)

7   Ibid., 21.

8   Wengart, *Human Freedom, Christian Righteousness,* 48.

9   See Plett, *Rhetoric and Renaissance Culture.*

10 See Frost, *Introduction to Classical Legal Rhetoric,* 67.

11 Serjeantson, "Proof and Persuasion," esp. 147–48.

12 See Grafton, *What Was History?,* 96–122.

13 Huijgen, *Divine Accommodation in John Calvin's Theology,* 68.

14 See, e.g., Prosperi, "Missionary." See also O'Malley et al., *Jesuits,* in particular part 4, "Encounters with the Other," and part 5, "Tradition, Innovation, Accommodation."

15 See Ginzberg, *History, Rhetoric, and Proof,* chap. 3.

16 Quoted in ibid., 78.

17 See, e.g., Sommerville, "'New Art of Lying.'"

18 See Serjeantson, "Proof and Persuasion."

19 Shapiro, *Culture of Fact.* See also Shapin, *Social History of Truth.*

20 Biagioli, *Galileo's Instruments of Credit,* 1.

21 For other analyses of Kepler's use of rhetorical techniques, see the discussion of the form of Kepler's *Apologia for Tycho against Ursus* in N. Jardine, *Birth of the History and Philosophy of Science;* see also the analysis of the rhetorical nature of the *New Astronomy* in Voelkel, *Composition of Kepler's*

"Astronomia Nova"; finally, see the investigation of Kepler as humanist in Grafton, *Defenders of the Text*, chap. 7.

22   *KGW* 13:91.

23   *KGW* 13:40.

24   *KGW* 13:58.

25   See Martens, *Kepler's Philosophy and the New Astronomy*.

26   *KGW* 13:64.

27   *KGW* 13:40.

28   Although Edward Rosen ("Galileo and Kepler," 264) argues that "Kepler sent Galileo a book because he had heard about him," Stillman Drake's (*Galileo Studies*, 123) assertion "that up to this time, Kepler had no more heard of Galileo than Galileo of Kepler" fits better with the textual evidence—in particular, with Kepler's letter to Maestlin in which he mentioned Galileo's letter and described him as "a Paduan mathematician, by the name of Galileo Galilei" (*KGW* 13:75). Although Rosen argues that Kepler was merely struck by the repetition of Galileo's full name in its Latin form (Galileus Galileus), it seems more likely that Kepler had simply not heard of Galileo before receiving his letter. Moreover, as Drake points out, Kepler's comment that he sent his books *"in Italiam"* seems to indicate that he sent them generally to Italy rather than to a specific location (Padua) or person (Galileo).

29   *KGW* 13:73.

30   Ibid. Although Galileo did not specify the particular natural effects whose causes he had used Copernican theory to determine, Kepler guessed that he referred to the theory of the tides—and noted in a letter to Herwart von Hohenburg that if this was the case, he believed that Galileo was wrong (*KGW* 13:91).

31   *KGW* 13:73.

32   *KGW* 13:76.

33   Ibid.

34   Ibid.

35   Ibid.

36   Ibid.

37   Along similar lines, Biagioli (*Galileo's Instruments of Credit*, 26) argues that Galileo relied upon distance as a crucial device by which his authority could be constructed; he posits "knowledge as constituted through a range of distance-based partial perceptions."

38   *KGW* 13:76.

39   *KGW* 13:30.

40   *KGW* 13:28. For a more detailed description of the goblet and Kepler's plans for it, see Mosley, "Objects of Knowledge."

41   *KGW* 13:31.

42   *KGW* 13:80.

43 *KGW* 13:85.

44 Interestingly, Galileo may not have been averse to Kepler's strategy of "deceptive" persuasion. Paul Feyerabend (*Against Method*, 65) argues that in the *Dialogue*, "Galileo . . . uses *psychological tricks* . . . insinuating that the new results which emerge are known and conceded by all, and need only be called to our attention to appear as the most obvious expression of the truth." Maurice Finocchiaro (*Galileo and the Art of Reasoning*, 22) likewise examines the rhetorical and propagandistic elements of the *Dialogue*, which he sees as "an attempt by verbal means and techniques to induce or increase adherence to Copernicanism."

45 Shortly thereafter, Kepler obtained a telescope and published his verification of Galileo's results, titled *Narratio de Jovis Satellitibus* (1611). For a translation and detailed analysis of the two texts, see Pantin, *Discussion avec "Le messager céleste."* For an English translation of the *Dissertatio*, with detailed notes, see Kepler, *Kepler's Conversation with Galileo's "Sidereal Messenger."*

46 For a different interpretation of Kepler's relationship with Galileo, and in particular of Kepler's support of Galileo's *Starry Messenger*, see Biagioli, *Galileo's Instruments of Credit*. My approach follows Biagioli's general depiction of Galileo by emphasizing Kepler's construction of authority. My disagreement with Biagioli's specific interpretation of Kepler's response to the *Starry Messenger* rests primarily on its emphasis on Galileo's prestige at the expense of Kepler's. According to Biagioli, Kepler supported Galileo because by attaching himself to Galileo he would advance his own credit and prestige. Yet as imperial mathematician, Kepler's prestige was at this point far greater than Galileo's. Moreover, Kepler viewed Italian science, and in particular Italian astronomy, as less advanced than the astronomy of Germany; he noted to Galileo, for instance, that he should not be surprised at opposition in Italy, a place "in which parallax—a most familiar matter, and proven matter to all astronomers—has attackers of a highly eminent position and celebrated reputation" (*KGW* 16:584).

47 Galileo, *Discoveries and Opinions of Galileo*, 23.

48 *KGW* 16:573.

49 *KGW* 4:285.26–27.

50 *KGW* 4:286.2–4.

51 *KGW* 4:286.28–30.

52 *KGW* 4:286.32–33.

53 *KGW* 4:290.13–15.

54 *KGW* 4:290.20–22. Kepler here referenced not only Galileo's telescope but also his own myopia.

55 Ibid.

56 *KGW* 4:290.25–27.

57 Kepler likely also had personal reasons for emphasizing Galileo's scholarly precedents. He was not pleased that Galileo had neglected to mention any

of Kepler's own achievements that had contributed to Galileo's work. He wrote to Magini that though he approved of Galileo's book and rejoiced in its claims, "Nevertheless I think (if you read attentively) that I take enough defensive action; and where I could I recalled the thing to its principles." Further, Kepler, like any good humanist, always felt it to be his duty to document his sources fully and precisely, and Galileo's failure to give credit where credit was due would have been seen as an omission on his part. Along these lines, Isabelle Pantin (*Discussion avec "Le messager céleste,"* cv) notes that Kepler's citation of Galileo's sources was intended to add back to the work the crucial element of dialogue with a textual tradition. Yet whatever critique Kepler hoped to convey in his citation of Galileo's sources, he meant it to be mild and subsumed into his larger goal of supporting Galileo's work and the Copernicanism that underpinned it. This is evident in his distress when his book was misunderstood, as we will see below, and in the lengths he took to correct the misunderstanding.

58 *KGW* 4:291.36–37.

59 *KGW* 4:291.37–38.

60 *KGW* 4:292.28–35.

61 *KGW* 4:303.34–35.

62 *KGW* 4:305.6–7.

63 As Elizabeth Spiller (*Science, Reading, and Renaissance Literature*, 119) notes, "Kepler's emphasis on established textual evidence clearly works to create a more recognizable scholarly context for Galileo's remarkable claims. If Galileo is anticipated by Pythagoras or supported by Maestlin, he is less likely to be wrong."

64 *KGW* 16:562.

65 *KGW* 16:565.

66 *KGW* 16:570.

67 *KGW* 16:580.

68 *KGW* 16:575.

69 *KGW* 16:592.

70 *KGW* 16:566.

71 As noted earlier, the misunderstanding may have had some basis in fact, as Kepler was certainly frustrated by Galileo's failure to cite his sources. Yet he expressed his allusions to Galileo's predecessors in ways that he hoped would bolster Galileo's credibility, rather than the reverse. Galileo certainly understood Kepler's text as praise and wrote to the Tuscan minister that Kepler had written approvingly of everything in his book (see *KGW* 16:572). Modern scholars have also overwhelmingly read Kepler's *Dissertatio* as supporting Galileo. See Rosen, "Galileo and Kepler," 264; and Field, "Cosmology in the Work of Kepler and Galileo," 207. Catherine Chevalley ("Kepler et Galilee dans la bataille du *Sidereus Nuncius*," 167) writes that "it seems certain" that Kepler immediately and wholeheartedly

agreed with Galileo's claims. Michele Camerota (*Galileo Galilei e la cultura scientifica nell'età della Controriforma*, 177–78), by contrast, notes the many conflicting threads in Kepler's *Dissertatio* and concludes that Kepler's true motivations and intentions are difficult to discern.

72 *KGW* 16:592.
73 *KGW* 16:584.
74 Ibid.
75 Ibid.
76 Ibid.
77 Ibid.
78 Ibid.
79 See Serjeantson, "Proof and Persuasion," 147–48.
80 *KGW* 16:584.
81 Ibid.
82 Ibid.
83 *KGW* 16:585.
84 *KGW* 16:597.
85 Ibid.
86 Ibid.
87 Ibid.
88 Ibid.
89 Ibid.
90 For examples of the ways that "impolite" or other unruly behavior altered the landscape of the otherwise civil Republic of Letters, see Mulsow, *Die unanständige Gelehrtenrepublik*.
91 *KGW* 16:597.
92 *KGW* 16:584.
93 *KGW* 16:587.
94 *KGW* 16:597.
95 Ibid.
96 *KGW* 16:604.
97 Biagioli, *Galileo's Instruments of Credit*, 81.
98 See ibid., 97–98.
99 Ibid., 119.
100 *KGW* 14:190.
101 Though it is unclear to what exactly Bruce alluded, Drake ("Galileo's 'Platonic' Cosmogony and Kepler's *Prodromus*") has speculated that it was likely Galileo's discussion of the relationship between planetary periods and orbital distances, which he made on the basis of the table from Kepler's *Mysterium Cosmographicum*.
102 *KGW* 14:268.
103 See Biagioli, *Galileo, Courtier*.
104 On the evolving connotations of the term "secrets of nature," see Eamon, *Science and the Secrets of Nature*.

105 See, e.g., Mosley, Jardine, and Tybjerg, "Epistolary Culture, Editorial Practices, and the Propriety of Tycho's Astronomical Letters." See also L. Jardine, *Erasmus, Man of Letters.*

106 See Voelkel, *Composition of Kepler's "Astronomia Nova."*

107 Augustine, *Expositions of the Psalms* 45.4.7, ed. Boulding and Rotelle, 315.

108 *Assayer* (1623), in *Discoveries and Opinions of Galileo*, 238.

109 See Blacketer, *School of God*, 59.

110 Ibid.

111 Kepler, *Kepler's Conversation with Galileo's "Sidereal Messenger,"* 39.

112 Vickers, "Recovery of Rhetoric," 30.

113 Ibid., 31.

114 See M. Spitzer, *Metaphor and Musical Thought.*

115 See Bonds, *Wordless Rhetoric*, 93.

116 See Bartel, *Musica Poetica*, 108.

117 *Harmony of the World*, 128.

118 Ibid., 147.

119 See Vickers, "Recovery of Rhetoric," 38.

120 Remer, *Humanism and the Rhetoric of Toleration*, 5–6.

121 *KGW* 13:107.

122 See N. Jardine, *Birth of the History and Philosophy of Science*, chap. 6.

123 See ibid., chap. 8.

124 See Schmidt-Biggemann, "New Structures of Knowledge," 507.

125 *KGW* 18:1072.

126 *KGW* 18:1083.

127 See also Methuen, *Science and Theology in the Reformation*, esp. chap. 6.

128 *KGW* 12:29.17–19.

129 See Barker, "Role of Religion in the Lutheran Response to Copernicus." Barker also ends his essay with the phrase "God was a Copernican."

CHAPTER 5

1 *KGW* 18:963.

2 On the term *politicus*, see Rubinstein, "History of the Word *Politicus* in Early-Modern Europe."

3 Plato, *Republic* 3.389b, trans. Grube, 64–65.

4 Ginzberg, *Wooden Eyes*, chap. 2, esp. 40.

5 Ibid., 36.

6 See Hankins, *Plato in the Italian Renaissance*; and Allen, *Mysteriously Meant.*

7 Ginzberg, *Wooden Eyes*, 36.

8 See Williams, "'Si Faut-il Voir Si Cette Belle Philosophie.'"

9 Kahn, *Future of Illusion*, 91.

10 See Bireley, *Counter-Reformation Prince*; and Smuts, "Court-Centered Politics and the Uses of Roman Historians."

11 See Malcolm, *Reason of State, Propaganda, and the Thirty Years' War*, 95.

12　See Grafton, "Commentary," 231.

13　Burke, "Tacitism, Skepticism and Reason of State," 484.

14　Smuts, "Court-Centered Politics and the Uses of Roman Historians," 25.

15　See Giglioni, "Philosophy according to Tacitus." Giglioni writes that Tacitus "could be used to foster republican liberty (Niccolò Machiavelli), to promote political realism (Giovanni Botero), to preserve a sphere of intellectual freedom in situations dominated by tyrannical rule (Justus Lipsius), and to claim a divine origin for monarchical regimes (King James I). He could show people how to live safely under tyranny and tyrants and how to secure their power in situations of political instability (Francesco Guicciardini)" (159).

16　Burke, "Tacitism, Skepticism and Reason of State," 485.

17　On early modern Tacitism and in particular the linkage between Tacitism and reason of state, see Soll, *Publishing "The Prince"*; Oestreich, *Neostoicism and the Early Modern State*; Tuck, *Philosophy and Government*, esp. chaps. 2–3; and Schelhase, *Tacitus in Renaissance Political Thought*. On Lipsius, see Morford, "Tacitean *Prudentia* and the Doctrines of Justus Lipsius."

18　Grafton, "Portrait of Justus Lipsius," 382.

19　Giglioni, "Philosophy according to Tacitus," 165.

20　See Stanciu, "Prudence in Lipsius's *Monita et Exempla Politica*."

21　See "Of the Interpretation of Nature," in *Works of Francis Bacon*, ed. Montagu, 2:549: "The morals of Aristotle and of Plato many admire; yet Tacitus breathes more living observations of manners."

22　See "The Wisdom of the Ancients," in ibid., 3:17.

23　Keller, "Mining Tacitus," 193.

24　Strauss, *Hobbes's Critique of Religion and Related Writings*, 113.

25　See Bouwsma, "Renaissance Discovery of Human Creativity."

26　Cited in ibid., 21. See Cervantes, *Adventures of Don Quixote*, 425–26.

27　See also Trinkhaus, *In Our Image and Likeness*.

28　See Rutkin and Charette, "Astrology," 86. See also Azzolini, *The Duke and the Stars*.

29　See Grafton, *Cardano's Cosmos*, 118.

30　See Hayton, "Astrology as Political Propaganda." See also Grafton, *Cardano's Cosmos*, 110.

31　See Evans, *Rudolf II and His World*, chap. 2. See also Hausenblasová, *Der Hof Kaiser Rudolfs II*; and Marshall, *Magic Circle of Rudolf II*.

32　Hayton, "Astrology as Political Propaganda," 63.

33　For a collection of Kepler's horoscopes, see Greenbaum, "Kepler's Astrology."

34　See Rutkin, "Various Uses of Horoscopes," 170.

35　I have used J. V. Field's translation of "On the More Certain Foundations of Astrology" included in her "A Lutheran Astrologer."

36　Ibid., 265.

37　"Judicium astrologicum de Hungaria," in *KOO* 8.2:335.

38 "Discurs von der grossen conjunction," in *KOO* 7:708.

39 See Bauer, "Die Rolle des Hofastrologen und Hofmathematicus als fürstlicher Berater." Much of my discussion here takes Bauer's article as its starting point.

40 *KGW* 16:612.

41 Ibid.

42 Ibid.

43 *KGW* 16:454.

44 See the commentary by Friederike Boockmann on the "Tacitus Übersetzung" in *KGW* 12:367–83.

45 Title page of "Tacitus Übersetzung," in *KGW* 12:103.

46 *KGW* 17:643.

47 Grafton, "Kepler as Reader," 567.

48 See Bauer, "Die Rolle des Hofastrologen und Hofmathematicus als fürstlicher Berater," 40.

49 *KGW* 4:161.32–39.

50 *KGW* 12:238.25–30.

51 *KGW* 12:239.2–4. Of course, critics of uroscopy took it as a primary example of a theoretically inconsistent and empirically invalid procedure, as Kepler no doubt knew. See Stolberg, "Decline of Uroscopy in Early Modern Learned Medicine."

52 See, e.g., Portuondo, *Secret Science.*

53 *KGW* 13:43.

54 *KGW* 13:44.

55 On the dedicatory letter and its role in early modern patronage relationships, see Glomski, "Careerism at Cracow."

56 *KGW* 1:6.11–19.

57 *KGW* 1:7.5–8.23–27.

58 See Dear, *Revolutionizing the Sciences,"* esp. chap. 3. See also P. Smith, *Body of the Artisan.*

59 See Findlen, *Possessing Nature.*

60 See also Biagioli, *Galileo, Courtier.*

61 As he wrote in a letter to Frederick, the Duke of Württemberg—to whom he dedicated the image of the goblet in the *Mysterium Cosmographicum*—he hoped to create an actual "credenza-goblet . . . with a real and actual image of the world and a model of the creation, [representing] the distance that human reason can reach, and the like of which has previously neither been seen nor heard by anyone" (*KGW* 13:28). For a more detailed description of the goblet and Kepler's plans for it, see Mosley, "Objects of Knowledge."

62 *KGW* 1:7.31–36.

63 When Ferdinand II, then Archduke of Styria, expelled all the Lutheran priests and teachers from Graz, Kepler left with them, because of his role as teacher of mathematics at the Lutheran school. He was soon allowed to

return, likely because of those very princely connections that his earlier work had cultivated.

64 *KGW* 14:132.

65 Ibid.

66 *KGW* 14:168.

67 Ibid.

68 Though others from the empire moved to England, Kepler himself had no desire to leave his homeland. When he was ultimately invited to James's court by Henry Wotton, he declined the invitation. As he wrote to Matthias Bernegger in 1620 from Austria, "Mr. Wotonius . . . told me to come to England. Yet I do not think I ought to leave this second home of mine, especially now, when it is suffering so much insult, unless I would be more ungrateful by becoming a burden to my country" (*KGW* 18:891). If Kepler did not hope to move to England, then reaching out to James— and dedicating a book to him in 1619—was a risky move. In 1613 James's daughter married Frederick, ruler of the Palatinate and leader of the German Protestants. Just as Kepler's *Harmony of the World* was set to appear in print in 1619, Frederick joined the cause of the Bohemians against the emperor by accepting their crown. Kepler thus ultimately dedicated his book to a monarch with close ties to the emperor's enemies.

69 The copy to King James, with this inscription, is currently housed at the British Library.

70 *KGW* 16:470.

71 Raz Chen-Morris drew my attention to Kepler's move from Charles V to James I as alternative models of kingship in his talk at the Scientiae conference in Toronto in May 2015.

72 *KGW* 17:799.

73 Patterson, *King James VI and I and the Reunion of Christendom*.

74 Ferrell, *Government by Polemic*, 136.

75 Evans, *Rudolf II and His World*, 82.

76 Patterson, *King James VI and I and the Reunion of Christendom*, 152.

77 *Harmony of the World*, 3–5.

78 See Lloyd, "Introduction," 5. See also Blair, "Authorial Strategies in Jean Bodin," for a list of the various editions of the text.

79 Quoted in Soll, "Empirical History," 300.

80 Quoted in Tentler, "Meaning of Prudence in Bodin," 370.

81 Soll, "Empirical History," 302.

82 Ong, *Ramus, Method, and the Decay of Dialogue*, 32.

83 See McRae, "Ramist Tendencies in the Thought of Jean Bodin," 309.

84 See McRae, "Postscript on Bodin's Connections with Ramism."

85 Bodin was also responding directly to François Hotman's *Francogallia* (1573), which used historical arguments to demonstrate the superiority of popular sovereignty in France. According to Hotman, the constitution of the ancient Francogallia supported a blend of democracy, aristocracy, and

monarchy, in which decisions were made by a common council with supreme authority—a council that Hotman explicitly linked to the modern-
day Estates General. To support his claims, Hotman relied on the musical
metaphor and in particular on the notion of harmony, arguing that in
both music and government, harmony required the coming together of
different voices—and consequently could not be achieved via one absolute sovereign alone. Bodin's discussion of mathematical harmony was
in part a response to Hotman's challenge; he argued not merely that Hotman's understanding of political harmony was flawed but that harmonic
theory supplied a mathematical foundation for absolute monarchy. See
van Orden, *Music, Discipline, and Arms in Early Modern France*, chap. 2.

86 Bodin, *Six Books of the Commonwealth*, 199–200.
87 For a detailed discussion of the proportions and their roots in Archytas, on which my explanation below rests, see Harvey, "Two Kinds of
Equality."
88 Quoted in ibid., 103–4.
89 Ibid., 104.
90 Plato, *Laws* 6.757, trans. Jowett, 121.
91 Aristotle, *Politics* 5.1301b29–1302a8, trans. Reeve, 136. See also Arena,
Libertas and the Practice of Politics in the Late Roman Republic, 104–6.
92 Bodin, *Six Books of the Commonwealth*, 205.
93 Ibid., 206.
94 Harvey ("Two Kinds of Equality") discusses the contention that Plato, in
his *Republic*, suggested the harmonic proportion as the ideal form, but he
dismisses this suggestion as a misunderstanding, especially in light of the
many places where Plato explicitly highlights the geometric proportion.
95 Here, Bodin's definition differed from the one in Archytas. According
to Bodin's conception, 3, 4, 6, 8, 12, 16 was a harmonic series because
the arithmetic mean was inserted between each of the terms of a
geometric set.
96 Bodin, *Six Books of the Commonwealth*, 206.
97 Ibid., 211. On Bodin's discussion of harmonic justice and the linkages
between his understanding of political harmony and the harmonic
tradition more broadly, see also Kouskoff, "Justice arithmetique, justice
geometrique, justice harmonique."
98 Bodin, *Six Books of the Commonwealth*, 212.
99 Harvey ("Two Kinds of Equality," 126) finds its "last echo" in the twelfth-
century grammarian Johannes Tzetzes.
100 On the linkages between Bodin's theory of harmony and his theory of
justice—and in particular, the connections to mathematics and Ramism—
see also Desan, "La justice mathematique de Jean Bodin."
101 *Harmony of the World*, 256.
102 Ibid.
103 *"Institutio oratoria" of Quintilian*, 4.3.14.

104 Ibid., 4.3.1–2.
105 *Harmony of the World*, 257.
106 Ibid., 268.
107 Ibid.
108 Bodin derived this story from Xenophon, *Cyropaedia* 1.3.
109 *Harmony of the World*, 261.
110 Ibid., 267.
111 Ibid., 268.
112 Ibid., 260.
113 Ibid., 275.
114 Ibid.
115 Ibid.
116 As an example of this, Kepler cited with distaste Bodin's description of
the ways that judges assigned fees for their cases. Often, Bodin had noted,
cases that required the most effort were the least profitable; therefore,
harmonic justice allowed judges to assess their cases and demand appropriate fees from the litigants, particularly in cases where the fee from
the republic was insufficient for the effort required by the case. "I leave
this harmonic part-song to its author Bodin as a Frenchman," Kepler
responded. "Among us Germans justice in the chief states and provinces is
kept far away from meanness of that kind, and it is not lawful to demand
anything beyond what is prescribed by law" (ibid., 272).
117 Ibid., 275.
118 Ibid., 276.
119 In fact, it is in the realm of religious harmony that Kepler and Bodin's
views are most similar. Kepler, as we've seen already, argued for a religious
unity that embraced diverse views within it, and he used the language
of harmony in order to do so. Similarly, in Bodin's *Colloquium Heptaplo-
meres*, a religious dialogue between seven men of differing religious views,
harmony figures centrally. As in music, "tones in unison would take away
all sweetness of harmony," so in religion, Bodin's dialogue suggests. See
Bodin, *Colloquium of the Seven about Secrets of the Sublime*, 146. Over the
course of the dialogue, the participants reach no definitive conclusion
about which religion is the truest but end simply by citing the principle of
musical harmony. This ending without resolution suggested that "truth—
especially religious truth—is complex, and each speaker represents a different facet of that multifaceted truth" (Remer, "Dialogues of Toleration,"
308). It suggested as well that religious tolerance was necessary; as truth
was multiple, it was "not a monopoly maintained by force, but a shared
commodity divided between different, mutually tolerant persuasions"
(309). This position has much in common with Kepler's, as we saw in
chapters 2 and 3 when Kepler called for a reunified church that contained
elements of all three major confessions, no one of which had a monopoly
on the truth.

120 Nitschke, "Keplers Staats- und Rechtslehre."

121 *Harmony of the World*, 147. This is also the period in which military music came to be used as a way to keep soldiers marching in unison. See Mc-Neill, *Keeping Together in Time*, 86.

122 In his *Theater of Nature*, Bodin had portrayed harmony in terms that Kepler would have heartily embraced: "far from causing the ruin of the world, the contrariety of the elements would by its absence destroy the world, just as harmony is altogether lost if the intermediates between the high and low voices are removed." See Blair, *Theater of Nature*, 126–42.

123 *Harmony of the World*, 278.

124 Ibid., 304.

125 Hobbes, *Leviathan*, 24.

126 The frontispiece of *Leviathan* reads, "There is no power on earth which compares to him." In Job itself—though not in *Leviathan*'s frontispiece—the text continues: "where were you when I laid the earth's foundations? Who settled its dimensions? . . . Who stretched this measuring line over it?" See Strong, "How to Write Scripture."

127 Hobbes, *On the Citizen*, 4.

128 *Harmony of the World*, 304.

129 See, e.g., Tuck, *Philosophy and Government*, 302.

130 See Pettit, *Made with Words*.

131 Cited in Hull, *Hobbes and the Making of Modern Political Thought*, 70.

132 Cited in Pettit, *Made with Words*, 20.

133 *Harmony of the World*, 139.

134 Cited in Pettit, *Made with Words*, 20.

135 Ibid.

136 Hobbes, *Leviathan*, 35.

137 Quoted in Hull, *Hobbes and the Making of Modern Political Thought*, 21.

138 Hobbes, *Elements of Law*, 75. See also Remer, "Hobbes, the Rhetorical Tradition, and Toleration."

139 See Shapin and Schaffer, *Leviathan and the Air-Pump*, esp. 152.

140 "For I doubt not," Hobbes (*Leviathan*, 78) wrote, "but if it had been a thing contrary to any man's right of dominion . . . that the three angles of a triangle, should be equal to two angles of a square, that doctrine should have been, if not disputed, yet by the burning of all books of geometry, suppressed."

141 Ibid., 21.

142 Hobbes, *On the Citizen*, 5.

143 See, e.g., Probst, "Infinity and Creation."

144 On Hobbes as breaking from Aristotle, see Strauss, *Political Philosophy of Hobbes*, esp. 136–42.

145 See Skinner, "Hobbes and the Purely Artificial Person of the State," esp. 5.

146 To Jakob Bartsch, November 6, 1629, in Caspar and von Dyck, *Kepler in Seinem Briefe*, 2:308.

CHAPTER 6

1   *KGW* 13:93.
2   *KGW* 17:829.
3   *KGW* 21.1:354.4–13.
4   Porter, *Trust in Numbers*, 4.
5   Ibid.
6   Daston, "Objectivity and the Escape from Perspective," 598. See also Daston and Galison, *Objectivity*.
7   Daston, "Objectivity and the Escape from Perspective," 609.
8   Daston, "Moral Economy of Science," 10.
9   See Nagel, *View from Nowhere*.
10 See Murphy and Traninger, "Introduction," in *The Emergence of Impartiality*, 4. Much of my discussion of impartiality in this introductory section is indebted to this book.
11 Ibid., 26.
12 See Berns, *"Partheylichkeit* and the Periodical Press," 128.
13 See Murphy and Traninger, "Introduction," 2.
14 See, e.g., Forhan, "Respect, Interdependence, Virtue," 79–80.
15 See, e.g., Henry Cockeram's dictionary of 1623, which defines neutrality as "a wretchless Being on neither side." See also the text of the British Solemn League and Covenant of 1643: "We . . . shall not . . . give ourselves to a detestable indifference or neutrality in this cause, which so much concerneth the glory of God, the good of the kingdoms, and the honour of the king" (Cressy and Ferrell, *Religion and Society in Early Modern England*, 214). See also Craig's ("Neutrality in the Nineteenth Century," 143–44) description of Gustavus Adolphus's attitude toward the elector of Brandenburg and his supposed neutrality in 1630: "God and the Devil are fighting here. . . . *Was ist doch das für ein Ding: Neutralität?* I don't understand it."
16 Scholar, "Reasons for Holding Back in Two Essays of Montaigne."
17 "Of Husbanding Your Will," in *Complete Essays of Montaigne*, 774.
18 See Murphy and Traninger, "Introduction," 22. See also Parker, "To the Attentive, Nonpartisan Reader," esp. 76–77.
19 Murphy and Traninger, "Introduction," 24.
20 See Hanegraaf, *Esotericism and the Academy*, 124.
21 Berns, *"Partheylichkeit* and the Periodical Press," 98.
22 Murphy and Traninger, "Introduction," 12.
23 See Comeanu, *Regimens of the Mind*, 11.
24 *KGW* 12:29.17–19.
25 On Gregorian calendar reform, see Kaltenbrunner, "Die Vorgeschichte der Gregorianischen Kalenderreform"; and Coyne, Hoskin, and Pedersen, *Gregorian Reform of the Calendar*. See also Kaltenbrunner, "Die Polemik"; and Stieve, *Der Kalenderstreit*.

26 See, e.g., Dixon, "Urban Order and Religious Coexistence in the German Imperial City."

27 Ibid., 12.

28 On the particular reactions of Heerbrand and Maestlin to the Gregorian reforms, see Methuen, *Science and Theology in the Reformation*, chap. 5.

29 Coyne, Hoskin, and Pedersen, *Gregorian Reform of the Calendar*, 260.

30 Quoted in "Kepler and His Discoveries," 341.

31 On Maestlin's approach to the Gregorian calendar reforms, see Jarrell, "Life and Scientific Work of the Tübingen Astronomer Michael Maestlin," 46–64.

32 Quoted in ibid., 55.

33 Ibid., 59.

34 See Coyne, Hoskin, and Pedersen, *Gregorian Reform of the Calendar*, part 6; Whitrow, *Time in History*, 118; and Janssen, *History of the German People at the Close of the Middle Ages*, 55. Indeed, Kepler's preference for the Gregorian calendar is even cited—wrongly—as the basis for his excommunication in Linz (see, e.g., Lamont, "Reform of the Julian Calendar").

35 *KGW* 13:63.

36 Another notable Protestant who supported the new calendar was Tycho Brahe, Kepler's soon-to-be employer in Prague. Tycho believed that the issues surrounding the calendar reform were largely political rather than theological; further, he argued that its introduction by the pope was warranted, since only the pope had the kind of authority necessary for such a universal change. Tycho likewise emphasized that there would always be problems inherent in any calendar, for no calendar could perfectly represent the movements of the heavens. See Kaltenbrunner, "Die Polemik," 574.

37 A nail, or a reference to Clavius, the calendar's chief promoter.

38 *KGW* 13:64.

39 Ibid.

40 Ibid.

41 Ibid.

42 Ibid.

43 Ibid.

44 *KGW* 13:112.

45 *KGW* 13:76.

46 *KGW* 16:584.

47 *KGW* 15:400. Interestingly, Christoph Scheiner lodged the same complaint against the Holy Roman Empire that Kepler lodged against Italy, citing Kepler himself as the only exception to the sad state of mathematics he found there. Writing about his visit, he noted that "the muses of mathematicians mourn on the shores here: there is no mathematician of any name in Prague, none in Bohemia, none in Moravia, no one in Austria (excepting Kepler, living in Linz, whom I will visit in my ascent)"

(*KGW* 17:755). Yet while for Scheiner this was simply a comment about the problems found in a given location, for Kepler it seems to have been more about national character, as he even included Clavius among the Germans.

48 *KGW* 17:761.

49 *KGW* 13:64.

50 For a discussion of the dating of the dialogue, see Friederike Boockmann's commentary on Kepler's calendar manuscripts in *KGW* 21.1:642–67, esp. 649. Kepler mentions a recent Reichstag in the dialogue, which Christian Frisch and others have assumed to be the Reichstag held in Regensburg in 1613, during the reign of Rudolf II. Yet Boockmann notes that Kepler refers to his time with Tycho as relatively recent, and Tycho died in 1601. Likewise, Kepler refers to works written through 1603 and relies upon datings of the equinox through 1604. Finally, he writes that it has been "around twenty years" since the empire began operating with two different calendars; the Catholic states of the empire adopted the calendar in 1583. Thus, the later drafts of "Ein Gespräch von der Reformation des alten Calenders" seem to date to 1604–9.

51 *KGW* 17:629.

52 *KGW* 21.1:351.27–29.

53 *KGW* 21.1:352.1–3.

54 *KGW* 21.1:352.9–10.

55 *KGW* 21.1:353.41–43.

56 *KGW* 21.1:354.1–2.

57 *KGW* 21.1:354.4–13.

58 *KGW* 21.1:354.14–15.

59 *KGW* 21.1:354.23–27.

60 *KGW* 21.1:354.28–29.

61 *KGW* 21.1:354.31–35.

62 *KGW* 21.1:354.43–45.

63 *KGW* 21.1:355.6–7.

64 *KGW* 21.1:355.15–16.

65 *KGW* 21.1:355.33–34.

66 *KGW* 21.1:355.37–41.

67 *KGW* 21.1:356.10–15.

68 *KGW* 21.1:356.27–29.

69 *KGW* 21.1:356.29–44.

70 *KGW* 21.1:377.35–36.

71 *KGW* 21.1:387.1–2.

72 See N. Jardine, *Birth of the History and Philosophy of Science*, chaps. 6–7.

73 *KGW* 21.1:395.34–35.

74 *KGW* 21.1:395.37.

75 *KGW* 21.1:397.27–28.

76 *KGW* 21.1:397.28–32.

77 *KGW* 15:417.

78 *KGW* 21.1:399.18–20.

79 See Westman, "Astronomer's Role in the Sixteenth Century."

80 *On the Revolutions*, in *Complete Works, Nikolaus Kopernikus*, 7.

81 Ibid., 5.

82 Ash, "Introduction," 8.

83 Leibniz similarly argued for mathematicians as social mediators; see Mercer, *Leibniz's Metaphysics*, 49–59. See also M. Jones, *The Good Life in the Scientific Revolution*, 246–47.

84 Proclus, commentary on book 1 of the *Elements* of Euclid, cited in *Harmony of the World*, preface to book 3, 127.

85 See Gingerich, "Civil Reception of the Gregorian Calendar."

86 Vallée, "Fellowship of the Book," 43.

87 Lim, "Christians, Dialogues, and Patterns of Sociability," 170.

88 "Est enim . . . epistola absentium amicorum quasi mutuus sermo." Quoted in L. Jardine, *Erasmus, Man of Letters*, 150.

89 Grafton, *Worlds Made by Words*; and Goldgar, *Impolite Learning*.

90 See, e.g., Mulsow, *Die unanständige Gelehrtenrepublik*.

91 Chordas, "Dialogue, Utopia, and the Agencies of Fiction," 27.

92 Ibid., 31.

93 See, e.g., Wilson, *Incomplete Fictions*; and Houston, *Renaissance Utopia*.

94 See "Apology for Raymond Sebond," in *Complete Essays of Montaigne*, 377.

95 See Rigolot, "Problematizing Renaissance Exemplarity," 13–16; Bénouis, *Le dialogue philosophique de la littérature française de seizième siècle*, 66; Frame, "Montaigne and the Problem of Consistency"; and Good, *Observing Self*, 1–6.

96 Bakhtin, *Problems of Dostoevsky's Poetics*, 40.

97 Ibid., 6.

98 Ibid., 49.

99 Ibid., 6.

CONCLUSION

1 Crary, *Techniques of the Observer*, 27.

2 See Henry Wotton's 1620 letter to Francis Bacon, describing Kepler's camera obscura (*Reliquiae Wottoniae*, 413–14).

3 See Della Porta, *Natural Magick*, 364–65.

4 *Optics*, 209. Though the analogy itself delighted Kepler, as Della Porta had hoped, Kepler's claims of pleasure were made with some irony, as he proceeds to explain why Della Porta was wrong in the specifics of his explanation.

5 Ibid., 173. Even before the publication of his *Optics*, he referred to the pupil as the "aperture" of the eye in a letter to Michael Maestlin of 1601.

6   See Lindberg, *Theories of Vision from Al-Kindi to Kepler*, for the first posi-
    tion, and Straker, "Kepler's *Optics*," for the second.

7   See Crary, *Techniques of the Observer*. His larger argument is that the cam-
    era obscura changed from being a metaphor for objectivity and knowl-
    edge in the early modern period to a "model for procedures and forces
    that conceal, invert, and mystify truth" (29) in later invocations by Marx,
    Bergson, and Freud, for example.

8   Kitao, "*Imago* and *Pictura*," 509.

9   Ibid., 510.

10 See Alpers, *Art of Describing*, 33–34.

11 Chen-Morris and Gal, *Baroque Science*, 24.

12 Ibid., 25.

13 *Optics*, 236, quoted in Chen-Morris and Gal, *Baroque Science*, 25.

14 For a far more detailed discussion of Kepler's *Optics* and its epistemological
    implications, see Chen-Morris, *Measuring Shadows*.

15 See, for example, the discussion in chapter 4 of this book about Kepler
    taking issue with Galileo's own presentation of his telescopic discoveries
    precisely because Galileo has severed these communal ties and pretended
    to speak as an isolated observer.

16 Jean Pelerin, *De Artificiali Perspectiva*, 1505: "Les quantitez et les distances,
    Ont concordables differences."

17 *Optics*, 56.

18 Ibid., 59.

19 In his *Optics*, Kepler labeled the inverted retinal image a *pictura*, a paint-
    ing, in contrast to the *imago*, the image of the world outside the eye.

20 Panofsky, *Perspective as Symbolic Form*, 67.

21 *Optics*, 336.

22 *KGW* 7:277.20–27.

23 *Harmony of the World*, 291.

24 Ibid., 314.

25 Ibid., 496. On this theme, see also Hallyn, *Poetic Structure of the World*,
    243–46.

26 *KGW* 1:253.15–17.

27 Field, *Kepler's Geometrical Cosmology*, 18. See also the discussion by
    Heller-Roazen in *Fifth Hammer*, 140.

28 Claessens, "Imagination as Self-Knowledge," 197.

29 *New Astronomy*, 60.

30 *KGW* 6:416.27–34. Translated by Boner in *Kepler's Cosmological
    Synthesis*, 161.

31 Arendt, *Human Condition*, 263.

32 Westman, *Copernican Question*, 399.

33 *KGW* 11.2:345.36–41.

34 See also Spiller, *Science, Reading, and Renaissance Literature*, esp. 119–36.

35 *KGW* 11.2:333.7–15.

36   *KGW* 11.2:350.53–54.
37   *KGW* 11.2:354.23–27.
38   Chen-Morris, "Shadows of Instruction," 241.
39   Reiss, *Discourse of Modernism*, 144–47.
40   Galileo, *Dialogue concerning the Two Chief World Systems*, 536.
41   Reiss, *Discourse of Modernism*, 166.
42   Arendt, *Human Condition*, 286.
43   *KGW* 1:6.7–10.
44   Alberti, *On the Art of Building in Ten Books*, 305, 301.
45   Ibid., 24.
46   Ibid., 303.
47   Ibid., 24.
48   Ibid.
49   *Vitruvius, On Architecture*, 105. See also Payne, "Creativity and Bricolage in Architectural Literature of the Renaissance."
50   Alberti, *On the Art of Building in Ten Books*, 310.
51   See Trachtenberg, "Building outside Time."
52   See also Joost-Gaugier, *Pythagoras and Renaissance Europe*, 162–80.
53   Alberti, *On the Art of Building in Ten Books*, 156.
54   Ibid., 310.
55   Ibid., 318–19.
56   Trachtenberg, "Building outside Time," 126.
57   See Hon and Goldstein, *From* Summetria *to* Symmetry, esp. 157–77. See also Westman, "Proof, Poetics, and Patronage." Martin Kemp likewise explicitly links the Vitruvian ideal to both Alberti and Copernicus in "Temples of the Body and Temples of the Cosmos."
58   Copernicus, *On the Revolutions*, in *Complete Works, Nikolaus Kopernikus*, 4.
59   Ibid.
60   Ibid.
61   Ibid., 22.
62   *KGW* 8:32.42, 33.1–4.
63   *KGW* 8:33.8–9.
64   *KGW* 1:62.4–7, 14–19.
65   *Harmony of the World*, 488.
66   *KGW* 1:60.14–15.
67   For more on the image and its relationship to Hebenstreit's poem, see N. Jardine, Leedham-Green, and Lewis, "Johann Baptist Hebenstreit's *Idyll on the Temple of Urania*."
68   Gattei, "Engraved Frontispiece of Kepler's *Tabulae Rudolphinae*." On the frontispiece, see also Arnulf, "Das Titelbild der Tabulae Rudolphinae des Johannes Kepler."
69   Gattei, "Engraved Frontispiece of Kepler's *Tabulae Rudolphinae*," 346.
70   Ibid., 365.
71   See M. Jones, *The Good Life in the Scientific Revolution*, 169–266.

72 Ibid., 212.
73 Ibid.
74 Ibid., 246.
75 *KGW* 12:27.16–18.
76 *KGW* 16:600.
77 Appiah, *Cosmopolitanism*, xiv.
78 See McMahon, "Fear and Trembling," esp. 8.
79 See Plutarch, *On the Fortune of Alexander* 326b, quoted in Pagden, "Stoicism, Cosmopolitanism," 3.
80 Cicero, *On the Nature of the Gods* 2.61.154–56 (*Treatises of M. T. Cicero*, 101).
81 Quoted in McMahon, "Fear and Trembling," 6 (*Tusculan Disputations* 5.37).
82 Aurelius, *Meditations*, 24. See also A. Long, "Concept of the Cosmopolitan in Greek and Roman Thought."
83 Aurelius, *Meditations*, 29.
84 Pagden, "Stoicism, Cosmopolitanism," 5.
85 Galatians 3:28.
86 Cicero, *On the Nature of the Gods* 2.61.154–56 (*Treatises of M. T. Cicero*, 101).
87 See also Beercroft, *Ecology of World Literature*, esp. 137.
88 Laertius, *Lives of Eminent Philosophers*, 2:145.
89 See Erasmus's 1521 "Complaint of Peace," quoted in McMahon, "Fear and Trembling," 13. McMahon quotes the subsequent two claims by Montaigne and Lipsius that I cite as well, and I am indebted to his account.
90 McMahon, "Fear and Trembling," 13.
91 "On Vanity," in *Complete Essays of Montaigne*, 743.
92 Lipsius, "First Book on Constancy," 404.
93 Toulmin, *Cosmopolis*, 55.
94 As I've argued here, Kepler, though temporally in the later part of the story, aligns more with Erasmus and Montaigne than he does with the Descartes or Hobbes of Toulmin's narrative.
95 *Harmony of the World*, 252.
96 *KGW* 1:349.11–13, 29–30.
97 *KGW* 1:330.7–8, 348.8–9.
98 *KGW* 1:349.39–41.
99 Appiah, *Cosmopolitanism*, xv.
100 Pagden, "Stoicism, Cosmopolitanism," 20.
101 *KGW* 18:1111.
102 *Harmony of the World*, 263.
103 See Eden, *Friends Hold All Things in Common*.
104 Nussbaum, "Patriotism and Cosmopolitanism," condensed in *For Love of Country*, 5.
105 Ibid., 15.
106 Ibid.
107 Ibid., 9.

108 Nussbaum, "Towards a Globally Sensitive Patriotism," 80.
109 Ibid.
110 Arendt, *Human Condition*, 57.
111 *Harmony of the World*, 308.
112 Ibid., 311.
113 *KGW* 1:60.14–15.

Bibliography

Alberti, Leon Battista. *On the Art of Building in Ten Books*. Translated by Joseph Rykwert, Neil Leach, and Robert Tavernor. Cambridge, MA: MIT Press, 1988.

Allen, Don Cameron. *Mysteriously Meant: The Rediscovery of Pagan Symbolism and Allegorical Interpretation in the Renaissance*. Baltimore: Johns Hopkins University Press, 1970.

Alpers, Svetlana. *The Art of Describing: Dutch Art in the Seventeenth Century*. Chicago: University of Chicago Press, 1983.

Andreae, Johann Valentin. *Selbstbiographie*. Translated by David Christoph Seybold. Winterthur: Steiner, 1799.

Andritsch, Johann. "Gelehrtenkreise um Johannes Kepler in Graz." In *Johannes Kepler: 1571–1971*, edited by Paul Urban and Berthold Sutter, 159–95. Graz: Leykam-Verlag, 1975.

Antognazza, Maria Rosa. "Leibniz and the Post-Copernican Universe." *Studies in the History and Philosophy of Science* 34 (2003): 309–29.

———. *Leibniz on the Trinity and the Incarnation*. New Haven, CT: Yale University Press, 2008.

Appiah, Kwame Anthony. *Cosmopolitanism: Ethics in a World of Strangers*. New York: W. W. Norton, 2006.

Aquinas, Thomas. *Aquinas: Political Writings*. Translated by R. W. Dyson. Cambridge: Cambridge University Press, 2002.

Arena, Valentina. *Libertas and the Practice of Politics in the Late Roman Republic*. Cambridge: Cambridge University Press, 2013.

Arendt, Hannah. *The Human Condition*. 2nd ed. Chicago: University of Chicago Press, 1998.

Ariew, Roger. *Descartes and the Last Scholastics*. Ithaca, NY: Cornell University Press, 1999.

Aristotle. *Metaphysics*. Translated by Hugh Tredennick. Cambridge, MA: Harvard University Press, 1989.

———. *Politics*. Translated by C. D. C. Reeve. Indianapolis: Hackett, 1997.

Arnulf, Arwed. "Das Titelbild der Tabulae Rudolphinae des Johannes Kepler." *Zeitschrift des deutschen Vereins für Kulturwissenschaft* 54/55 (2000): 176–98.

Ash, Eric H. "Introduction: Expertise and the Early Modern State." In *Expertise: Practical Knowledge and the Early Modern State*, edited by Eric H. Ash, 1–24. Chicago: University of Chicago Press, 2010.

Augustine. *Expositions of the Psalms*. Translated by Maria Boulding and John Rotelle. New York: New City Press, 2000.

Aurelius, Marcus. *Meditations: With Selected Correspondence*. Translated by Robin Hard. Oxford: Oxford University Press, 2011.

Azzolini, Monica. *The Duke and the Stars: Astrology and Politics in Renaissance Milan*. Cambridge, MA: Harvard University Press, 2013.

Bacon, Francis. *The Wisedome of the Ancients*. Translated by Sir Arthur Gorges. London, 1619.

———. *The Works of Francis Bacon: Philosophical Works*. Edited by James Spedding, Robert Leslie Ellis, and Douglas Denon Heath. New York: Hurd and Houghton, 1864.

———. *The Works of Francis Bacon, Lord Chancellor of England*. Edited by Basil Montagu. London: William Pickering, 1825.

Bainton, Roland. "The Development and Consistency of Luther's Attitude to Religious Liberty." *Harvard Theological Review* 22, no. 2 (1929): 107–49.

———. "Luther's Attitudes on Religious Liberty." In *Studies on the Reformation*, by Roland Bainton, 20–45. Boston: Beacon Press, 1963.

———. *The Travails of Religious Liberty: Nine Biographical Studies*. Eugene: Wipf and Stock, 2008.

Bakhtin, Mikhail. *Problems of Dostoevsky's Poetics*. Translated by Caryl Emerson. Minneapolis: University of Minnesota Press, 1984.

Baldini, Ugo. "The Roman Inquisition's Condemnation of Astrology: Antecedents, Reasons, and Consequences." In *Church, Censorship, and Culture in Early Modern Italy*, edited by Gigliola Fragnito, translated by Adrian Belto, 79–110. Cambridge: Cambridge University Press, 2001.

Baldwin, Martha. "Pious Ambition: Natural Philosophy and the Jesuit Quest for the Patronage of Printed Books in the Seventeenth Century." In *Jesuit Science and the Republic of Letters*, edited by Mordechai Feingold, 285–330. Cambridge, MA: MIT Press, 2003.

Barker, Peter. "Kepler's Epistemology." In *Method and Order in Renaissance Philosophy*, edited by Daniel A. DiLiscia, Eckhard Kessler, and Charlotte Methuen, 355–68. Burlington, VT: Ashgate, 1997.

———. "The Role of Religion in the Lutheran Response to Copernicus." In *Rethinking the Scientific Revolution*, edited by Margaret J. Osler, 59–88. Cambridge: Cambridge University Press, 2000.

Barker, Peter, and Bernard Goldstein. "Realism and Instrumentalism in Sixteenth-Century Astronomy: A Reappraisal." *Perspectives on Science* 6, no. 3 (1998): 232–58.

Barnes, Robin B. *Astrology and Reformation*. Oxford: Oxford University Press, 2015.

Bartel, Dietrich. *Musica Poetica: Musical-Rhetorical Figures in German Baroque Music*. Lincoln: University of Nebraska Press, 1997.

Bauer, Barbara. "Die Rolle des Hofastrologen und Hofmathematicus als fürstlicher Berater." In *Hoefischer Humanismus*, edited by A. Buck, 93–117. Weinheim: VCH, 1989.

Beercroft, Alexander. *An Ecology of World Literature: From Antiquity to the Present Day*. London: Verso, 2015.

Benin, Stephen D. "The 'Cunning of God' and Divine Accommodation." *Journal of the History of Ideas* 45, no. 2 (1984): 179–91.

Bénouis, M. K. *Le dialogue philosophique de la littérature française de seizième siècle*. Berlin: Mouton, 1976.

Berns, J. J. "*Partheylichkeit* and the Periodical Press." In *The Emergence of Impartiality*, edited by Kathryn Murphy and Anita Traninger, translated by Pamela E. Selwyn, 85–139. Leiden: Brill, 2013.

Biagioli, Mario. *Galileo, Courtier: The Practice of Science in the Culture of Absolutism*. Chicago: University of Chicago Press, 1994.

———. *Galileo's Instruments of Credit: Telescopes, Images, Secrecy*. Chicago: University of Chicago Press, 2006.

Bireley, Robert. *The Counter-Reformation Prince: Anti-Machiavellianism or Catholic Statecraft in Early Modern Europe*. Chapel Hill: University of North Carolina Press, 1990.

Blacketer, Raymond Andrew. *The School of God: Pedagogy and Rhetoric in Calvin's Interpretation of Deuteronomy*. Berlin: Springer Science and Business Media, 2007.

Blackwell, Richard J. *Galileo, Bellarmine, and the Bible*. Notre Dame, IN: University of Notre Dame Press, 1991.

Blair, Ann. "Authorial Strategies in Jean Bodin." In *The Reception of Bodin*, edited by Howell A. Lloyd, 137–56. Leiden: Brill, 2013.

———. *The Theater of Nature: Jean Bodin and the Renaissance*. Princeton, NJ: Princeton University Press, 1997.

Bodin, Jean. *Colloquium of the Seven about Secrets of the Sublime*. Edited and translated by Marion Leathers Daniels Kuntz. Princeton, NJ: Princeton University Press, 1975.

———. *The Six Books of a Common-weale*. Translated by Richard Knolles. London: Adam Islip, 1606.

———. *Six Books of the Commonwealth*. Translated by M. J. Tooley. Oxford: Blackwell, 1955.

Bonds, Marc Evan. *Wordless Rhetoric: Musical Form and the Metaphor of the Oration*. Cambridge, MA: Harvard University Press, 1991.

Boner, Patrick J. *Kepler's Cosmological Synthesis: Astrology, Mechanism, and the Soul*. Leiden: Brill, 2013.

———. "Kepler v. the Epicureans: Causality, Coincidence, and the Origins of the New Star of 1604." *Journal for the History of Astronomy* 38 (2007): 207–21.

———. "Statesman and Scholar: Herwart von Hohenburg as Patron and Author in the Republic of Letters." *History of Science* 52, no. 1 (2014): 29–51.

Bono, James J. *The Word of God and the Languages of Man: Interpreting Nature in Early Modern Science and Medicine*. Madison: University of Wisconsin Press, 1995.

Bouwsma, William J. "The Renaissance Discovery of Human Creativity." In *Humanity and Divinity in Renaissance and Reformation*, edited by John O'Malley, Thomas M. Izbicki, and Gerald Christianson, 17–34. Leiden: Brill, 1993.

Burke, Peter. "Tacitism, Skepticism and Reason of State." In *The Cambridge History of Political Thought, 1450–1700*, edited by J. H. Burns, 479–98. Cambridge: Cambridge University Press, 1994.

Burke-Gaffney, M. W. *Kepler and the Jesuits*. Milwaukee, WI: Bruce, 1944.

Calvin, John. *Institutes of the Christian Religion*. Edited by John T. McNeill. Translated by Ford Lewis Battles. Louisville, KY: Westminster John Knox Press, 1960.

Cameron, Euan. *Interpreting Christian History: The Challenge of the Churches' Past*. Oxford: Blackwell, 2005.

Camerota, Michele. *Galileo Galilei e la cultura scientifica nell'età della Controriforma*. Rome: Salerno, 2004.

Carabine, Deirdre. *The Unknown God: Negative Theology in the Platonic Tradition; Plato to Eriugena*. Grand Rapids, MI: Eerdmans, 1995.

Cardano, Girolamo. *Writings on Music*. Edited by Clement Albin Miller. Middleton: American Institute of Musicology, 1973.

Cary, Phillip. *Augustine's Invention of the Inner Self: The Legacy of a Christian Platonist*. Oxford: Oxford University Press, 2000.

Caspar, Max. *Kepler*. Edited and translated by C. Doris Hellman. New York: Dover, 1993.

Caspar, Max, and Walter von Dyck. *Kepler in Seinem Briefe*. Berlin: Oldenburg, 1930.

Cervantes, Miguel de. *The Adventures of Don Quixote*. Translated by J. M. Cohen. London: Penguin Books, 1963.

Chen-Morris, Raz. *Measuring Shadows: Kepler's Optics of Invisibility*. University Park: Pennsylvania State University Press, 2016.

———. "Shadows of Instruction: Optics and Classical Authorities in Kepler's *Somnium*." *Journal of the History of Ideas* 66, no. 2 (2005): 223–43.

Chen-Morris, Raz, and Ofer Gal. *Baroque Science*. Chicago: University of Chicago Press, 2013.

Chevalley, Catherine. "Kepler et Galilee dans la bataille du *Sidereus Nuncius*." In *Novità celesti e crisi del sapere: Atti del Convegno internationale di studi Galileiani*, edited by Paolo Galluzzi, 165–75. Florence: Giunti Barbera, 1984.

Chordas, Nina. "Dialogue, Utopia, and the Agencies of Fiction." In *Printed Voices: The Renaissance Culture of Dialogue*, edited by Dorothea B. Heitsch and Jean-François Vallée, 27–41. Toronto: University of Toronto Press, 2004.

Cicero, M. T. *The Treatises of M. T. Cicero on the Nature of the Gods; On Divination; On Fate; On the Republic; On the Laws; and on Standing for the Consulship*. Translated by C. D. Yonge. London: Bohn, 1853.

Claessens, Guy. "Imagination as Self-Knowledge: Kepler on Proclus' *Commentary on the First Book of Euclid's 'Elements.'*" *Early Science and Medicine* 16 (2011): 179–99.

Comeanu, Sorana. *Regimens of the Mind: Boyle, Locke, and the Early Modern Cultura Animi Tradition*. Chicago: University of Chicago Press, 2012.

Conley, Thomas. "Vituperation in Early Seventeenth-Century Historical Studies." *Rhetorica: A Journal of the History of Rhetoric* 22, no. 2 (2004): 169–82.

Cook, Harold J. "Body and Passions: Materialism and the Early Modern State." *Osiris* 17 (2002): 25–48.

Copernicus, Nicholas. *The Complete Works, Nikolaus Kopernikus*. Edited by Jerzy Dobrzycki and Edward Rosen. Translated by Edward Rosen. Vol. 2. London: Macmillan, 1978.

Coudert, Allison. *Religion, Magic, and Science in Early Modern Europe and America*. Santa Barbara, CA: Praeger, 2011.

Coudert, Allison, and Taylor Corse. *Franciscus Mercurius van Helmont: The Alphabet of Nature*. Leiden: Brill, 2007.

Coyne, G. V., M. A. Hoskin, and O. Pedersen. *Gregorian Reform of the Calendar: Proceedings of the Vatican Conference to Commemorate Its 400th Anniversary, 1582–1992*. Vatican City: Pontifical Academy of Sciences, Specolo Vaticano, 1983.

Craig, Gordon. "Neutrality in the Nineteenth Century." In *War, Politics, and Diplomacy: Selected Essays*, 143–52. Munich: Frederick A. Praeger, 1966.

Crary, Jonathan. *Techniques of the Observer: On Vision and Modernity in the Nineteenth Century*. Cambridge, MA: MIT Press, 1992.

Creese, David. *The Monochord in Ancient Greek Harmonic Science*. Cambridge: Cambridge University Press, 2010.

Cressy, David, and Lori Anne Ferrell, eds. *Religion and Society in Early Modern England: A Sourcebook*. London: Routledge, 2007.

Cusa, Nicholas of. *Nicholas of Cusa's "De Pace Fideo" and "Cribratio Alkorani": Translation and Analysis*. Edited by Jasper Hopkins. Minneapolis, MN: A. J. Banning Press, 1994.

———. *Of Learned Ignorance*. Translated by Germain Heron. Eugene, OR: Wipf and Stock, 2007.

Daly, James. *Cosmic Harmony and Political Thinking in Early Stuart England*. Philadelphia: American Philosophical Society, 1979.

Daston, Lorraine. "The Moral Economy of Science." *Osiris* 10 (1995): 2–24.

———. "Objectivity and the Escape from Perspective." *Social Studies of Science* 22, no. 4 (1992): 597–618.

Daston, Lorraine, and Peter Galison. *Objectivity.* New York: Zone Books, 2010.

Daston, Lorraine, and Katharine Park. *Wonders and the Order of Nature, 1150–750.* New York: Zone Books, 2001.

Dear, Peter. *Revolutionizing the Sciences: European Knowledge and Its Ambitions, 1500–1700.* Princeton, NJ: Princeton University Press, 2001.

De Dominis, Marcus Antonius. *De Republica Ecclesiastica.* London, 1617.

Della Porta, Giovanni Battista. *Natural Magick in Twenty Books.* Edited by Thomas Young and Samuel Speed. London, 1658.

Desan, P. "La justice mathematique de Jean Bodin." *Corpus* 4 (1987): 19–29.

Dickreiter, Michael. *Der Musiktheoretiker Johannes Kepler.* Munich: Francke, 1973.

Dickson, Donald R. "Johann Valentin Andreae's Utopian Brotherhoods." *Renaissance Quarterly* 49, no. 4 (1996): 760–802.

Dixon, C. Scott. "Popular Astrology and Lutheran Propaganda in Reformation Germany." *History* 84 (1999): 403–18.

———. "Urban Order and Religious Coexistence in the German Imperial City: Augsburg and Donauwoerth, 1548–1608." *Central European History* 40 (2007): 1–33.

Donahue, William. "Astronomy." In *The Cambridge History of Science*, vol. 3, *Early Modern Science*, edited by Katherine Park and Lorraine Daston, 562–95. Cambridge: Cambridge University Press, 2006.

Donne, John. *John Donne: The Complete English Poems.* Edited by A. J. Smith. London: Penguin Books, 1971.

Drake, Stillman. "Galileo's 'Platonic' Cosmogony and Kepler's *Prodromus.*" In *Essays on Galileo and the History and Philosophy of Science*, vol. 1, edited by Noel M. Swerdlow and Trevor Harvey Levere, 367–74. Toronto: University of Toronto Press, 1999.

———. *Galileo Studies: Personality, Tradition, and Revolution.* Ann Arbor: University of Michigan Press, 1970.

Dryden, John. *The Works of John Dryden.* Edited by Walter Scott. Vol. 11. London: James Ballantyne, 1808.

Duffy, Eamon. *The Stripping of the Altars: Traditional Religion in England, c. 1400–c. 1580.* New Haven, CT: Yale University Press, 1992.

Eamon, William. *Science and the Secrets of Nature: Books of Secrets in Medieval and Early Modern Culture.* Princeton, NJ: Princeton University Press, 1996.

Eden, Kathy. *Friends Hold All Things in Common: Tradition, Intellectual Property, and the Adages of Erasmus.* New Haven, CT: Yale University Press, 2001.

Edwards, Jonathan. "Calvin and Hobbes: Trinity, Authority, and Community." *Philosophy and Rhetoric* 42, no. 2 (2009): 115–33.

Elert, Werner. *The Structure of Lutheranism.* Translated by Walter A. Hansen. Vol. 1. St. Louis, MO: Concordia, 1962.

Elwood, Christopher. *The Body Broken: The Calvinist Doctrine of the Eucharist and the Symbolization of Power in Sixteenth-Century France.* Oxford: Oxford University Press, 1998.

Ermarth, Elizabeth. *History in the Discursive Condition: Reconsidering the Tools of Thought.* New York: Routledge, 2011.

Evans, R. J. *Rudolf II and His World: A Study in Intellectual History, 1576–1612.* Oxford: Oxford University Press, 1984.

Falcom, Andrea. *Aristotle and the Science of Nature: Unity without Uniformity.* Cambridge: Cambridge University Press, 2005.

Feingold, Mordechai. "Jesuits: Savants." In *Jesuit Science and the Republic of Letters,* edited by Mordechai Feingold, 1–46. Cambridge, MA: MIT Press, 2003.

Ferrell, Lori Anne. *Government by Polemic: James I, the King's Preachers, and the Rhetorics of Conformity.* Redwood City, CA: Stanford University Press, 1998.

Feyerabend, Paul. *Against Method.* London: Verso, 1993.

Field, J. V. "Cosmology in the Work of Kepler and Galileo." In *Novità celesti e crisi del sapere: Atti del Convegno internationale di studi Galileiani,* edited by Paolo Galluzzi, 207–15. Florence: Giunti Barbera, 1984.

———. *Kepler's Geometrical Cosmology.* Chicago: University of Chicago Press, 1988.

———. "A Lutheran Astrologer: Johannes Kepler." *Archive for History of Exact Sciences* 31, no. 3 (1983): 189–272.

Findlen, Paula. *Possessing Nature: Museums, Collecting, and Scientific Culture in Early Modern Italy.* Berkeley: University of California Press, 1994.

Finocchiaro, Maurice. *Galileo and the Art of Reasoning: Rhetorical Foundations of Logic and Scientific Method.* Dordrecht: Springer, 1980.

Forhan, Kate Langdon. "Respect, Interdependence, Virtue." In *Difference and Dissent: Theories of Toleration in Medieval and Early Modern Europe,* edited by Cary J. Nederman and John Christian Laursen, 67–82. Lanham, MD: Rowan and Littlefield, 1996.

Frame, Donald M. "Montaigne and the Problem of Consistency." *Kentucky Romance Quarterly* 21, no. 2 (1974): 157–72.

Frost, Michael H. *Introduction to Classical Legal Rhetoric: A Lost Heritage.* Burlington, VT: Ashgate, 2005.

Funkenstein, Amos. *Theology and the Scientific Imagination: From the Middle Ages to the Seventeenth Century.* Princeton, NJ: Princeton University Press, 1986.

Galilei, Galileo. *Dialogue concerning the Two Chief World Systems: Ptolemaic and Copernican.* Translated by Stillman Drake. New York: Modern Library Edition, 2001.

———. *Discoveries and Opinions of Galileo.* Edited by Stillman Drake. New York: Doubleday, 1957.

Gattei, Stephen. "The Engraved Frontispiece of Kepler's *Tabulae Rudolphinae* (1627): A Preliminary Study." *Nuncius* 24 (2009): 341–65.

Gerlach, Walther. "Humor und Witz in Schriften von Johannes Kepler." In *Sitzungsberichte der Bayerische Akademie der Wissenschaften, mathematisch-naturwissenschaftliche Klasse, Jahrgang 1968,* 13–29. Munich: Bayerische Akademie der Wissenschaften, 1969.

Giglioni, Guido. "Philosophy according to Tacitus: Francis Bacon and the Inquiry into the Limits of Human Self-Delusion." *Perspectives on Science* 20, no. 2 (2012): 159–82.

Gingerich, Owen. "The Civil Reception of the Gregorian Calendar." In *Gregorian Reform of the Calendar: Proceedings of the Vatican Conference to Commemorate Its 400th Anniversary, 1582–1992*, edited by G. V. Coyne, M. A. Hoskin, and O. Pedersen, 265–79. Vatican City: Pontifical Academy of Sciences, Specolo Vaticano, 1983.

Ginzberg, Carlo. *History, Rhetoric, and Proof.* Lebanon, NH: University Press of New England, 1999.

———. *Wooden Eyes: Nine Reflections on Distance.* New York: Columbia University Press, 2001.

Glomski, Jacqueline. "Careerism at Cracow: The Dedicatory Letters of Rudolf Agricola Junior, Valentin Eck, and Leonard Cox." In *Self-Presentation and Social Identification: The Rhetoric and Pragmatics of Letter Writing in Early Modern Times*, edited by Toon Van Houdt et al., 165–82. Leuven: Leuven University Press, 2002.

Goldgar, Ann. *Impolite Learning: Conduct and Community in the Republic of Letters, 1680–1750.* New Haven, CT: Yale University Press, 1995.

Good, Graham. *The Observing Self: Rediscovering the Essay.* London: Routledge, 2014.

Gorman, Michael John. "From 'The Eyes of All' to 'Usefull Quarries in Philosophy and Good Literature': Consuming Jesuit Science, 1600–1665." In *The Jesuits: Cultures, Sciences, and the Arts, 1540–1773*, edited by John W. O'Malley et al., 170–89. Toronto: University of Toronto Press, 1999.

Gouk, Penelope. *Music, Science, and Natural Magic in Seventeenth-Century England.* New Haven, CT: Yale University Press, 1999.

Grafton, Anthony. *Cardano's Cosmos: The Worlds and Works of a Renaissance Astrologer.* Cambridge, MA: Harvard University Press, 1999.

———. "Commentary." In *The Classical Tradition*, edited by Anthony Grafton, Glenn W. Most, and Salvatore Settis, 225–33. Cambridge, MA: Harvard University Press, 2010.

———. *Defenders of the Text: The Traditions of Scholarship in an Age of Science, 1450–1800.* Cambridge, MA: Harvard University Press, 1991.

———. "Kepler as Reader." *Journal of the History of Ideas* 53, no. 4 (1992): 561–72.

———. "Portrait of Justus Lipsius." *American Scholar* 56, no. 3 (1987): 382–90.

———. *What Was History? The Art of History in Early Modern Europe.* Cambridge, MA: Harvard University Press, 2007.

———. *Worlds Made by Words: Scholarship and Community in the Modern West.* Cambridge, MA: Harvard University Press, 2009.

Granada, Miguel A. "After the Nova of 1604: Roeslin and Kepler's Discussion on the Significance of the Celestial Novelties (1607–1613)." *Journal for the History of Astronomy* 42, no. 3 (2011): 353–90.

Graves, Neil D. "Milton and the Theory of Accommodation." *Studies in Philology* 98, no. 2 (2001): 251–72.

Green, Lowell C. *The Formula of Concord: An Historiographical and Bibliographical Guide.* St. Louis, MO: Center for Reformation Research, 1977.

Greenbaum, Dorian Gieseler, ed. "Kepler's Astrology." Special double issue, *Culture and Cosmos* 14, nos. 1–2 (2010).

Haar, James. "*Musica Mundana*: Variations on a Pythagorean Theme." PhD diss., Harvard University, 1961.

Hallyn, Fernand. *The Poetic Structure of the World: Copernicus and Kepler.* Translated by Donald M. Leslie. New York: Zone Books, 1993.

Hanegraaf, Wouter J. *Esotericism and the Academy: Rejected Knowledge in Western Culture.* Cambridge: Cambridge University Press, 2012.

Hankins, James. *Plato in the Italian Renaissance.* Leiden: Brill, 1990.

Harrison, Peter. *The Bible, Protestantism, and the Rise of Natural Science.* Cambridge: Cambridge University Press, 2001.

Harvey, F. D. "Two Kinds of Equality." *Classica et Mediaevalia* 26 (1965): 101–29.

Hausenblasová, Jaroslava. *Der Hof Kaiser Rudolfs II: Eine Edition der Hofstaatsverzeichnisse, 1576–1612.* Prague: Artefactum, 2002.

Hayton, Darin. "Astrology as Political Propaganda: Humanist Responses to the Turkish Threat in Early Sixteenth-Century Vienna." *Austrian History Yearbook* 38 (2007): 61–91.

Heller-Roazen, Daniel. *The Fifth Hammer: Pythagoras and the Disharmony of the World.* New York: Zone Books, 2011.

Hobbes, Thomas. *Elements of Law.* Translated by John Charles Addison Gaskin. Oxford: Oxford University Press, 1999.

———. *Leviathan; or, The Matter, Forme, and Power of a Commonwealth, Ecclesiasticall and Civil.* Edited by Michael Oakeshott. New York: Simon and Schuster, 1962.

———. *On the Citizen.* Edited by Richard Tuck and Michael Silverthorne. Cambridge: Cambridge University Press, 1998.

Hollander, John. *The Untuning of the Sky: Ideas of Music in English Poetry, 1500–1700.* Princeton, NJ: Princeton University Press, 1961.

Hon, Giora, and Bernard Goldstein. *From* Summetria *to* Symmetry: The Making of a Revolutionary Scientific Concept. New York: Springer, 2008.

Honeygosky, Stephen R. *Milton's House of God: The Invisible and Visible Church.* Columbia: University of Missouri Press, 1993.

Hotson, Howard. "Irenicism in the Confessional Age: The Holy Roman Empire, 1563–1648." In *Conciliation and Confession: The Struggle for Unity in the Age of Reform, 1415–1648,* edited by Howard P. Louthan and Randall C. Zachman, 228–85. Notre Dame, IN: University of Notre Dame Press, 2004.

Houston, Chloe. *The Renaissance Utopia: Dialogue, Travel, and the Ideal Society.* Burlington, VT: Ashgate, 2014.

Howell, Kenneth J. *God's Two Books: Copernican Cosmology and Biblical Interpretation in Early Modern Science.* Notre Dame, IN: University of Notre Dame Press, 1992.

Hsia, R. Po-Chia. "Monstrous Births, Propaganda, and the German Reformation." In *Monstrous Bodies / Political Monstrosities in Early Modern Europe,* edited by Laura Lunger Knoppers and Joan B. Landes, 67–92. Ithaca, NY: Cornell University Press, 2004.

Hübner, Jürgen. *Die Theologie Johannes Keplers zwischen Orthodoxie und Naturwissenschaft.* Tübingen: Mohr, 1975.

Huijgen, Arnold. *Divine Accommodation in John Calvin's Theology.* Göttingen: Vandenhoeck und Ruprecht, 2011.

Hull, Gordon. *Hobbes and the Making of Modern Political Thought.* London: Bloomsbury, 2011.

Hunter, Ian. *Rival Enlightenments: Civil and Metaphysical Philosophy in Early Modern Germany.* Cambridge: Cambridge University Press, 2001.

Hunter, Michael. *Establishing the New Science: The Experience of the Early Royal Society.* Woodbridge, Suffolk: Boydell Press, 1995.

Ignatius of Loyola. *Ignatius of Loyola: The Spiritual Exercises and Selected Works.* Edited by George E. Ganss. New York: Paulist Press, 1991.

Ishiguro, Hide. *Leibniz's Philosophy of Logic and Language.* Cambridge: Cambridge University Press, 1990.

James I. *The Political Works of James I.* Edited by Charles Howard McIlwain. Cambridge, MA: Harvard University Press, 1918.

Janssen, Johannes. *History of the German People at the Close of the Middle Ages.* Translated by A. M. Christie and M. A. Mitchell. London: K. Paul, Trench, Truebner, 1905.

Jaquette, James L. *Discerning What Counts: The Function of the Adiaphora Topos in Paul's Letters.* Atlanta, GA: Scholars Press, 1995.

Jardine, Lisa. *Erasmus, Man of Letters: The Construction of Charisma in Print.* Princeton, NJ: Princeton University Press, 1993.

Jardine, Nicholas. *The Birth of the History and Philosophy of Science: Kepler's "A Defence of Tycho against Ursus," with Essays on Its Provenance and Significance.* Cambridge: Cambridge University Press, 1984.

———. "God's 'Ideal Reader': Kepler and His Serious Jokes." In *Johannes Kepler: From Tübingen to Żagań,* edited by Richard L. Kremer and Jarosław Włodarczyk, 41–51. Studia Copernicana. Warsaw: Institute for the History of Science, Polish Academy of Sciences, 2009.

———. "The Places of Astronomy in Early-Modern Culture." *Journal for the History of Astronomy* 29 (1998): 49–62.

Jardine, Nicholas, Elisabeth Leedham-Green, and Christopher Lewis. "Johann Baptist Hebenstreit's *Idyll on the Temple of Urania*, the Frontispiece Image of Kepler's *Rudolphine Tables*, Part 1: Context and Significance." *Journal for the History of Astronomy* 45 (2014): 1–19.

Jarrell, Richard A. "The Life and Scientific Work of the Tübingen Astronomer Michael Maestlin." PhD diss., University of Toronto, 1971.

John of Salisbury. *Policraticus*. Edited by Cary J. Nederman. Cambridge: Cambridge University Press, 1990.

Johnson, Carina L. "Stone Gods and Counter-Reformation Knowledges." In *Making Knowledge in Early Modern Europe: Practices, Objects, and Texts, 1400–1800*, edited by Pamela Schmidt and Benjamin Smith, 233–47. Chicago: University of Chicago Press, 2007.

Jones, Matthew. *The Good Life in the Scientific Revolution: Descartes, Pascal, Leibniz and the Cultivation of Virtue*. Chicago: University of Chicago Press, 2006.

Jones, Serene. *Calvin and the Rhetoric of Piety*. Louisville, KY: Westminster John Knox Press, 1995.

Joost-Gaugier, Christiane L. *Pythagoras and Renaissance Europe: Finding Heaven*. Cambridge: Cambridge University Press, 2014.

Kahn, Victoria. *The Future of Illusion: Political Theology and Early Modern Texts*. Chicago: University of Chicago Press, 2014.

Kaltenbrunner, Ferdinand. "Die Polemik über die Gregorianische Kalender-Reform." *Sitzungsberichte der Kaiserlichen Akademie von Wissenschaften* 87 (1877): 485–586.

———. "Die Vorgeschichte der Gregorianischen Kalenderreform." *Sitzungsberichte der Kaiserlichen Akademie von Wissenschaften* 82 (1876): 289–414.

Kant, Immanuel. *Kant: Political Writings*. Translated by H. B. Nisbet. Cambridge: Cambridge University Press, 1991.

Kaplan, Benjamin J. *Divided by Faith: Religious Conflict and the Practice of Toleration in Early Modern Europe*. Cambridge, MA: Harvard University Press, 2010.

Keller, Vera. "Mining Tacitus: Secrets of Empire, Nature, and Art in the Reason of State." *British Journal for the History of Science* 45, no. 2 (2012): 189–212.

Kemp, Martin. "Temples of the Body and Temples of the Cosmos: Vision and Visualization in the Vesalian and Copernican Revolutions." In *Picturing Knowledge: Historical and Philosophical Problems concerning the Use of Art in Science*, edited by Brian Scott Baigrie, 40–79. Toronto: University of Toronto Press, 1996.

Kepler, Johannes. *Kepler's Conversation with Galileo's "Sidereal Messenger."* Edited and translated by Edward Rosen. New York: Johnson Reprint, 1965.

"Kepler and His Discoveries." In *National Quarterly Review*, vol. 8, edited by Edward Isidore Sears, 335–57. New York: Pudney and Russell, 1864.

Kitao, T. Kaori. "*Imago* and *Pictura*: Perspective, Camera Obscura, and Kepler's Optics." In *La Prospettiva Rinascimentale*, edited by Marissa Dalai Emiliani, 499–510. Florence: Centro Di, 1980.

Knoppers, Laura Lunger, and Joan B. Landes. *Monstrous Bodies / Political Monstrosities in Early Modern Europe*. Ithaca, NY: Cornell University Press, 2004.

Kolb, Robert, ed. *Luther's Heirs Define His Legacy: Studies on Lutheran Confessionalization*. London: Variorum, 1996.

Kolb, Robert, and James Arne Nestingen. *Sources and Contexts of the Book of Concord*. Minneapolis, MN: Fortress Press, 2001.

Kouskoff, Georges. "Justice arithmetique, justice geometrique, justice harmonique." In *Jean Bodin: Actes du colloque interdisciplinaire d'Angers, 24 au 27 mai 1984*, 327–36. Angers: Presses de l'Université d'Angers, 1985.

Kraemer, Hendrik. *A Theology of the Laity*. London: Westminster, 1958.

Kusukawa, Sachiko. *The Transformation of Natural Philosophy: The Case of Philip Melanchthon*. Cambridge: Cambridge University Press, 1995.

Laertius, Diogenes. *Lives of Eminent Philosophers*. Translated by Robert Drew Hicks. London: Heinemann, 1925.

Lamont, Roscoe. "The Reform of the Julian Calendar." *Popular Astronomy* 28 (1920): 18–32.

Laursen, John Christian, and Cary J. Nederman. *Beyond the Persecuting Society: Religious Toleration before the Enlightenment*. Philadelphia: University of Pennsylvania Press, 2011.

———. *Difference and Dissent: Theories of Toleration in Medieval and Early Modern Europe*. New York: Rowman and Littlefield, 1996.

Leitão, Henrique. "Entering Dangerous Ground: Jesuits Teaching Astrology and Chiromancy in Lisbon." In *The Jesuits II: Cultures, Sciences, and the Arts, 1540–1773*, edited by John W. O'Malley et al., 371–404. Toronto: University of Toronto Press, 2006.

Le Roy, Louis. *Aristotle's Politiques or Discourses of Government*. Translated by I. D. London: Adam Islip, 1598.

Lim, Richard. "Christians, Dialogues, and Patterns of Sociability." In *The End of Dialogue in Antiquity*, edited by Simon Goldhill, 151–72. Cambridge: Cambridge University Press, 2008.

Lindberg, David C. *Theories of Vision from Al-Kindi to Kepler*. Chicago: University of Chicago Press, 1981.

Lipmann, Edward A. "Hellenic Conceptions of Harmony." *Journal of the American Musicological Society* 16, no. 1 (1963): 3–35.

Lipsius, Justus. "The First Book on Constancy." Edited by John Stradling. *Philosophical Forum* 37, no. 4 (2006): 398–426.

Lloyd, Howell A. "Constitutionalism." In *The Cambridge History of Political Thought, 1450–1700*, edited by J. H. Burns, 254–97. Cambridge: Cambridge University Press, 1994.

———. Introduction to *The Reception of Bodin*, edited by Howell A. Lloyd, 1–20. Leiden: Brill, 2013.

Locke, John. *An Essay concerning Human Understanding*. Edited by Alexander Campbell Fraser. Vol. 1. New York: Dover, 1959.

———. *A Letter concerning Toleration*. Edited by James H. Tully. Indianapolis: Hackett, 1983.

Long, A. A. "The Concept of the Cosmopolitan in Greek and Roman Thought." *Daedalus* 137, no. 3 (2008): 50–58.

Long, Kathleen P. "Odd Bodies: Reviewing Corporeal Difference in Early Modern Alchemy." In *Gender and Scientific Discourse in Early Modern Culture*, edited by Kathleen P. Long, 63–86. Burlington, VT: Ashgate, 2010.

Luther, Martin. *D. Martin Luthers Werke: Kritische Gesamtausgabe (Weimarer Ausgabe)*. 56 vols. Weimar: Hermann Boehlaus Nachfolger, 1883–1929.

Malcolm, Noel. *De Dominis, 1560–1624: Venetian, Anglican, Ecumenist, and Relapsed Heretic*. London: Strickland and Scott Academic Publications, 1994.

———. *Reason of State, Propaganda, and the Thirty Years' War: An Unknown Translation by Thomas Hobbes*. Oxford: Oxford University Press, 2007.

Manschreck, Clyde L. *Melanchthon: The Quiet Reformer*. Eugene, OR: Wipf and Stock, 2009.

———. "The Role of Melanchthon in the Adiaphora Controversy." *Archive for Reformation History* 48 (1958): 165–87.

Marion, Jean-Luc. "The Idea of God." In *The Cambridge History of Seventeenth-Century Philosophy*, vol. 2, edited by Daniel Ayers and Michael Garber, 265–304. Cambridge: Cambridge University Press, 2003.

Marshall, Peter. *The Magic Circle of Rudolf II: Alchemy and Astrology in Renaissance Prague*. New York: Bloomsbury Publishing USA, 2006.

Martens, Rhonda. *Kepler's Philosophy and the New Astronomy*. Princeton, NJ: Princeton University Press, 2000.

McColley, Diane Kelsey. *Poetry and Music in Seventeenth-Century England*. Cambridge: Cambridge University Press, 1997.

McDonough, Jeffrey K. "Leibniz's Conciliatory Account of Substance." *Philosophers' Imprint* 13, no. 6 (2013): 1–23.

McMahon, Darrin M. "Fear and Trembling: Strangers and Strange Lands." *Daedalus* 137, no. 3 (2008): 5–17.

McMullin, Ernan. "Galileo on Science and Scripture." In *The Cambridge Companion to Galileo*, edited by Peter K. Machamer, 271–347. Cambridge: Cambridge University Press, 1998.

McNeill, William H. *Keeping Together in Time: Dance and Drill in Human History*. Cambridge, MA: Harvard University Press, 1997.

McRae, Kenneth D. "A Postscript on Bodin's Connections with Ramism." *Journal of the History of Ideas* 24, no. 4 (1963): 569–71.

———. "Ramist Tendencies in the Thought of Jean Bodin." *Journal of the History of Ideas* 16, no. 3 (1955): 306–23.

Melanchthon, Philip. *A Melanchthon Reader*. Edited and translated by Ralph Keen. Bern: P. Lang, 1988.

Mercer, Christia. *Leibniz's Metaphysics: Its Origins and Development*. Cambridge: Cambridge University Press, 2001.

Methuen, Charlotte. *Kepler's Tübingen: Stimulus to a Theological Mathematics*. Aldershot: Ashgate, 1998.

———. *Science and Theology in the Reformation: Studies in Interpretation of Astronomical Observation in Sixteenth Century Germany*. London: T&T Clark International, 2008.

Mitchell, Margaret M. *Paul and the Rhetoric of Reconciliation*. Louisville, KY: Westminster John Knox Press, 1991.

Montaigne, Michel de. *The Complete Essays of Montaigne*. Edited by Donald M. Frame. Stanford, CA: Stanford University Press, 1958.

Morford, Mark. "Tacitean *Prudentia* and the Doctrines of Justus Lipsius." In *Tacitus and the Tacitean Tradition*, edited by T. J. Luce and A. J. Woodman, 129–51. Princeton, NJ: Princeton University Press, 1993.

Mosley, Adam. "Objects of Knowledge: Mathematics and Models in Sixteenth-Century Cosmology and Astronomy." In *Transmitting Knowledge: Words, Images, and Instruments in Early Modern Europe*, edited by Sachiko Kusukawa and Ian Maclean, 193–216. Oxford: Oxford University Press, 2006.

Mosley, Adam, Nicholas Jardine, and Karin Tybjerg. "Epistolary Culture, Editorial Practices, and the Propriety of Tycho's Astronomical Letters." *Journal for the History of Astronomy* 34 (2003): 419–51.

Moss, Jean Dietz. *Novelties in the Heavens: Rhetoric and Science in the Copernican Controversy*. Chicago: University of Chicago Press, 1993.

Muir, Edward. *Ritual in Early Modern Europe*. Cambridge: Cambridge University Press, 1997.

Mulsow, Martin. *Die unanständige Gelehrtenrepublik*. Stuttgart: J. B. Metzler, 2007.

Murphy, Kathryn, and Anita Traninger. "Introduction: Instances of Impartiality." In *The Emergence of Impartiality*, edited by Kathryn Murphy and Anita Traninger, 1–30. Leiden: Brill, 2013.

Nadler, Steven. "Arnauld, Descartes, and Transubstantiation: Reconciling Cartesian Metaphysics and Real Presence." *Journal of the History of Ideas* 49 (1988): 229–46.

Nagel, Thomas. *The View from Nowhere*. Oxford: Oxford University Press, 1989.

Nelson, Eric. "The Jesuit Legend: Superstition and Myth-Making." In *Religion and Superstition in Reformation Europe*, edited by Helen L. Parish and William G. Naphy, 94–132. Manchester: Manchester University Press, 2002.

Nitschke, August. "Keplers Staats- und Rechtslehre." In *Internationales Kepler-Symposium*, vol. 1, *Weil der Stadt, 1971*, 409–24. Hildesheim: HA Gerstenberg, 1973.

Novosel, Filip. "From Periphery to Centre: The Role of Marcus Antonius De Dominis in the 16th and 17th Century Scientific and Intellectual Movements." Master's thesis, Central European University, 2012.

Nussbaum, Martha. "Patriotism and Cosmopolitanism." In *For Love of Country: Debating the Limits of Patriotism*, edited by Joshua Cohen, 3–17. Boston: Beacon, 1996.

———. "Towards a Globally Sensitive Patriotism." *Daedalus* 137, no. 3 (2008): 78–93.

Oberman, Heiko. *Luther: Man between God and the Devil.* New Haven, CT: Yale University Press, 2006.

Oestreich, Gerhard. *Neostoicism and the Early Modern State.* Cambridge: Cambridge University Press, 2008.

O'Malley, John W., Gauvin Alexander Bailey, Steven J. Harris, and T. Frank Kennedy, eds. *The Jesuits: Cultures, Sciences, and the Arts, 1540–1773.* Toronto: University of Toronto Press, 1999.

Ong, Walter J. *Ramus, Method, and the Decay of Dialogue.* Chicago: University of Chicago Press, 2005.

Osler, Margaret J. *Reconfiguring the World: Nature, God, and Human Understanding from the Middle Ages to Early Modern Europe.* Baltimore: Johns Hopkins University Press, 2010.

Pagden, Anthony. "Stoicism, Cosmopolitanism, and the Legacy of European Imperialism." *Constellations* 7, no. 1 (2000): 3–22.

Panofsky, Erwin. *Perspective as Symbolic Form.* Translated by Christopher S. Wood. New York: Zone Books, 1991.

Pantin, Isabelle. *Discussion avec "Le messager céleste": Rapport sur l'observation des satellites de Jupiter.* Paris: Belles Lettres, 1993.

Parker, Charles H. "To the Attentive, Nonpartisan Reader: The Appeal to History and National Identity in the Religious Disputes of the Seventeenth-Century Netherlands." *Sixteenth Century Journal* 28, no. 1 (1997): 57–78.

Patterson, W. B. *King James VI and I and the Reunion of Christendom.* Cambridge: Cambridge University Press, 1998.

Payne, Alina A. "Creativity and Bricolage in Architectural Literature of the Renaissance." *RES: Anthropology and Aesthetics* 34 (1998): 20–38.

Pesic, Peter. *Music and the Making of Modern Science.* Cambridge, MA: MIT Press, 2014.

Pettit, Philip. *Made with Words: Hobbes on Language, Mind, and Politics.* Princeton, NJ: Princeton University Press, 2009.

Plato. *Laws.* Translated by Benjamin Jowett. New York: Cosimo Classics, 2008.

———. *The Republic.* Translated by G. M. A. Grube. Indianapolis: Hackett, 1992.

———. *Timaeus and Critias.* Translated by Desmond Lee. London: Penguin Books, 1971.

Plett, Heinrich F. *Rhetoric and Renaissance Culture.* Berlin: Walter de Gruyter, 2004.

Pope, Alexander. *The Works of Alexander Pope.* Vol. 5. Edinburgh: J. Balfour, 1764.

Porter, Theodore M. *Trust in Numbers: The Pursuit of Objectivity in Science and Public Life.* Princeton, NJ: Princeton University Press, 1996.

Portuondo, Maria M. *Secret Science: Spanish Cosmography and the New World.* Chicago: University of Chicago Press, 2009.

Preus, Robert. *The Inspiration of Scripture: A Study of the Theology of the Seventeenth-Century Lutheran Dogmaticians.* St. Louis, MO: Concordia, 2003.

Prins, Jacomien. "Harmony." In *Internet Encyclopedia of Philosophy*, edited by Marco Sgarbi. Springer Online, forthcoming.

Probst, Siegmund. "Infinity and Creation: The Origin of the Controversy between Thomas Hobbes and the Savilian Professors Seth Ward and John Wallis." *British Journal for the History of Science* 26 (1993): 271–79.

Prosperi, Adriano. "The Missionary." In *Baroque Personae*, edited by Rosario Villari, 160–94. Chicago: University of Chicago Press, 1995.

Pufendorf, Samuel. *The Present State of Germany*. London: Richard Chiswell, 1690.

Quintilian. *The "Institutio oratoria" of Quintilian*. Translated by H. E. Butler. Cambridge, MA: Harvard University Press, 1980.

Rabin, Sheila J. "Kepler's Attitude toward Pico and the Anti-astrology Polemic." *Renaissance Quarterly* 50, no. 3 (1997): 750–70.

Reich, Karin, and Eberhard Knoblock. "Die Kreisquadratur Matthias Hafenreffers." *Acta Historica Astronomiae* 17 (2002): 157–83.

Reiss, Timothy. *The Discourse of Modernism*. Ithaca, NY: Cornell University Press, 1982.

Remer, Gary. "Dialogues of Toleration: Erasmus and Bodin." *Review of Politics* 56, no. 2 (1994): 305–36.

———. "Hobbes, the Rhetorical Tradition, and Toleration." *Review of Politics* 54, no. 1 (1992): 5–33.

———. *Humanism and the Rhetoric of Toleration*. University Park: Pennsylvania State University Press, 1996.

Rigolot, François. "Problematizing Renaissance Exemplarity: The Inward Turn of Dialogue from Petrarch to Montaigne." In *Printed Voices: The Renaissance Culture of Dialogue*, edited by Dorothea B. Heitsch and Jean-François Vallée, 3–26. Toronto: University of Toronto Press, 2004.

Rosen, Edward. "Galileo and Kepler: Their First Two Contacts." *Isis* 57 (1966): 262–64.

Rubinstein, Nicolai. "The History of the Word *Politicus* in Early-Modern Europe." In *The Languages of Political Theory in Early-Modern Europe*, edited by Anthony Pagden, 41–56. Cambridge: Cambridge University Press, 1990.

Rublack, Ulinka. *The Astronomer and The Witch: Johannes Kepler's Fight for His Mother*. Oxford: Oxford University Press, 2015.

Rummel, Erika. "Erasmus and the Restoration of Unity in the Church." In *Conciliation and Confession: The Struggle for Unity in the Age of Reform, 1415–1648*, edited by Howard P. Louthan and Randall C. Zachman, 62–72. Notre Dame, IN: University of Notre Dame Press, 2004.

Rutkin, H. Darrel. "Various Uses of Horoscopes: Astrological Practices in Early Modern Europe." In *Horoscopes and Public Spheres: Essays on the History of Astrology*, edited by Günther Oestermann, H. Darrel Rutkin, and Kocku von Stuckard, 167–82. Berlin: Walter de Gruyter, 2005.

Rutkin, H. Darrel, and François Charette. "Astrology." In *The Classical Tradition*, edited by Anthony Grafton, Glenn W. Most, and Salvatore Settis, 84–89. Cambridge, MA: Harvard University Press, 2010.

Ryle, Gilbert. "Letters and Syllables in Plato." *Philosophical Review* 69, no. 4 (1960): 431–51.

Salmon, Vivian. *The Study of Language in 17th-Century England*. Amsterdam: John Benjamins, 1988.

Schäfer, Volker. *Blätter für Württembergische Kirchengeschichte*, June 18, 2009, 437–38.

Schaff, Philip. *Bibliotheca Symbolica Ecclesiae Universalis*. Vol. 1. New York: Harper and Brothers, 1878.

Schelhase, Kenneth C. *Tacitus in Renaissance Political Thought*. Chicago: University of Chicago Press, 1976.

Schinkel, Anders. *Conscience and Conscientious Objections*. Amsterdam: Amsterdam University Press, 2007.

Schmidt-Biggemann, Wilhelm. "New Structures of Knowledge." In *A History of the University in Europe: Universities in Early Modern Europe*, vol. 2, edited by Hilde De Ridder-Symoens, 489–530. Cambridge: Cambridge University Press, 1996.

Scholar, Richard. "Reasons for Holding Back in Two Essays of Montaigne." In *The Emergence of Impartiality*, edited by Kathryn Murphy and Anita Traninger, 65–83. Leiden: Brill, 2013.

Scribner, Robert W. *For the Sake of Simple Folk: Popular Propaganda for the German Reformation*. Cambridge: Cambridge University Press, 1981.

Serjeantson, Richard W. "Proof and Persuasion." In *The Cambridge History of Science*, vol. 3, *Early Modern Science*, edited by Katherine Park and Lorraine Daston, 132–75. Cambridge: Cambridge University Press, 2006.

Shakespeare, William. *Henry V*. Edited by Gary Taylor. Oxford: Oxford University Press, 1982.

Shapin, Steven. *A Social History of Truth: Civility and Science in Seventeenth-Century England*. Chicago: University of Chicago Press, 1994.

Shapin, Steven, and Simon Schaffer. *Leviathan and the Air-Pump: Hobbes, Boyle and the Experimental Life*. Princeton, NJ: Princeton University Press, 1985.

Shapiro, Barbara J. *A Culture of Fact: England, 1550–1720*. Ithaca, NY: Cornell University Press, 2002.

Skinner, Quentin. "Hobbes and the Purely Artificial Person of the State." *Journal of Political Philosophy* 7, no. 1 (1999): 1–29.

———. *Reason and Rhetoric in the Philosophy of Hobbes*. Cambridge: Cambridge University Press, 1996.

Smith, Barry D. *The Indescribable God: Divine Otherness in Christian Theology*. Eugene, OR: Wipf and Stock, 2012.

Smith, Pamela H. *The Body of the Artisan: Art and Experience in the Scientific Revolution*. Chicago: University of Chicago Press, 2004.

Smuts, Malcolm. "Court-Centered Politics and the Uses of Roman Historians, c. 1590–1630." In *Culture and Politics in Early Stuart England*, edited by K. Sharpe and P. Lake, 21–43. London: Macmillan, 1994.

Soll, Jacob. "Empirical History and the Transformation of Political Criticism in France from Bodin to Bayle." *Journal of the History of Ideas* 64, no. 2 (2003): 297–316.

———. *Publishing "The Prince": History, Reading, and the Birth of Political Criticism.* Ann Arbor: University of Michigan Press, 2005.

Sommerville, Johann P. "The 'New Art of Lying': Equivocation, Mental Reservation, and Casuistry." In *Conscience and Casuistry in Early Modern Europe*, edited by Edmund Leites, 159–84. Cambridge: Cambridge University Press, 2002.

Sparn, Walter. *Wiederkehr der Metaphysik.* Stuttgart: Calwer Verlag, 1976.

Spiller, Elizabeth. *Science, Reading, and Renaissance Literature: The Art of Making Knowledge, 1580–1670.* Cambridge: Cambridge University Press, 2004.

Spitzer, Leo. *Classical and Christian Ideas of World Harmony: Prolegomena to an Interpretation of the Word "Stimmung."* Baltimore: Johns Hopkins University Press, 1963.

Spitzer, Michael. *Metaphor and Musical Thought.* Chicago: University of Chicago Press, 2004.

Stahl, William Harris. *Commentary on the "Dream of Scipio" by Macrobius.* New York: Columbia University Press, 1990.

Stanciu, Diana. "Prudence in Lipsius's *Monita et Exempla Politica.*" In *(Un)masking the Realities of Power: Justus Lipsius and the Dynamics of Political Writing in Early Modern Europe*, edited by Erick Born, Marijke Janssens, and Toon van Houdt, 233–62. Leiden: Brill, 2010.

Stephenson, Bruce. *Kepler's Physical Astronomy.* Princeton, NJ: Princeton University Press, 1994.

Stieve, Felix. *Der Kalenderstreit des sechzehnten Jahrhunderts in Deutschland.* Munich: Verlag der k. Akademie, 1880.

Stolberg, M. "The Decline of Uroscopy in Early Modern Learned Medicine (1500–1650)." *Early Science and Medicine* 12, no. 3 (2007): 313–36.

Straker, Stephen. "Kepler's *Optics.*" PhD diss., Indiana University, 1970.

STRANGE NEWES of a prodigious Monster, borne in the Towneship of Adlington in the Parish of Standish in the Countie of Lancaster, the 17. day of Aprill last, 1613. Testified by the Reuerend Diuine Mr. W. LEIGH, Bachelor of Diuinitie, and Preacher of Gods. London, 1613.

Strauss, Leo. *Hobbes's Critique of Religion and Related Writings.* Chicago: University of Chicago Press, 2011.

———. *The Political Philosophy of Hobbes: Its Basis and Its Genesis.* Chicago: University of Chicago Press, 1952.

Strong, Tracy B. "How to Write Scripture: Words, Authority, and Politics in Thomas Hobbes." *Critical Inquiry* 20, no. 1 (1993): 128–59.

Tentler, Thomas N. "The Meaning of Prudence in Bodin." *Traditio* 15 (1959): 365–84.

Thorndike, Lynn. *The History of Magic and Experimental Science.* Vol. 6. New York: Columbia University Press, 1953.

Tomlinson, Gary. *Music in Renaissance Magic: Towards a Historiography of Others.* Chicago: University of Chicago Press, 1994.

Toulmin, Stephen. *Cosmopolis: The Hidden Agenda of Modernity.* Chicago: University of Chicago Press, 1990.

Trachtenberg, Marvin. "Building outside Time in Alberti's *De re aedificatoria.*" *RES: Anthropology and Aesthetics* 48 (2005): 123–34.

Trinkhaus, Charles. *In Our Image and Likeness: Humanity and Divinity in Italian Humanist Thought.* Chicago: University of Chicago Press, 1970.

Tuck, Richard. *Philosophy and Government, 1572–1651.* Cambridge: Cambridge University Press, 1993.

Vallée, Jean-François. "The Fellowship of the Book: Printed Voices and Written Friendships in More's *Utopia.*" In *Printed Voices: The Renaissance Culture of Dialogue,* edited by Dorothea B. Heitsch and Jean-François Vallée, 42–62. Toronto: University of Toronto Press, 2004.

van Orden, Kate. *Music, Discipline, and Arms in Early Modern France.* Chicago: University of Chicago Press, 2005.

Verkamp, Bernhard J. "The Limits upon Adiaphoristic Freedom: Luther and Melanchthon." *Theological Studies* 36, no. 1 (1975): 52–76.

Vickers, Brian. "The Recovery of Rhetoric: Petrarch, Erasmus, and Perelman." In *The Recovery of Rhetoric: Persuasive Discourse and Disciplinarity in the Human Sciences,* edited by Richard H. Roberts and James M. M. Good, 25–48. Charlottesville: University of Virginia Press, 1993.

Vitruvius, On Architecture. Vol. 1, *Books I–V..* Translated by Frank Granger. Loeb Classical Library 251. Cambridge, MA: Harvard University Press, 1996.

Voelkel, James. *The Composition of Kepler's "Astronomia Nova."* Princeton, NJ: Princeton University Press, 2001.

Wagner, Tobias. "Memoria Rediviva, Viri admodum Reverendi & Amplissimi, Dn. Danielis Hitzleri." 1660. 4 Diss 575, Staatsbibliothek München.

Walker, D. P. "Kepler's Celestial Music." *Journal of the Warburg and Courtauld Institutes* 30 (1967): 228–50.

Wandel, Lee Palmer. *The Eucharist in the Reformation: Incarnation and Liturgy.* Cambridge: Cambridge University Press, 2006.

Watson, Richard A. *The Breakdown of Cartesian Metaphysics.* Indianapolis: Hackett, 1998.

———. "Transubstantiation among the Cartesians." In *Problems of Cartesianism,* edited by Thomas M. Lennon, John M. Nicholas, and John Whitney Davis, 127–48. Kingston: McGill-Queen's University Press, 1982.

Wengart, Timothy J. *Human Freedom, Christian Righteousness: Philip Melanchthon's Exegetical Dispute with Erasmus of Rotterdam.* Oxford: Oxford University Press, 1997.

Westman, Robert S. "The Astronomer's Role in the Sixteenth Century: A Preliminary Survey." *History of Science* 18 (1980): 105–47.

———. *The Copernican Question: Prognostication, Skepticism, and Celestial Order.* Berkeley: University of California Press, 2011.

———. "The Melanchthon Circle, Rheticus, and the Wittenberg Interpretation of the Copernican Theory." *Isis* 66 (1975): 165–93.

———. "Proof, Poetics, and Patronage: Copernicus's Preface to the *De revolutionibus.*" In *Reappraisals of the Scientific Revolution,* edited by R. S. Westman and D. C. Lindberg, 167–205. Cambridge: Cambridge University Press, 1990.

Whitman, James Q. *The Origins of Reasonable Doubt: Theological Roots of the Criminal Trial.* New Haven, CT: Yale University Press, 2008.

Whitrow, Gerald J. *Time in History: Views of Time from Prehistory to the Present Day.* Oxford: Oxford University Press, 1990.

Wilhelmi, Thomas. *Die griechischen Handschriften der Universitätsbibliothek Tübingen: Sonderband Martin Crusius; Handschriftenverzeichnis und Bibliographie.* Weisbaden: O. Harrassowitz, 2002.

Williams, Wes. "'Si Faut-il Voir Si Cette Belle Philosophie . . .': The Language of Fiction in Montaigne, Corneille, and Pascal." In *Fiction and the Frontiers of Knowledge in Europe, 1500–1800,* edited by Richard Scholar and Alexis Tadié, 31–52. Burlington, VT: Ashgate, 2010.

Wilson, Kenneth J. *Incomplete Fictions: The Formation of English Renaissance Dialogue.* Washington, DC: Catholic University of America Press, 1985.

Wotton, Henry. *The Life and Letters of Henry Wotton.* Edited by Logan Persall Smith. Vol. 2. Oxford: Clarendon Press, 1907.

———. *Reliquiae Wottoniae.* London, 1651.

Ziebart, K. Meredith. *Nicolaus Cusanus on Faith and the Intellect: A Case Study in 15th-Century Fides-Ratio Controversy.* Leiden: Brill, 2013.

Ziolkowski, John E. "The Bow and the Lyre: Harmonizing Duos in Plato's Symposium." *Classical Journal* 95, no. 1 (1999): 19–36.

Index

Page numbers in italics refer to figures.